Scientific Realism and the Quantum

Scientific Realism and the Quantum

EDITED BY
Steven French
and Juha Saatsi

OXFORD
UNIVERSITY PRESS

OXFORD
UNIVERSITY PRESS

Great Clarendon Street, Oxford, OX2 6DP,
United Kingdom

Oxford University Press is a department of the University of Oxford.
It furthers the University's objective of excellence in research, scholarship,
and education by publishing worldwide. Oxford is a registered trade mark of
Oxford University Press in the UK and in certain other countries

First Edition published in 2020

Published in the United States of America by Oxford University Press
198 Madison Avenue, New York, NY 10016, United States of America

British Library Cataloguing in Publication Data
Data available

Library of Congress Control Number: 2019951277

ISBN 978–0–19–881497–9

Printed and bound in Great Britain by
Clays Ltd, Elcograf S.p.A.

Contents

Part IV Wavefunction and Quantum State Realism

Part V Scientific Realism and Quantum Field Theory

List of Illustrations

List of Contributors

VALIA ALLORI Northern Illinois University

ALISA BOKULICH Boston University

CRAIG CALLENDER University of California at San Diego

DOREEN FRASER University of Waterloo

JAMES D. FRASER Durham University

STEVEN FRENCH University of Leeds

RICHARD HEALEY University of Arizona

CARL HOEFER University of Barcelona and Institució Catalana de Recerca i Estudis Avançats

LINA JANSSON University of Nottingham

PETER J. LEWIS Dartmouth College

WAYNE C. MYRVOLD Western University

LAURA RUETSCHE University of Michigan

JUHA SAATSI University of Leeds

DAVID WALLACE University of Pittsburgh

J. E. WOLFF University of Edinburgh

Acknowledgements

We'd like to thank the Arts and Humanities Research Council for their generous funding of the *Scientific Realism and the Quantum* project. We'd also like to thank the participants in the series of workshops and conferences that it supported, including the contributors to this volume, as well as: Pablo Acuña, Guido Bacciagaluppi, Elena Castellani, Kevin Coffey, Elise Crull, Dennis Dieks, Mauro Dorato, Matthias Egg, Simon Friederich, Ruth Kastner, Adrian Kent, Claus Kiefer, Raffael Krismer, Soazig LeBihan, Diego Romero Maltrana, Michela Massimi, Owen Maroney, George Musser, Joshua Rosaler, Chris Timpson, and Karim Thébault. We'd especially like to thank the members of our reading group in philosophy of physics, including Fabio Ceravolo, Callum Duguid, Douglas Earl, Simon Newell, and Nahuel Sznajderhaus.

1

Introduction

Steven French and Juha Saatsi

Quantum physics represents our best and most fundamental physical science. Our understanding of numerous physical phenomena and our knowledge of the nature of light, electricity, solid matter, elementary particles, and even parts of chemistry, is rooted in quantum physics. But exactly what kind of knowledge does it provide us? This question gains significance from weighty epistemological issues that forcefully arise in this context—issues that are also at the heart of a more general debate on 'scientific realism' in the philosophy of science. This volume aims to advance both the realism debate as well as our understanding of the nature of quantum physics by bringing the two together in a productive dialogue.

Scientific realism was famously announced dead already back in 1984 by Arthur Fine, an American philosopher of science. He explained that its demise "was hastened by the debates over the interpretation of quantum theory, where Bohr's nonrealist philosophy was seen to win out over Einstein's passionate realism," and that its death "was certified, finally, as the last two generations of physical scientists turned their backs on realism and have managed, nevertheless, to do science successfully without it" (1984, p. 261). Fine's diagnosis appears flawed, however, as more than thirty-five years later realism doesn't just linger on, but thrives in discussions about quantum physics! But debates over the interpretation of quantum theory have not become any calmer in the hundred or so years after its inception, even though Bohr's ideas have been debunked many times over (Becker 2018). If anything, it is currently even harder to find a consensus about critical interpretive issues, as the range of seriously considered alternatives has steadily increased amongst broadly realist approaches to quantum physics. In this state of affairs there is considerable pressure to better articulate not just what "realism about the quantum" amounts to, but also what *justifies* a realist perspective over alternatives that rescind one or another realist theses. This volume collects together new work from the cutting edge of this active area of current research.

As indicated, the core realist intuitions about the physical sciences are resilient and hard to deny, but what exactly does "realism" stand for? As Richard Healey explains in his contribution to this volume, this term of philosophical trade has many meanings. In the extensive philosophical and foundational literature on quantum theory, realism has most typically signified the notion that we can specify what the mind-independent physical world could be like so as to render quantum theory (approximately) true. Painting a picture of reality compatible with the truth of quantum theory is a business

Steven French and Juha Saatsi, *Introduction* In: *Scientific Realism and the Quantum*. Edited by: Steven French and Juha Saatsi, Oxford University Press (2020). © Steven French and Juha Saatsi.
DOI: 10.1093/oso/9780198814979.003.0001

of *interpreting* the theory, its mathematical formalism and models. Realists like Einstein traditionally placed intuitive constraints on plausible pictures of reality compatible with quantum theory—in particular, that they should conform to a principle of locality according to which physical systems have determinate local properties not influenced by action at a distance.

Today's realist interpretations are a much more motley bunch, filling more of the space of logically coherent possible ways the world could be to make true one or another formal variant of quantum theory. These interpretations have emerged over decades of work by both philosophers and physicists engaged in foundational research. This work has largely aimed to tease out quantum theory's metaphysical and ontological implications, but hitherto much less attention has been paid to the concomitant epistemological issues. It is these latter issues that form the primary focal point of the present volume, which aims to engage more directly with the relevant epistemological questions that are also debated within general philosophy of science, concerning the status of our best scientific theories as a source of knowledge about unobservable reality or as furnishing representations of it.

The realist attitude towards well-established scientific theories is widely shared, seemingly common-sensical, and presupposed by the broadly accepted idea that such theories indeed do provide us genuine scientific understanding of natural phenomena through explanations in terms of how the unobservable world is structured and how it "works". Considerations in favour of realism tend to capitalize on the empirical success of science, variously manifested in triumphant theoretical predictions and the way science ever-increasingly supports our ability to manipulate the world to our liking through powerful interventions and applications that put to concrete use quantum theoretical notions such as 'spin'.

But while realists proclaim optimism about science's ability to tell us how things stand behind the veil of observable appearances, a very long tradition steadily opposes any such optimism on the basis of varied considerations regarding science at large. Two sources of scepticism have been particularly pervasive. First, there are historically driven concerns about the status of our current scientific theories' as 'approximately true', based on the historical track-record of radical and unexpected (r)evolutions in foundational scientific theorizing. Secondly, there are general "underdetermination" worries about the possibility of there being empirically indistinguishable—either in principle or for all practical purposes—theories that represent the world in radically different ways. These two broad sets of concerns have been raised time and again against scientific realism in various specific ways, sometimes individually, sometimes in unison.

Like much of general philosophy of science, the debates surrounding such concerns have been traditionally largely conducted in rather broad and abstract terms, quite independently of specific scientific detail. In a significant recent trend philosophers have become increasingly troubled about the potential limitations of sweeping, general arguments for or against realism, due to the variability of evidential, methodological, and explanatory contexts and practices that seem relevant for the assessment and outcomes of these arguments. As a result, there has been increasing emphasis within the realism debate of the importance of discipline or domain specific scientific details. In this spirit more 'local' analyses of the key issues animating this debate have been undertaken in relation to disciplines such as, e.g. cosmology, economics, geology, molecular biology, paleontology (see, e.g. contributions to Saatsi 2018, Part IV).

Quantum physics of course also provides a very natural locus for such a 'local' analysis, as the contributions in this volume nicely demonstrate. On the one hand, realism towards quantum physics is very easy to motivate in the light of its truly outstanding empirical successes. On the other hand, the theory is well known for its exceptional interpretational challenges and the resulting bifurcation regarding what it is taken to tell us about reality. This bifurcation powerfully brings to life the kind of underdetermination that many anti-realists have tended to worry about in the abstract. Relatedly, many classic philosophical questions concerning the relationship between science and metaphysics—the latter being deeply problematic according to some prominent anti-realist philosophers—are also nicely brought into focus in this context. In addition to throwing new light on such well-known issues, there are also entirely new ideas to be considered that have recently emerged specifically in the context of the philosophy of quantum physics, such as quantum pragmatism (advocated by, e.g. Richard Healey), and quantum Bayesianism (advocated by, e.g. Christopher Fuchs). The fourteen chapters that follow engage with all these issues and many, many more.

* * *

Let's now turn to the contributions to this volume. A theme running through many of them is to respond to the above problems by changing the terms in which realism is articulated. **PART I** presents two proposals for accomplishing this by explicitly rethinking what scientific realism amounts to.

Carl Hoefer articulates and defends 'Tautological Scientific Realism' (TSR). It eschews standard ways of defending and delineating realist commitments with regard to Inference to the Best Explanation and considerations of what might be preserved across theory change. Nevertheless, TSR, much like standard realism, maintains that our current scientific picture of the world is to a significant extent correct and will be retained through future changes at the theoretical level. But such a realist stance is only appropriate, Hoefer argues, with respect to those areas of current science for which we simply cannot seriously imagine future developments that would show that we are seriously mistaken in our current ontological commitments—in the way that we were with regard to phlogiston, for example. These 'safe' areas of science embrace the core properties of atoms and the way they combine, as well as our knowledge of electronics, for example, but not, crucially, quantum physics. Hence, Hoefer argues—more contentiously—for a new way of delineating realist commitments, according to which our current '*fundamental*' theories, such as quantum mechanics and quantum field theory, are specifically excluded from the scope of TSR. The grounds for this are two-fold: first, quantum physics is subject to the kind of underdetermination indicated above (and as discussed in one way or another by most of the papers in this collection); and secondly, it is expected to be replaced by a theory capable of unifying quantum theory and general relativity. Thus, Hoefer argues that the appropriate attitude towards quantum physics is one of anti-realism: agnosticism about its ontology, coupled with instrumentalism about its theories.

Juha Saatsi also proposes an alternative articulation of realism, focusing his discussion on the exemplary quantum property of spin. As is well-known, spin has no classical analogue and, as Saatsi notes, it not only lies at the heart of many quantum theoretic explanations, but has come to be understood as increasingly manipulable in a way that allows it to feature in a number of exciting new technological developments (e.g. 'spintronics'). These features strongly motivate a realist stance towards spin in a way that is, Saatsi argues, analogous to the motivations behind Hoefer's TSR, thereby questioning the latter's exclusion of quantum physics from its scope. Yet spin also lies at the heart of the 'interpretational conundrums' of quantum theory and spelling out what spin *is* involves 'deep' metaphysical commitments that go beyond what is necessary to account for any theory's explanatory and predictive success. Here the underdetermination that bedevils realism raises its ugly head again: even the comparatively straightforward setup of silver atoms passing through a Stern–Gerlach apparatus arguably comes to be characterized in radically different ways depending on one's theoretical approach. Yet, these different ways seem to add nothing to the epistemic virtues of the theory—hence, Saatsi argues, they involve 'deep' metaphysics that remains unjustifiable by the realist's lights.

However, rather than abandoning a realist stance towards spin altogether, Saatsi argues that we should step away from such deep metaphysics and modify our realism accordingly. Thus he suggests we should give up the epistemic ambitions of what he calls 'truth-content' realism, grounded as it is in notions of reference and approximate truth. Instead we should accept and articulate a form of 'progress' realism which, in the case of spin, does not commit to claims about what spin is like, but nevertheless acknowledges that the relevant models function as representations of reality and to that extent can be considered to 'latch onto' the world in ways that ground the empirical success of the theory. This maintains a representational role for these models and, in naturalistic terms, allows for physics theorizing itself to explain the success of spintronics, for example. It also constrains future theorizing by pointing to those well-known exemplars of inter-theoretic relationships that motivate claims of theoretical progress and emphasizing that this is how physics can make sense of its own empirical success.

PART II contains three chapters that further explore the challenges faced by realism in the quantum context, focusing on the interconnected issues of underdetermination and interpretation. As already noted, there are well-known alternative realist approaches to quantum theory, such as Bohmian ('hidden variables'), Everettian ('many worlds'), and dynamical collapse formulations, as well as specific interpretations, such as Quantum Bayesianism, the 'transactional' approach, and myriad others. Which, if any, should a realist embrace, and on what grounds? Or does the problem of underdetermination completely undermine scientific realism in relation to quantum theory?

Craig Callender voices a degree of pessimism about realists' prospects for quarantining the blight of underdetermination. He helpfully characterizes the different foundational/interpretational approaches in terms of Lakatosian 'research programmes', delineated by their hard cores and negative and positive heuristics. As he also points out, given the flexibility inherent in each programme, no crucial experimental test between them is likely in the near future. Furthermore, as Callender goes on to argue, the underdetermination here cannot be dismissed as artificial (with the various interpretations construed as philosophers' toys or mere notational variants); hence, the realist faces a genuine problem.

A natural move for the realist is to try to find some common ground between these different programmes in which she can plant her flag. Unfortunately, as Callender spells out, all we seem able to find are small disconnected 'islands of reprieve'. Following a suggestion by Alberto Cordero (2001), Callender looks at basic 'textbook level' cases of quantum models, to see what common ground can be found. He concludes that insofar as quantum models of, e.g. the water molecule underwrite incontestable claims, these claims are not distinctly quantum in nature. And much of the distinctly quantum theoretic content of models of, e.g. quantum tunnelling or the hydrogen atom, turns out to differ between different variants of quantum mechanics (e.g. the Bohmian theory vs. standard textbook presentations)—a point that chimes with Saatsi's claim about different accounts of the Stern–Gerlach apparatus. Even relaxing what one means by 'quantum' and shifting to the semi-classical level offers little hope for the realist, as Callender shows that most of what we say about the quantum realm is dependent on the chosen foundational perspective. And unless the realist substantially reins in her ambitions, along the lines suggested by Hoefer and Saatsi, for example, it seems she must make such a choice. But which? As Callender notes, it's not just a matter of balancing competing theoretical virtues, but of reconciling different attitudes towards such virtues and their relationship to empirical confirmation. In a sense, he concludes, we have a kind of philosophical gridlock.

David Wallace locates the disagreement between the different realist approaches to quantum theory at a more fundamental level: what is the theory that one is trying to interpret in the first place? In other words, when it comes to the issue of identifying the 'best interpretation of quantum theory', it is not just a matter of debating theoretical virtues, or what we mean by an 'interpretation', but how we identify quantum theory itself. Wallace argues that 'abstract' quantum mechanics, with its formalism of Hilbert space and self-adjoint operators representing observables and so on, should not be conceived of as a scientific theory at all, but as a theoretical 'framework' within which concrete quantum theories—e.g. quantum particle mechanics, quantum field theory, or quantum cosmology—can be expressed. The latter theories have limited applicability, depending on the type of system, the energy level involved, and so on. (This is analogous to classical mechanics, according to Wallace, where an overarching framework, provided by Hamiltonian or Lagrangian mechanics, encompasses a wide range of concrete theories.) These concrete theories are inter-related in various ways, and Wallace argues that in the quantum case these inter-relationships should be seen not as establishing a hierarchy, but something more akin to a patchwork (although not necessarily a disunified one). Given this understanding of quantum theory, he asks: what should we expect from an interpretation thereof?

The answer, Wallace argues, is an interpretive *recipe* that tells us how to understand any specific quantum theory, in a manner that is compatible with the relevant inter-theory relations. (Again, this is analogous to classical mechanics.) Such an interpretive recipe is arguably provided by the Everett interpretation. Other interpretational approaches, he claims, fail to similarly make sense of the relationships between specific, concrete quantum theories. Dynamical collapse and hidden variable approaches, for example, typically focus only on non-relativistic particle mechanics and, further, under the fiction that it is a fundamental and universal theory. The theoretical commitments of such approaches that in one way or another modify 'standard' quantum mechanics are disconnected from the actual practice of physics, and incapable of accounting for the successes of quantum theories as *non-fundamental*, effective theories applicable in a given domain or energy regime. In the same spirit Wallace argues that interpretational strategies that take the quantum state to be non-representational (e.g. Richard Healey's quantum pragmatism, see Ch. 7) fail to make sense of how quantum physics captures a quark-gluon plasma, for example, involving an interplay of many concrete quantum theories and their relationships.

J. E. Wolff adopts a broader perspective on the question of what it is to interpret quantum theory. She contrasts a 'naturalistic', science-driven philosophical stance towards theories with that of the more principled, 'empiricist' stance, as represented by van Fraassen. Regarding the former, the Everettian interpretation favoured by Wallace is a clear example of an attempt to naturalistically 'read off' ontology from the theory. However, van Fraassen raises a challenge for the naturalistic stance. Following Maddy (2007), a 'naturalistic native' is someone so deeply immersed in scientific practice that she approaches *all* questions, including interpretive ones, from within that practice. Van Fraassen questions the idea that a naturalistic philosopher can consistently regard a paradigmatic participant of current science as such a naturalistic native. If such a native is incapable of adopting a distinctly philosophical interpretation of her own scientific practice, how will she cope with situations where scientists are more or less forced to step back and reflect on their aims and methodologies? Paradigmatic cases of such a situation were involved in the development of quantum mechanics, and hence a naturalist philosopher must here face van Fraassen's challenge: on the one hand, if the naturalistic native cannot engage in such reflection, then she cannot function as the paradigmatic participant in science, as she will be unable to handle crisis situations; on the other hand, if she is allowed to step back and reflect, then she cannot be really characterized as a 'naturalistic native'. Thus, van Fraassen insists (from within his empiricist stance) that interpreting theories necessarily requires stepping outside of science itself, since interpreting a theory like quantum mechanics involves considering how the world could possibly be the way the theory says it is and that involves investigating alternatives.

As Wolff suggests, the naturalist, in response, might question whether interpretation necessarily involves stepping outside of science in this way. To this end she identifies three different 'moments of interpretation' that arose in the development of quantum mechanics: interpretive questions that featured in that very development; the presentation of alternative views in competition with the 'orthodox' interpretation; and articulating what the world would be like were the theory to be true. With regard to the first, Wolff argues that this did not require scientists at the time to step back from their scientific practice or engage in particularly philosophical reflection, so there is no issue for naturalism here. When it comes to the second 'moment', the naturalist could maintain that the hidden-variables and the dynamical collapse approaches are actually different theories rather than different interpretations per se. As for the third project of interpretation, which is the one van Fraassen primarily has in mind, this does seem to present a problem for the naturalist, insofar as it invites metaphysical speculation. One option for the naturalistic philosopher identified by Wolff is to deny that there is a plurality of such speculative interpretations worth engaging with. Thus she could follow Wallace, for example, in adopting a 'literalist' line and arguing that the Everett interpretation is the only one that takes the theory literally, thereby rejecting the basis of van Fraassen's challenge.

However, Wolff continues, the risk then is that the naturalist might be unable to accommodate the way that such interpretive projects aid our *understanding* of the theory. After drawing a distinction between 'symbolic', 'objectual', and 'explanatory' forms of understanding, she argues that the last, in particular, is not closed off to the naturalist. Focusing on de Regt and Dieks' contextualist approach to this form of understanding, Wolff notes that by characterizing it as an epistemic aim of science—something the empiricist would reject, of course—this approach would surely look attractive to the naturalist. And in the quantum context, both the hidden-variables and dynamical collapse approaches, for example, can be viewed as offering such explanatory understanding. This underwrites them as appropriate for the naturalist's consideration, and thus the naturalistic native is ultimately not precluded from engaging in various forms of interpretive endeavour.

PART III comprises three chapters that focus on pragmatism about quantum theory, representing a step further beyond traditional conceptions of scientific realism, but without embracing traditional anti-realism either.

Richard Healey is a key advocate of such a pragmatist interpretation, at the heart of which lies the rejection of the 'representationalist' assumption that a scientific theory can give us a literally true account of what the world is like only by faithfully representing the world. As Healey notes, those parts of quantum physics that are actually used in incredibly successful technological developments such as spintronics, for instance, are independent of foundational and interpretational issues. If we think carefully about how quantum mechanics is actually applied to physical systems, Healey argues, we should see that the name of the game is not the representation of quantum reality but rather to give us, the users of the theory, appropriate information about the significance and credibility of claims regarding non-quantum physical features associated with those systems. Thus, for example, according to Healey the primary role of the notion of the interpretationally troublesome 'quantum state' is not to represent, or describe, some system, but rather to prescribe how we should determine the probabilities associated with various measurable eventualities (by applying the Born rule).

However, Healey insists, this is not a form of instrumentalism or empiricism (of the sort advocated by van Fraassen, for instance), since it does not rely on any distinction between 'observable' and 'unobservable' features of the world; rather, various claims about unobservable physical magnitudes are significant and true or false, depending on how the world is. It is the function of (non-quantum) magnitude claims to represent the relevant features of reality and ultimately it is the truth or falsity of such claims that we care about. This is still compatible with a 'thin' version of the correspondence theory of truth, in the sense of one that eschews some form of causal account of reference, so that we're not misled into thinking that terms appearing in magnitude claims refer to their subject matter via some form of causal connection. Indeed, Healey maintains, recent arguments regarding a form of the Bell inequalities put paid to such thinking. Nevertheless, we can still accept the existence of a physical world that is independent of our thinking about it. What the pragmatist adds to this conception is a broader perspective on how we gain knowledge of that world: this is achieved not via representation per se but, in effect, by taking the theory's advice on how the world might be meaningfully represented and what the likelihood is of such meaningful representations being true. Furthermore, from this pragmatist perspective quantum theory still helps us to explain a range of otherwise puzzling phenomena by showing that they were to be expected and also what they depend on.

The following two chapters focus on critical issues about the pragmatist attempt to construct a 'middle road' between realism and instrumentalism.

Lina Jansson focuses on the issue of explanation and the close ties that it has with scientific realism and argues that from this perspective, Healey's pragmatist interpretation comes with significant costs. She begins by inviting us to consider the putative truism that genuine explanations posit true explanans. This intuitive idea has to be immediately qualified, however, due to well-known challenges arising from the roles of idealizations, distortions, and fictions within scientific explanations. A realist can try to hang onto the gist of the putative truism by appropriately distinguishing the explanatory from the non-explanatory roles played by different aspects of scientific explanations, in such a way that the latter's ontological commitments are tracked. However, such moves are not open to Healey who rejects, as we have seen, the claim that quantum models explain by virtue of representing quantum reality, arguing instead that they explain by virtue of telling us what to expect regarding non-quantum magnitudes, together with what such magnitudes depend upon. A crucial issue is how to make sense of this explanatory dependence by the pragmatist's own lights.

As Jansson suggests, one possibility is to adopt a popular counterfactual approach based on 'what-if-things-had-been-different' questions (in the spirit of James Woodward), while also allowing for non-causal dependencies. However, without causation to rely on, there is no straightforward way of distinguishing the explanatory theoretical posits from the non-explanatory roles played by idealizations and the like. The way to proceed, she avers, is to carefully distinguish different kinds of dependence within the epistemic dependence approach to explanation and, in particular, to look to what it is that allows us to make the relevant inferences about the counterfactually robust connections between the initial input of the explanans and the explanandum. Idealizations, distortions, and fictions can serve to do that, without acting as the relevant input into the explanans. In Healey's account, since the quantum state is not taken to represent the system in question, it cannot serve as such an initial input but it may nevertheless be indispensable to us in offering the appropriate explanations. As a result, Jansson argues, crucial information about the physical grounds for the appropriate assignment of such states has to be effectively 'black-boxed', a feature that she highlights as one of the costs of adopting this form of pragmatist stance.

Peter Lewis also examines the costs of pragmatist approaches—here taken to embrace also Simon Friederich's (2015) account—not only with regard to explanation but also when it comes to our understanding of the content of propositions. In articulating their position the pragmatists appeal to an inferentialist account of meaning according to which the meaning of a proposition lies with the material inferences that it supports, rather than in its representational content. It arguably follows from such an account that claims concerning, e.g. quantum states, spin etc., are best viewed not as describing physical systems, but rather as *prescribing* degrees of belief in non-quantum magnitude claims that do have descriptive content. Lewis illustrates this by reference to a quantum state associated with a particular molecule. This quantum state can licence appropriate probabilistic inferences regarding, e.g. the molecule's location upon encountering a silicon surface through the application of the Born rule (underwritten by decoherence). A claim concerning the molecule's location on the surface is an example of a non-quantum magnitude claim that has descriptive, empirical content that is worth asserting, since it supports material inferences about, e.g. image formation in an electron microscope. The quantum state itself allegedly does not have such content; any claim about the molecule's location at a diffraction grating, for example, would lead to erroneous inferences concerning which slit the particle is going to go through, for instance. Hence, apart from prescribing probabilistic inferences supported by the Born rule in situations where the quantum state decoheres, arguably the state ascription has no content, especially when the Born rule is inapplicable.

As Lewis notes, one might worry that the distinction between prescriptive and descriptive content is not supported by the inferentialist account itself and here perhaps appeals to further elements of the pragmatist toolbox must be made. More acutely perhaps, Lewis raises the issue that it is not clear how counterfactual inferences are to be treated on the pragmatist approach: if a diffraction grating were to be replaced by a silicon surface, we would shift from a situation in which no credences regarding location can be assigned, the relevant claims being taken to be devoid of content, to one where definite probabilistic prescriptions can be made, the relevant claims being contentful. But given that latter point, if counterfactuals contribute to the content of quantum state attributions, then, Lewis argues, the former claims should also be understood as having at least some content, contrary to what the pragmatists assert. One possible response to this worry would be to reconsider the role of decoherence with regard to this shift in context—rather than delimiting the range over which claims have content it should be understood as delimiting the range over which our material inferences can unproblematically draw on our classical intuitions. Resolving these sorts of worries, together with those concerning explanation, Lewis concludes, will crucially determine whether this sort of pragmatist approach has enduring advantages over its realist rivals.

PART IV comprises three chapters that focus on various issues concerning the nature of the quantum state, standardly taken to be—in contrast to the pragmatist approach—represented by the wave function. Indeed, advocates of so-called wave function realism argue that this representational role should place the wave function at the centre of the scientific realist endeavour.

Alisa Bokulich challenges this view and the 'hegemony of the wave function' in general by presenting a formulation of quantum mechanics that doesn't make use of it. This is 'Lagrangian quantum hydrodynamics' according to which the state is represented via the displacement function of a continuum of interacting 'particles' following trajectories in spacetime. Schrödinger's equation is then recast as a second order Newtonian law governing such trajectories. Although this formulation is helpfully motivated by classical hydrodynamics, Bokulich is at pains to emphasize that it does not require commitment to some notion of a 'quantum fluid'. Instead the fundamental state entity via which one can understand the time evolution of the system is given by the congruence of the trajectories.

As she goes on to note, the centrality of these trajectories in this formulation suggests an obvious comparison with Bohmian mechanics. However, there are crucial differences, most notably with regard to the role played by the wave function in the latter. Furthermore, Bokulich insists, Bohmian mechanics is an *interpretation*, whereas Lagrangian quantum hydrodynamics is a *formulation*, and as such has entirely different interpretive ambitions (with regard to the measurement problem, for example). Interestingly, as Bokulich outlines, this alternative formulation reveals a previously obscured symmetry of quantum mechanics, associated with the infinite-parameter particle relabelling group, which implies the conservation of quantum forms of circulation, density, and current. Controversially, perhaps, when transposed to the relativistic context, the conservation law allows for the definition of global simultaneity manifolds. More significantly, it is partly because it allows the articulation of this relabelling symmetry that the Lagrangian formulation should be regarded as more fundamental than the apparently equivalent Eulerian formulation of quantum hydrodynamics (associated with Erwin Madelung), which retains the wave function representation of the quantum state.

What does this imply for the various realist projects adopted and pursued in the context of quantum theory? As Bokulich notes, the formulation not only challenges the hegemony of the wave function, but also offers a new perspective on experiments—such as those involving protective measurements—that are invoked as evidence for its reality. More generally, the existence of the Lagrangian formulation encourages us to be cautious in reading off our realist commitments from features of the standard mathematical presentation of quantum mechanics. Finally, one could also adopt a realist stance towards this formulation itself. Here Bokulich identifies three possible ways forward. One is to render it an interpretation of the theory, as in the 'Many Interacting Worlds' or 'Newtonian QM' views. Another is to adopt a 'duality' line towards the quantum state, with the wave functional- and trajectory-based aspects regarded as a new take on the (in)famous wave-particle duality. The third approach is what Bokulich calls 'inferential realism' which urges a shift in realist focus from asking 'what is the world like?' to 'what true things can we learn?' instead. Drawing on Ernan McMullin's emphasis on the role of metaphors in the realist enterprise, Bokulich insists that inferential realism is more about developing a plurality of fertile interpretations than finding the one true picture of the world, and both the trajectory-based and wave function-conceptions of state feature in this plurality.

Valia Allori similarly seeks to decentre the wave function in realist approaches to the theory. Like Bokulich she argues that we should not simply read off our realist commitments from a given formulation, but instead start the interpretive project with a 'primitive ontology' and construct our interpretation around that. In Newtonian mechanics the primitive ontology is that of particles, for example, represented by points in three-dimensional space and our understanding of the theory is grounded in this. Shifting to the quantum domain, Allori argues that we should retain the same approach, dropping the representational role of the wave function, not least because of the issue of how to understand superpositions. Instead, she maintains, we should begin with a primitive ontology located in space-time, and select an appropriate law of evolution for the relevant entities and aim to understand the theory on that basis. Different such primitive ontologies can then be combined with different laws of evolution, and Allori considers three kinds of the former: particles, matter fields and 'flashes'. This array of alternatives can accommodate a whole slew of theories, as she sets out. Within this interpretive framework the role of the wave function is to help implement the law that governs the spatio-temporal evolution of whatever primitive ontology has been chosen. Thus the wave function can be regarded as having a nomological character, a suggestion that appears more palatable to many if understood from a Humean perspective, according to which law statements are simply the axioms and theorems of our best theoretical system, representing regularities found in the world. Given the choice of modifying our conception of what counts as ontology or that of what counts as a law in the quantum context, Allori prefers the latter.

This general approach meshes well, she argues, with selective realism about the 'working' posits of the theory that are responsible for its explanations and predictions. Here the primitive ontology would supply the working posits, the wave function counting as a merely 'presuppositional' auxiliary that is necessary for the theory's mathematical formulation (however, see Bokulich above) but not to be understood realistically. Nevertheless, in some of the theories canvassed here, there remains a kind of dependence of the primitive ontology on the wave function and Allori suggests that this yields a useful way of categorizing solutions to the measurement problem: in theories of type 1 the primitive ontology and the wave function are independent, as in particle theories; in theories of type 2 the two are co-dependent and these include flash and matter density theories. Armed with this distinction Allori goes on to explore how such theories differ with respect to their super-empirical virtues (e.g. empirical coherence, simplicity, and relativistic invariance), arguing that Bohmian interpretations with particles as their primitive ontology and GRW approaches with a flash ontology should be viewed as the leading contenders, with the former, according to Allori, just nosing ahead. More importantly and generally, she concludes that once we get the wave function off centre stage we can more easily explore the different ways quantum mechanics can be made compatible with realism.

Wayne Myrvold considers a broader set of reasons for denying that quantum states represent something physically real and argues that at best these provide grounds for pursuing theories in whose ontologies quantum states don't appear. Such reasons may draw on certain classical 'toy' models in the context of which apparently quantum phenomena can be reproduced, such as the existence of pure states that cannot reliably be distinguished. However, Myrvold notes, these phenomena are at best only 'weakly' non-classical and such models cannot capture the Bell inequalities, for example, which are regarded as exemplars of quantum behaviour. Likewise, he argues, the fact that quantum mechanics exhibits classical behaviour under certain restrictions is better regarded as a prerequisite for taking the theory as comprehensive in the first place, rather than as evidence that quantum states are not real.

Myrvold then goes on to give positive reasons for an ontic construal of quantum states, within the context of the information theoretic 'ontological models framework'.

From this perspective, he explores the importance of two theorems that constrain the set of possible theories that could account for quantum phenomena. The first is due to Barrett, Cavalcanti, Lal, and Maroney and shows that quantum states cannot be construed as probability distributions over an underlying state space in such a way that the operational indistinguishability of such states can be accounted for in terms of overlap of the corresponding probability distributions. The motivation for constructing an interpretation under which the quantum states are not ontic is thus stymied. The second theorem, due to Pusey, Barrett, and Rudolph, demonstrates that distinct pure quantum states are ontologically distinct. Crucially this assumes the so-called Preparation Independence Postulate (PIP), which has to do with independent preparations performed on distinct systems. Myrvold problematizes PIP in relation to quantum field theory, and proposes that it be replaced with what he calls the Preparation Uninformativeness Condition (PUC), which, he argues, suffices to show that distinct quantum states must be ontologically distinct.

Given these results, he concludes, the project—which goes back to Einstein, of course—of understanding the quantum state in epistemic terms must be abandoned. Myrvold's argument hinges on the requirement that the ontological lessons we draw from a theory should rely only on premises that could reasonably be expected to be preserved when we shift to the successor theory, in this case quantum field theory. This raises the further question: how does realism fare when we move to consider quantum physics beyond the realm of non-relativistic quantum mechanics?

PART V examines various responses to this question, specifically in the context of quantum field theory.

Doreen Fraser focuses on the example of the Higgs boson as exemplifying the use of certain formal analogies holding between mathematical structures in the absence of any physical similarity between the relevant models. This, she suggests, represents a major challenge to the support that is typically adduced in favour of scientific realism. The construction of the Higgs model proceeded by drawing formal analogies with the relevant order parameters in certain models of superconductivity: with regard to the latter, it is the effective collective wave function of the superconducting electrons, which distinguishes the normal state of the metal from the superconducting state, and in the case of the Higgs model, this is the complex scalar quantum field associated with the Higgs boson.

Fraser argues that the physical dissimilarities of the various elements means that these analogies must be regarded as purely formal. Thus, for example, the internal relationships these elements enter into in each model are quite different: the transition to a superconducting state, involving spontaneous symmetry breaking, is a temporal process, but there is no analogue of this in the Higgs model.

What then is the explanation for the successful application of such analogies? Fraser argues that they opened up the space of mathematically conceivable models by showing that it was possible to incorporate spontaneous symmetry breaking into one's model accompanied by massive bosons. Furthermore, because of the physical disanalogies, crucial features of the superconductivity model were open to experimental investigation, allowing the formal analogies to play a heuristically useful role.

This then presents a fundamental challenge to realism, as the purely formal reasoning used to construct the Higgs model was instrumentally successful, yet, Fraser insists, the truth or falsity of the theoretical statements asserting the appropriate causal connections cannot be relevant to explaining its success because there is no plausible physical analogy underpinning them. Thus we seem to have an example of scientific success that cannot be accommodated in realist terms.

Fraser concludes by noting that realists and anti-realists alike tend to draw on diachronic sequences of theories in defence of their opposing claims. However, she argues, what the Higgs case study demonstrates is that when it comes to the development of specific quantum theories (and here we might recall Wallace's point above), it is also *synchronic* relationships that need to be considered, involving new sets of challenges.

James Fraser is more sanguine, insisting that despite the challenges, we can give a realist reading of quantum field theory. However, he argues, restricting our attention to perturbative or axiomatic treatments is unhelpful in that regard. It is the former approach that lies behind the striking empirical predictions of the theory, including many of those tested at the Large Hadron Collider, for example. Yet the underlying strategy is famously problematic and has been widely regarded as lacking in mathematical rigour, depending as it does on the removal of certain infinities in a suspiciously ad hoc manner. Indeed, it leaves the realist unable to specify what the theory says about the world. In desperation, perhaps, one might turn to the so-called axiomatic approach that at least gives a clear set of theoretical principles for the realist to work with. Unfortunately, as is well known, these principles can only be used as the framework for certain physically unrealistic 'toy' models.

All is not lost, however. As Fraser notes, developments in renormalization theory offer a way forward and here he sketches the core features of the momentum space approach—in particular the way in which certain coarse-grained transformations induce a 'flow' on the space of possible theories which offers information on the behaviour of systems at different scales. These systems, modelled by QFT, display a feature known as 'universality', whereby models that display very different behaviour at high energies manifest very similar physics at lower energy levels. What this means is that if the high energy degrees of freedom are removed, as in a 'cut-off', this will leave the lower energy behaviour more or less unaffected. This in turn allows the realist to 'bracket off' what the world is like at the fundamental level, while still accurately modelling its lower energy properties.

This then helps to justify the various steps of the perturbative renormalization procedure. Thus, for example, it justifies the absorption of the physics beyond a certain 'cut-off' point into an effective action and reveals that what this procedure is really about is ensuring the right kind of scaling behaviour exhibited by the system in question. Finally it pragmatically justifies taking the cut-off to infinity, which yields significant computational benefits.

Given this procedure, Fraser argues that the renormalization group offers a way of developing a selective realist reading of QFT, according to which we should be realist about those constituents that underwrite a theory's predictive success. In particular, it reveals that certain features play no role in that success and can be set aside as far as the realist is concerned. However, it also helps the realist articulate the relevant positive theoretical commitments, with regard to relatively large-scale, non-fundamental aspects of the world. It shows, for example, that they are largely insensitive to the details that obtain at high energies and hence can be considered 'robust' and thereby worthy of realist commitment.

Nevertheless there remain challenges. Thus, for example, even granted the robustness of low-energy features of the relevant models, it remains unclear what aspects of the world they are latching onto. Here, Fraser argues, the realist needs to pay further attention to such claims about the non-fundamental and consider more carefully the terms in which they are characterized. This is in addition to the more well-known concerns regarding what the world is like according to quantum theory, as canvassed in this volume, as well as the additional puzzle posed in the context of QFT by the existence of unitarily inequivalent Hilbert space representations. As Fraser concludes, such puzzles and concerns highlight the need for a comprehensive re-examination of realist strategies in general.

Laura Ruetsche agrees that the development of the Standard Model and the quantum field theory that underpins it present a range of new challenges to realism. She focuses on one of the strategies indicated by James Fraser, namely adopting a selective attitude as embodied in what she calls 'effective realism'. As she notes, this takes seriously the point that our best current theories are merely 'effective' in the sense that they're not true across all energy regimes and uses the renormalization group as a means of motivating the core 'divide et impere' move of such an attitude. Unfortunately, she points out, the action of the renormalization group is defined on a specific space of theories and whatever the virtues are of our best current models, if the true, final theory lies outwith that space, then, as she puts it, all bets are off.

Even more worryingly, Ruetsche notes that it is not clear how realist 'effective realism' is! Through an examination of various features of effective theories, she concludes that in order to distance herself from the empiricist, the effective realist must approach such features in the light of certain interpretive projects. So, for example, the effective realist might endorse particles corresponding to fields that are robustly present in the relevant Lagrangian at a certain length scale, but to do so she must engage in interpretive manoeuvres that are typically articulated in terms of a theory's truth conditions and which she supposedly repudiates.

However, Ruetsche suggests, the effective realist needn't disavow such interpretive work per se, as long as she is mindful of the distinction between asking what the world is like according to a given theory, and asking why that theory is so successful. According to the view Ruetsche labels 'fundamentalism', the answer to the second question is given in terms of the answer to the first: the theory is so successful because it accurately describes how the world is. When it comes to effective theories, however, she argues that this intertwining of the answers is a mistake because what explains the success of an effective theory is something exogenous to it. Recognizing that and rejecting fundamentalism then brings effective realism closer to what she calls a 'humble' form of empiricism that explains a theory's success in terms of its approximation to the predictions of some final theory within experimentally accessible regimes. The humble empiricist accepts that we can give an explanation of a given theory's success, just not in terms of its truth and that we can entertain the possibility of some true, final theory but that we should adopt an agnostic stance towards a given effective theory's set of unobservables.

To conclude, Ruetsche suggests that whether the commitments of effective realism actually count as realist or not depends on how they're understood. It is better, she maintains, to embrace a stance of humble empiricism that has the resources to accommodate the myriad ontological subtleties of quantum physics.

* * *

This collection of chapters is thus 'book-ended' by attempts to shift realism away from the traditional conception in the face of the multiple problems posed by quantum physics. The nature and extent of that shift varies from author to author—in some cases it involves a move away from the foundational, in others dropping the emphasis on the truth-content of theories, and in yet others it requires some form of non-literal 'reading' of quantum theory. There are a variety of options 'on the table' and what realists in general need to do now is not just take quantum physics seriously but to continue articulating, defending and contrasting these epistemic alternatives along the lines presented in this volume, which we hope will come to be seen as a significant step in the right direction.

References

Becker, A. (2018). *What Is Real?: The Unfinished Quest for the Meaning of Quantum Physics.* New York: Basic Books.

Cordero, A. (2001). Realism and underdetermination: some clues from the practices-up. *Philosophy of Science 68*, S301–S312.

Fine, A. (1984). The natural ontological attitude. In J. Leplin (Ed.), *Scientific Realism*, pp. 83–107. Los Angeles: University of California Press.

Friederich, S. (2015). *Interpreting Quantum Theory: A Therapeutic Approach.* London: Palgrave Macmillan.

Maddy, P. (2007). *Second Philosophy.* New York: Oxford University Press.

Saatsi, J. (Ed.) (2018). *The Routledge Handbook of Scientific Realism.* London: Routledge.

PART I

Rethinking Scientific Realism

2

Scientific Realism without the Quantum

Carl Hoefer

2.1 Introduction

The purpose of this chapter is to handle one part of a larger project proposing and defending a (relatively) new and untried form of Scientific Realism (SR). SR is the family of philosophical views that assert that we have strong reasons to believe in the truth or approximate truth of much of the content of our best current scientific theories. The specific form of SR that I defend will appear to some to be a black sheep in the SR family, since I reject many tenets central to earlier SR variants. For example, I claim that inference to the best explanation (IBE) is not the ground on which SR stands, nor is it a defensible inference rule in general. And I do not aim to look back into the history of science and argue that SR was defensible prior to the twentieth century, or that we can find a crucial type of content that has been preserved for centuries across apparently dramatic changes in theory. Instead, I will defend only the claim that a large proportion of our *current* scientific picture of the world is correct, and that we have ample reason to believe that it will never be overthrown and replaced in the way that important theories from earlier centuries (Newtonian gravity, phlogiston theory, Lamarckian evolution theory) were overthrown and replaced. So the SR I defend is 'selective' in the sense that it restricts its claims to fairly recent science (from roughly the mid-twentieth century onward).

It is selective in another important sense: it *excludes our current fundamental theories of physics*. That is this chapter's main purpose: to argue for the proposition that whatever the exact scope of the realist's claim may be, it should exclude the most fundamental physical theories we have right now. By this I mean: Quantum Mechanics (QM), the various quantum field theories that form parts of the Standard Model, General Relativity (GR), all the speculative theories of quantum gravity and quantum cosmology (string theories, AdS/CFT theories, etc.), and their related models (Standard Model of particle physics, Λ-CDM cosmological models, etc.). In the case of GR, which I will not discuss below, the case for exclusion is simple and is part of conventional wisdom in theoretical physics: GR treats matter only in a phenomenological sense, treats both matter and spacetime as classical continuum fields, and seems not to be quantizable. It also gives rise quite generically to infinities

Carl Hoefer, *Scientific Realism without the Quantum* In: *Scientific Realism and the Quantum*.
Edited by: Steven French and Juha Saatsi, Oxford University Press (2020). © Carl Hoefer.
DOI: 10.1093/oso/9780198814979.003.0002

(curvature singularities) that are presumed to be unphysical; and even if GR is accurate for low-energy gravitational phenomena, it is assumed that at very high energies (corresponding to the Planck scale), its accuracy must break down and a quantum gravity theory will be required to correctly describe the phenomena. So GR is assumed to be *empirically* adequate for a wide range of phenomena, but not to give a correct/true description of the fundamental nature of matter or spacetime.

In this chapter I will focus on quantum theories, and argue that we have ample reasons to exclude them from the domain of scientific knowledge that we can regard as established, true knowledge. As part of explaining why this is the case, of course, I will have to say quite a bit about my proposed new version of SR: enough for us to see clearly why large portions of foundational physics must still be excluded, and to see which bits of our current quantum picture of matter can already be included. Briefly laying out my version of SR will be the task in the next section. But I will have to leave a full exposition and defense of the view to future work.

2.2 Scientific Realism, Revived

In my career as a philosopher of science I have mostly explored topics other than realism/anti-realism, principally metaphysical issues such as the nature of causation, of objective probability, or of spacetime. But I have been aware of the shifting tides of debates about SR, and often have had to think about my own stance, in the classroom or in reading groups. That stance was quite variable and unstable. I would tend to lean strongly toward anti-realism after reading certain historical works, or when I was immersed in foundations of physics discussions or looking at the plethora of approaches to quantum gravity. But I would tend to revert to sympathy with scientific realism when not focused on these sources of doubt, or when I reflected on the stunning achievements of biomedicine and engineering of the last fifty years or so. And I didn't need to make up my mind or take sides (other than for fun, during discussions). But in the course of a year-long reading group on SR, a few years back, I ended up seeing a way to isolate what seemed good and perennially attractive about realism, and separate it from the problematic aspects that make it struggle to withstand careful historical and philosophical scrutiny. As a way of approaching my current view, here are some lessons that I believe I've learned over the years:

1. *Against method.* The idea that there is a thing called 'the scientific method' which is applied in all (or most) sciences and which leads us inevitably to the truth (or closer and closer to the truth) is one that I gave up a long time ago. There are many different methods that can count as scientific, and different ones apply at different times, both within a single scientific domain and when we look across more than one domain. Success is possible—when the world happens to be amenable to our methods of thinking, exploring, and testing. But complete success may *not* be possible; that depends on how the world happens to be. So there is no way to argue, ahead of the game, that science can or will 'track truth.'

2. *Don't go back in time.* In order to evade Laudan's (1981) pessimistic induction,[1] some philosophers of science, perhaps partly under the spell of the notion of the inevitable success of the scientific method, try to find a way to convert many of Laudan's examples of radical theory overthrow into cases of merely partial theory overthrow, with preservation and accumulation over time of the stuff that the old dead theories 'got right.' I think this just doesn't work. When one finds something that (arguably) did get preserved from theory A to (very different) successor B, it often is not the same kind of thing that got preserved in the replacement of theory C with theory D. Frequently, the content allegedly preserved also (to my eyes) looks a bit thin and superficial, or very close to the purely observational. This being the case, why bother taking on this burden of tracing threads of truth back into the past? Some realists have felt they had no choice but to do so, but I will argue that this is not correct.
3. *IBE isn't your friend.* The anti-realists are just right about *inference to the best explanation* (IBE): it is a terrible inference rule. It will lead one astray if one happens to possess only a 'bad lot' of possible explanations—as was typically the case in science of the nineteenth century or earlier. Or if one has poor taste in explanations. Or if the world happens to be such that the *right* story is one that doesn't feel very explanatory to us, while also being such that a false story that feels very explanatory to us is empirically adequate. So leave IBE out, because, as we will see, you don't need it.
4. *Focus on why you can't help being a realist in daily life, after you stop reading anti-realist papers.* The core intuition behind SR is a feeling that it is absolutely *crazy* to not believe in viruses, DNA, atoms, molecules, tectonic plates, etc.; and in the correctness of at least many of the things we say about them. By focusing on why and in what ways it is indeed crazy to doubt certain parts of our *current* science, we will arrive directly at a justification for believing in them.
5. *Leave fundamental physics out of the scope of SR's claim.* As many philosophers have noted, in various ways, physics is different. It aims to describe the fundaments and the behavior of *everything* in the physical world, not just give us a good story about some important but limited domain. So physics is special in the scope of its aims. And the way physics has gone, in our world, has led to the postulation of ever-more-severely-unobservable entities and properties. These two factors lead to a special vulnerability to underdetermination, i.e., to the existence of two or more possible stories about the unobservable substructures of the world that give rise to the behaviors of material stuff that we can observe. And quantum theories seem to illustrate precisely such underdetermination. Moreover, physicists working toward a theory of quantum gravity seem to be *counting* on the possibility of coming up with an entirely different fundamental story that explains all that GR plus current quantum theories already get right. So despite their impressive successes, these theories have to be placed outside the

[1] Laudan reviews a number of episodes in the history of science and shows that they confound the realist belief that science has been bringing us closer and closer to the truth. If anything, the correct induction to make from the history of science would be to conclude that our current theories will one day be overthrown and viewed, by our successors centuries hence, as having been very seriously mistaken.

scope of what we can confidently endorse as true knowledge. (We will discuss all this in more detail in the next section.)

Putting these bits of wisdom, if such they are, together in the right way, we arrive at a form of realism that is novel to the literature, and which I think is stronger than previous variants.[2] I still don't know what to call it, but for now the name that seems best to me is *tautological scientific realism* (TSR).

Key points of TSR arise from lessons 2 and 4 above. There is a large swath of established scientific knowledge that we now possess which includes significant parts of microbiology, chemistry, electricity and electronics (understood as not fundamental), geology, natural history (the fact of evolution by natural selection and much coarse-grained knowledge of the history of living things on Earth), and so forth. It seems crazy to think that any of this lore could be *entirely mistaken*, *radically wrong* in the way that phlogiston theories and theories of the solid mechanical aether were wrong. But at what point did accepted scientific lore really start to have this indubitable character? Not, I submit, until the twentieth century, and in some cases well into the twentieth century. So TSR is a philosophical doctrine about our *current* best scientific theories (parts of them, that is), and makes no claim that a strongly realist attitude toward any parts of science going beyond the directly observational was ever warranted in the nineteenth century or earlier.

The claim of TSR is not that our current science is so very successful that we should *infer* that it is probably true. That would be IBE, and would be wrong. The claim of TSR is rather that certain parts of our current scientific lore are such that we can't conceive of any way that they could be seriously mistaken, other than by appealing to scenarios involving *radical* skepticism. That is, if we try to conceive of how we could come to think we were simply, drastically wrong in believing some chunk of the relevant lore, we find we can't do it by any of the normal routes of correction and discovery that led to radical theory changes and entity-rejections in earlier centuries.

Consider the case of phlogiston theory, rejected because as more and more distinct experiments were conducted a coherent set of properties and behaviors for phlogiston became almost impossible to configure, and because (later on) a more successful and explanatory chemical theory was developed that didn't need phlogiston and offered testable explanations of everything that originally was thought to be explained by phlogiston (and much more).[3] Compare that with current chemistry: today, the

[2] In any area as widely explored as SR, complete novelty of a philosophical view is of course impossible. Among the authors who have defended views with some significant overlap with my proposal, I should mention McMullin (1984), Miller (1987), Roush (2010), and Doppelt (2014). Here I will not discuss the similarities and differences; suffice it to say that, to my knowledge, nobody has ever put together the package of ideas that I describe in this section.

[3] Here I mean something like: chemistry by the end of the nineteenth century. Hasok Chang (2012) has argued that the rejection of phlogiston theory in favor of Lavoisierian chemistry was premature, given the latter's many problems and anomalies. Even if Chang is correct about this, it does not mean that a coherent, improved phlogistonic theory could have developed in some nearby possible world, without evolving into something that would have been substantially the same as our chemistry (but somehow preserving the word 'phlogiston' in some way or another). Chang notes that phlogiston played a role similar to that assigned, later, to (negative) electricity, and (in other contexts) had a role similar to that of chemical potential energy. This means that we can imagine how science could have developed in a way such that stuff

properties and behaviors of the atoms and molecules are understood in great detail and there are no *large* gaps in our understanding, nor important empirical anomalies yet to be explained. How can we imagine our current lore to be seriously wrong?

Well, we can't do this by imagining something parallel to what *actually* happened when phlogiston was rejected: the accumulation of new experimental data that puts great new constraints on what atoms could be like, or molecular bonds, or solutions. We can't imagine this because, basically, *the experiments have all been done already, and current theory is confirmed by them*. I am not, of course, claiming that literally every interesting experiment that could be done in chemistry has actually already been done. But the terrain of possibilities has been explored quite substantially, and in particular, all or most of the *types* of experiments that chemists can think of as being possible places to look for important anomalies or unexpected results, have already been done. Nothing remotely similar could be said in the late eighteenth century. There just are not the kind of huge gaps in our experimental knowledge or in our ability to offer theoretical stories and explanations about what goes on that there were in the eighteenth and most of the nineteenth centuries, that would allow us to imagine future experiments that prove our current theory (seriously) wrong.

So, if we want to imagine our current chemistry one day being overthrown, we are reduced to doing one of three things. (A) We can imagine that things start to behave *differently* than they have so far, in our labs, in ways that lead us to scrap our current theoretical beliefs. But this is to imagine the very laws of nature breaking down or changing significantly—a form of radical skeptical scenario familiar from debates on Hume's riddle of induction. (B) We can imagine some super-genius coming along and simply crafting a theory empirically near-equivalent to our current chemistry, from scratch, with seriously different ontology and different explanations of most phenomena. (Or perhaps we meet a much cleverer alien civilization, and they let us read *their* chemistry textbooks.) This is to imagine a 'possibility' that we have no reason to think is an actual logical possibility. Without fleshing out the story in detail, i.e., without actually producing this brilliant alt-chemistry, we have no more reason to take this as a genuine possibility than we do the daydreams of a mid-twentieth-century mathematician who imagines herself coming up with a counterexample to Fermat's last theorem. (C) We can imagine coming to believe that we are all something like Putnam's brains in a vat (Putnam, 1981), or Bostrom's computer simulations (Bostrom, 2003): beings radically deceived about the nature of the physical universe they *in fact* inhabit, whose properties can be as different as one likes to those of the fake universe we have actually been thinking was real.

In other words—and this is a key point of TSR—with respect to large swaths of current science, we simply cannot see a way to imagine things going such that we decide we have been *seriously* mistaken, in the way that we take phlogiston theory to have been seriously mistaken. All efforts we make to try such an imaginative exercise either have no claim to being taken as genuine logical possibilities, or wind up simply

with the properties and roles we ascribe to electricity (or chemical potential energy—but not both) might have been called 'phlogiston.' It does not mean we can imagine a science as empirically successful as our current chemistry in which phlogiston, *roughly as understood by Priestly and others in the late eighteenth century*, survives and plays a non-trivial role.

positing one or another form of *radical skeptical* scenario. Now, I have no problem with philosophers thinking and working on the threats of radical skepticism. But that is epistemology, not philosophy of science. As philosophers of science we are entitled (and, I would say, obliged) to set aside radical skeptical doubts. Or to put it another way: once the scientific realist forces the anti-realist into positing radical skeptical scenarios in order to keep her anti-realist doubts alive, the game is over and SR has won.[4] Hence 'Tautological' Scientific Realism: We ought to believe in the truth of those parts of contemporary science that can no longer be *reasonably* doubted—where 'reasonably' means 'by being able to conceive, in a non-trivial sense, of how they might turn out to be wrong', setting aside radical skeptical moves, as we do at almost all times, as definitely 'unreasonable.' Here I am assuming that 'One ought to believe in the truth of that for which one has strong evidence and cannot see a reasonable way to doubt' is a tautology, or something near enough to one; it is, at any rate, not something that most philosophers will reject, whether they are realist or anti-realist.

The comparison to mathematical truths/falsehoods, and invocation of the notion of *logical* necessity here, may strike the reader as implausibly strong, but I do believe it to be justified. To be clear, my claim is not that I can see, myself, with Cartesian clarity and distinctness, that the idea of a radically different alt-chemistry yielding all the observable phenomena we have come to know (without resorting to radical skeptical scenarios) involves an inherent contradiction. What I am claiming is that the incredible variety of experimental and observational evidence we have accumulated, which meshes together in complex ways, makes the existence of such an alt-chemistry quite inconceivable for us, and thus the burden of proof lies on the philosopher who wants us to take it as a live possibility: show us how things in chemistry could be radically different.[5]

Above we looked at how impossible it is to imagine our current chemistry being overthrown and replaced for the same sorts of reasons that led to phlogiston theory's rejection: the accumulation of serious anomalies in new experiments gradually making a coherent theoretical story impossible to give, plus arrival on the scene of a clearly better replacement. But in fact, the difficulties of imagining a coherent replacement scenario are far greater than we've yet seen, because of the ways in which our current scientific theories and technologies are intertwined and entangled with each other. If you replace our current chemistry, you will have to revise many other areas of science which are connected with the theoretical posits of current chemistry. Exactly which other areas are affected depends on exactly *how* one tries to imagine that our current chemistry is mistaken, and what alternative story gets proposed to replace the mistaken parts. But let's think about a radical change, to make vivid the point about knock-on revisions. Suppose we try to imagine deciding that we were wrong

[4] Kyle Stanford (Stanford, 2006, ch. 1) has also discussed how some attempts to undermine SR devolve into radical skeptical scenarios, which he also takes to be impotent qua arguments against SR.

[5] These remarks situate my view with respect to Stanford (2006): I disagree strongly with his conclusion that unconceived alternatives are a serious threat to many branches of modern science, in particular to chemistry and biology; but I agree that the threat remains serious when it comes to fundamental physics, as we will see below.

in thinking that stable substances are made up mostly of stable tiny bits deserving the name 'atoms,' whose types and basic properties are codified in the periodic table of elements. We then need to rewrite all of chemistry from the ground up, no small feat. But we also would need to rewrite much of astrophysics (what is the process that happens in stars to produce energy?), and statistical mechanics (with its theory of heat as molecular motion). We would need to revise all of microbiology as well (stories about how living cells get nutrients and oxygen, for example), and our understanding of what goes on in electron-tunneling microscopes.

In earlier centuries, the sciences were not so intimately interconnected and mutually reliant. A theory in biology could overthrow an earlier one without any impact on physics or chemistry, and vice versa, because the theories and stories told about processes were so sketchy and incomplete. Things are entirely different nowadays. For many parts of our best sciences today, major revision concerning the basic ontology and/or theoretical posits would trigger a chain reaction of necessitation of revisions in other areas of science, whose scope would surely be very large, and possibly encompassing most or all of biology, chemistry, engineering and material sciences, neuroscience, and so forth. The point is not that the wholesale scrapping and replacement of our current scientific lore would be difficult. The point is that, given the vast swathes of experimental and observational evidence we now possess, we have no reason whatsoever to think it is even logically possible, other than by resorting to radical skeptical scenarios. The web of belief that Quine describes is incomparably more complex and its parts mutually connected by multiple threads, than it was in earlier centuries.[6]

Exactly which parts of contemporary scientific lore fall inside this domain of the not-reasonably-doubtable? This is an excellent question, and one that deserves a major interdisciplinary research program (bringing together philosophers and scientists) to answer it. I can't offer a definitive list of the parts of science that have achieved indubitable status, but I can offer a set of examples that seem to me excellent candidates, and which already show TSR to be a genuine realist position, with commitments far beyond what standard anti-realisms endorse.

- The existence and many basic properties of the atoms of the periodic table;
- The ways that atoms bind into molecules, and how collections of them behave (including interactions between distinct substance types) over a wide range of temperatures—hence, much of chemistry;
- Most or all of basic cell biology and microbiology (including the basics of our theories concerning DNA and RNA and reproduction);
- The existence of viruses and their roles in causing illnesses;
- Much of the core of astrophysics, including the processes of fusion, stellar life phases, novas and supernovas (but excluding black holes and cosmology theories);

[6] Another way to think of our current situation is to note that each significant chunk of the scientific lore I am talking about is a thing on which we have multiple *epistemic handles*. In a separate publication I will develop the notion of our epistemic handles on scientific truths, a topic in which both reference and manipulation or control play important roles.

- Most of our knowledge of electricity and electronics, taking electrical phenomena in a non-fundamental sense—committed to the existence of electrons and some of their key properties, but not to QM or QFT per se.

It's important to note here that in most of the above items, I am talking about 'how' these things occur, at a relatively crude level of description, and not about the stories about 'why' and 'how, in detail,' stories which in many cases rely on principles of QM and on things such as wavefunctions. It's the level of description typically found in high school or first year university courses that is, I claim, here to stay. For example, that water is comprised mainly of molecules whose compositional structure is H_2O; or that a carbon atom can bond with up to four other atoms, 'sharing' its outermost electrons with the bonded atoms. Deep structures or principles of quantum theory are not presupposed by statements like these, even though working scientists would appeal to such structures and principles if asked to *explain* these facts.

Two features of this list deserve to be noted, and discussed in detail in the next section. First, the absence of fundamental physics theories: they are not there, because we still can imagine, rather easily, coming to decide that they are mistaken in certain fundamental ways and need to be replaced. (In fact, many philosophers and physicists are already convinced of this.) And the imagined replacement would not provoke a chain reaction of revisions throughout other areas of science, because of the second feature I want to note: many parts of science on my list depend on and are linked with physics—but not *fundamental* or *foundational* physics. Instead, they are connected to a sort of meso-level physics, a mix of facts (like the mass and charge of electrons), coarse-grained or 'effective' descriptions (of, e.g., the sizes of atoms and how they move, or of light, including when to treat it using classical E-M equations and when to treat it as a ray and when to treat it as something like a particle), useful phenomenological equations, and more. This part of contemporary physics is very hard to describe exactly, but it definitely deserves a place on the list of the now-indubitable. How this can be, without dragging our quantum theories onto the list as well, we will discuss after we examine why those theories should not be on the list.

2.3 No Realism About Quantum Theory (Yet)

The case for excluding quantum theories from the ambit of the scientific realist's truth claims can be made fairly simply, and the reader can look at other chapters in this volume for related discussions (e.g., Ruetsche on QFTs). Juha Saatsi sums up the core of the issue nicely:

> Firstly, as I will discuss further below, interpretations of QM, when spelt out in the level of detail required for their defence, become deeply metaphysical due to indispensably involving ideas about quantum reality, and its relationship to observable features of the world, that are far removed from the actual scientific use of quantum theory to predict and explain empirical phenomena. Secondly, assuming that the different interpretations are underwritten by variants of QM that are all sufficiently empirically adequate, the realist faces a radical underdetermination of the metaphysical alternatives. (Saatsi, 2019, pp. 6–7)[7]

[7] Saatsi summarizes the problems for a realist stance regarding quantum mechanics, but then goes on to defend a limited form of realism for the quantum realm. The sort of realist claim that he offers is

The fundamental obstacle to realism about the quantum arises out of the measurement problem (MP) and the various ways of tackling it, which lead to serious threats of underdetermination (UD). Let's look at the threats briefly.

Underdetermination of theory by evidence arises when two theories appear to make exactly the same predictions for all possible observational situations (for some appropriate sense of 'possible'). UD is precisely what I have argued is not possible for the various parts of current science covered by TSR, unless one resorts to radical skeptical scenarios. The interconnectedness of the various branches of science and the technologies of observation and manipulation we have created in conjunction with them make it impossible to imagine (say) a rival theory to the virus theory of illnesses coming along that makes no different empirical predictions. (Notice that an imagined rival 'theory' would need to explain why the cells of sick or dead people are full of copies of certain viruses, even though they don't actually cause the illness; or perhaps explain how microscopes only seem to show us viruses, and we only seem to be able to manipulate and grow such small things in our labs, while in reality nothing of the sort exists; or. . . .) But the fundamental quantum realm is entirely different, precisely because it is fundamental and assumed to be the underlying story of all physical events. Every event that happens is, in some sense, a test case for quantum theory, and nothing can come 'from outside' so to speak, from another science or domain of phenomena, to put constraints on quantum theory's predictions. So given one quantum theory that seems empirically adequate to all phenomena so far observed, if one can construct a substantially different rival theory that (apparently) makes all the same empirical predictions, then UD is established and we have to admit that our first theory may after all not be the right one.

This seems to be how things stand for quantum theories. Most famously, Bohmian mechanics is a genuinely different theory when compared to any interpretation of standard non-relativistic QM, and the same goes for GRW collapse theories, but—for the range of phenomena in which QM is considered valid—these rivals make no predictions that we can experimentally distinguish from the predictions of QM.

One response to this situation would be to weaken the ontological claims associated with QM and advocate a realist stance toward what all the variants of QM (here including Bohm theory and GRW theories) have in common, that is, some core of the theory that is sufficient to yield the empirical predictions. For QM this is easy enough to specify: the Schrödinger equation and the Born rule (together with the lore concerning how certain operators correspond to certain measurable quantities, and the lore about how to construct appropriate Hamiltonians). But the quantum state appears in the Schrödinger equation, so one must at least specify whether that is to be regarded as a physically real thing, and if so, what is its nature. (If one *denies* that it represents anything real, then prima facie, all one is committed to are the empirical predictions of QM: precisely what anti-realists are ready to accept.)[8]

considerably weaker than that associated with the domains of TSR, in particular by eschewing claims about which, if any, of the mathematical elements of the theory represent existing entities.

[8] Healey (this volume Chapter 7; 2015) has, however, crafted a form of realism about QM that denies the reality of the quantum state but embraces the correctness of many quantum mechanical explanations of observable phenomena, thus committing to more than most anti-realist views are ready to accept. For reasons of space I cannot discuss Healey's viewpoint in this chapter.

One must also specify the quantum state's connection to ordinary objects in 3-D space, how it evolves over time, and whether or not there exists *something else* besides the quantum state. Answering these questions in diverse ways leads straight to the well-known variants of QM, such as GRW theories, Bohmian mechanics, or the Everett interpretation. For the domain of non-relativistic QM, these variants are demonstrably empirically equivalent, at least given currently available techniques and resources for experimental testing. Even if one thought that one of these must be approximately true, one would have to be agnostic about which one, which is to say, not make the scientific realist claim about any of them.

One response to the UD of standard QM is to dismiss Bohm theory and GRW as genuine live possibilities because of their apparent inability to be extended to the realm of quantum field theories.[9] This line of response seems promising because it is true that in the view of most physicists, the non-Lorentz-invariance of QM and its inability to fully handle light (photons and the E-M field) indicate its non-fundamental status compared to fully Lorentz invariant quantum field theories. From this perspective, standard QM is a low-energy effective theory derivable, in principle, from quantum field theories (even if the derivation is never explicitly carried out). That the non-fundamental, merely effective theory has empirically equivalent rivals is then no more than a curiosity. The scenario posited here would then be one in which a non-fundamental theory does have empirically equivalent rivals. But this is a distinct sense of 'non-fundamental'; ordinary QM would still count as a theory of 'fundamental physics,' albeit one that has been relegated to a status of mere approximate truth for certain purposes and in certain domains. (The same is already true of Newton's gravity theory, and will presumably be true of GR once a better replacement theory comes along.)

This response may appear adequate—unless, of course, QFTs themselves suffer problems of underdetermination, as seems to be the case. At least two types of UD regarding quantum field theories can be discerned as things stand currently.

1. Collapse/no-collapse. For those who are willing to contemplate the possible truth of quantum theories under an Everettian interpretation, there is a simple alternative that poses an UD problem: standard quantum theory with its assumed collapses of superpositions upon measurement. It is easy to forget this threat because in the foundations of physics community, the proposal that measurements in particular (unlike other physical interactions that might seem superficially similar) provoke a sudden change of state is considered beyond the pale. But whether it is considered beyond the pale or not by the foundations community, the standard collapse interpretation is one that has been taught to four generations of physicists so far and has seemed to work without running into anything that might count as counter-evidence.
2. Particles/fields-only. The way that physicists tend to think and speak about the theories comprising 'the Standard Model of particle physics' involves apparently

[9] There do exist efforts along these lines, with some successes; see for example Pearle (2015) and Struyve (2012). I think it is fair to say that the consensus among physicists, however, is that these extensions of GRW and Bohm are unlikely to become full-blown competitors to all the existing quantum field theories.

serious discussion about particles: their creation and destruction, their intrinsic properties, their effective sizes (cross sections), their trajectories in detectors, and so forth. And one way to interpret QFTs is precisely as theories about particles; the associated quantum fields being either assigned equal reality, or demoted to the status of mathematical tools not to be thought of as existing. But this is now a minority stance, a majority of physicists and philosophers of physics arguing that, strictly speaking, only the quantum fields are truly real, particles being mere aspects of them, 'excitations' of the fields.[10] The details of this debate are too complex and technical to even begin to present here, and for all I know perhaps one side has already definitively refuted the other. But at the moment there seems, to me, to be a live interpretive question remaining here concerning the actual ontological implications of QFTs. Even if it is one day considered resolved, it stands as a warning example of how fundamental theories tend to leave open issues concerning their real ontological implications, a phenomenon we should consider likely to arise again if or when one or more successful quantum gravity theories come onto the scene.

Finally, there is one further UD problem for QFTs that must be faced, even if one feels the above two concerns can be set aside: almost all theoretical physicists already believe that there is an empirically equivalent rival out there waiting to be discovered: the long-sought Quantum Gravity (QG) theory that will bring gravity into the domain of quantum physics. It is considered already fairly clear that gravity cannot be folded into quantum theory by some canonical quantization leading to a new field theory that captures gravity in the same sort of way that QED captured the electromagnetic interactions and QCD captured the strong force. So, what many physicists expect is that all existing QFTs will turn out to be merely 'effective field theories' in the domain of lower energies, derivable (in principle) from a non-effective fundamental theory that is valid at all energy regimes. But this is UD: two different theories, namely, our existing QFT and the future QG, make empirically indistinguishable predictions for all the phenomena covered by the former. Perhaps more to the point: to believe in the existence of such an as-yet-undiscovered QG theory that will be *better* than our current QFTs is to believe, already, that we have good reason to regard those QFTs as predictively accurate but, strictly speaking, false.

Now, whether this means that we will at some point consider our current QFTs to have been *not even approximately true*—the issue that matters for scientific realism—is not something that we are in a position to judge yet. To my mind, much depends on how/whether the future theory manages to solve the quantum measurement problem, and on how much the ontology of the QG theory resembles or fails to resemble the ontology of our QFTs (to the limited extent that the latter is even well-defined). But in any case, we do see here reason to think that an underdetermination problem *may* exist for our current QFTs, one whose gravity is not yet clear. Agnosticism is recommended.

[10] See, e.g., Fraser (2008), Baker (2009), and Ruetsche (2011).

2.4 The Parts We Know Are True

When I very briefly sketched, above, the sorts of knowledge that are part of our current sciences and should be held to be true, I mentioned some things that surely count as bits of physics: that atoms and electrons and protons and neutrons exist, for example. I would add that we know the charges and rest masses of these types of particle, and that they are constant over time. We know that electric phenomena involve flow of electrons from one place to another. We know that atoms are made up of combinations of electrons, protons, and neutrons, with the latter two types of particles being fairly well localized in a small core compared to the electrons, which in some sense not yet fully understood are to be found outside the nucleus. We know that certain combinations of electrons, protons, and neutrons compose stable, electrically neutral atoms, while other combinations display instability over time (namely, the radioactive elements and isotopes). We also know that light sometimes behaves in a particle-like way, and we know some ways of generating such particles (photons) and detecting them.

The above statements are mainly ontological: they express that we know that things of type *x* exist, and say a bit about the *x*'s. It is part of a quasi-classical picture of matter, the sort of thing one is taught in high school physics and first-year university courses, and there is no way we can really imagine this lore being rejected as false in some future time.[11] But it is very limited; what about all the laws and equations that accompany this ontological picture? For example, what about Maxwell's equations and the Lorentz force law, which apparently tell us so much about what large collections of electrons and protons do, how they affect each other's movements? Well, these equations are *field* equations, and the very existence of E-M fields, and their nature if they do exist, is still a matter of uncertainty due to the uncertainty of fundamental quantum physics. So a claim of literal truth (or even 'approximate truth') for such equations is not yet warranted. That said, such equations can be correctly embedded in an instrumentalist claim: 'Maxwell's equations and the Lorentz force law are reliable tools for making predictions concerning most observable electromagnetic phenomena.' This is just a way of stating that classical E-M theory is empirically adequate for a wide range of its intended phenomena, and is readily conceded by anti-realists, so although we can certainly add it to the stock of certain knowledge contained in our current science, it does not make the claims of TSR any stronger qua form of scientific realism.

Still, it is good that we can say that we know that electrons, protons, and neutrons exist, and some important facts about their natures and behaviors. Perhaps the class of particles for which realist truth claims are now secure includes further particles: positrons, mesons, perhaps even neutrinos; this is not the place to try to draw the boundary in detail. Instead, what we need to consider is whether it is really plausible to think that one can quarantine the quantum quagmire while preserving the pristine certitude of intro-level physics, basic nuclear physics, electronics, chemistry, and so

[11] In a separate paper with Genoveva Martí, in preparation, we will explain why we can be sure that future twists and turns of fundamental physical theory will not lead us to abandon these simple ontological claims of intro-level physics. The story has to do, naturally, with facts about how reference to natural kinds works.

forth—areas of science that seem to rely, *in some sense*, on the correctness of quantum theories.[12]

It seems clear to me that this can be done, by the simple expedient we saw just above in the case of E-M theory, which, like QM and QFTs, is used in many ways in the calculations and predictions of higher-level sciences such as chemistry or electronics. That is: by taking the fundamental physical theory as a predictively reliable and accurate tool for helping to calculate how certain things—e.g., electrons in a semiconductor—in whose existence we *do* believe, will behave. If the theory in question is ordinary QM, this means relying on the standard interpretation, 'with collapses' wherever the need to avoid too-large superpositions arises. But neither the ontology of that theory (wave functions, spinors, etc.), nor its equations, nor the collapse of superposition states need be taken as existing/true in a literal sense. They are just parts of a great big, extremely useful recipe for predicting the behaviors of things we *must* believe in: electrons, atoms, currents, ionic solutions, viruses, cell walls, and so on.

The entrenchment of QM in so many fields and applications of science may make some philosophers suspect that instrumentalism about QM can't really be a defensible stance once one starts looking at certain cases in detail. For example, at the September 2017 conference 'Realism and the Quantum' at Leeds University, I was pressed concerning how I could handle the successes of spintronics, the relatively new branch of electronics dealing with quantum spin-related effects that can be created in solid state devices. One might think that the way to go would be to add *spin*, i.e., intrinsic angular momentum, to the properties we can safely take electrons to possess, such as their mass and charge. But this probably will not work given the fact that all spin states in QM correspond to superpositions, and given the impossibility of giving a pure ignorance interpretation of such superposition states. (Mass and charge, by contrast, are stable intrinsic properties not subject to getting into superpositions of distinct values.) Instead, we can simply take QM as predictively useful when doing our electronics, including spintronics. Since QM tells us that electrons have this I-know-not-what sort of property with such-and-such upshots, we take this as *predictively reliable* information for calculating the kinds of behaviors electrons will display in our solid state devices—and nothing more. By being instrumentalists about quantum theories and other fundamental physics theories, we get the best of both realism and anti-realism: realism about the parts of science which are such that we really have no idea how they could be false, and agnosticism concerning the part of science which is such that we can all too easily imagine sweeping changes occurring in the future.

At this point the traditional scientific realist may be feeling agitated. How can one admit that quantum theories have this immense predictive utility, making truly surprising and novel predictions all over the place, and not infer that they are really onto something, really giving us at least *part* of the truth about the fundamental level? At the level of logic, the answer is simple: recall that IBE is a bad inference rule, period, and that predictive and even explanatory success is no guarantee of even approximate truth. At the level of psychology, it can be helpful to re-read Laudan (1981).

[12] See Craig Callender's 'Can We Quarantine the Quantum Blight?' in this volume, Chapter 4, for a somewhat contrary perspective.

One should also avoid the temptation of falling back on something structural being true in quantum theories, as is urged by structural realists. It may be that some identifiable structural feature of quantum theories will be present in the hoped-for future quantum gravity/grand unified theory, in a sense that does not just amount to empirical equivalence over a broad range of phenomena. But equally, it might not go that way, for all we can now see. The only sort of structure we can be sure will be carried over into future replacement theories is the structure of empirical predictions that quantum theories currently make.[13]

Since I am conscious that I have been, in this section, sounding somewhat like a scientific anti-realist, I will conclude by situating TSR in relation to the more prominent positions found in the literature of the last forty years, and let the reader decide for herself whether it deserves the label 'scientific realism.'

2.5 Comparisons

TSR surely has much in common with Hacking's 'entity realism', but at the same time I neither adopt his slogan ("If you can spray them, then they are real") nor do I focus on entities to the exclusion of theory. I demur from the slogan mainly because it seems nowadays that physicists make claims to being able to control 'sprays' of many sorts of fundamental particles, and I am not sure that a realist attitude is warranted with respect to all of these. Hacking's point, however, was that when we achieve the ability not merely to spray a putative entity, but control it sufficiently to be able to utilize it in the production and control of *other* sorts of phenomena, then realism is warranted. My stance concerning the reason for our being warranted in believing in entities such as atoms, electrons, perhaps even positrons and neutrinos, is close to Hacking's but not identical. Warrant for belief arises when the *epistemic handles*[14] that we have on an entity have grown to be numerous and diverse enough that it is no longer feasible to imagine a scenario where we decide that we were simply mistaken in thinking that such an entity exists. Being able (apparently) to skillfully manipulate an entity in the service of other ends is certainly one way of having epistemic handles on the entity, but not by itself sufficient to guarantee existence. Late nineteenth-century radio builders surely thought that they were creating and manipulating vibrations in the aether in order to produce a variety of further effects; but they were not.

Entity realism is also sometimes associated with Nancy Cartwright, who at one point endorsed a principle of *inference to the most probable cause*. I consider any such

[13] A proper discussion of the pros and cons of structural realism is beyond this chapter's scope, so I will simply quote (and endorse) Saatsi's (2019) response:

> A realist's account of what makes Newtonian gravity empirically successful can ultimately have little in common with her account of what makes Fresnel's ether theory empirically successful. In particular, I (for one) do not see any useful abstract characterisation of structure that furnishes a unified explanatory sense in which Newtonian gravity and Fresnel's ether theories can both be regarded as 'getting the structure right'. (p. 11)

[14] Epistemic handles include things like: direct or indirect observations; theoretical reasons for believing that an entity of such a type should exist; knowing how to produce and manipulate the entity; taking the entity to be the cause of a certain observable event or phenomenon (where this is not a case of observation), etc.

principle to be just a species of IBE, however, and to have all the weaknesses of its genus. Neither unobservable entities nor theoretical statements deserve belief simply because they do one explanatory job for you; the bar must be much higher.

It is already clear from all the above, I suppose, that TSR is quite different from Psillos' selective realism (Psillos, 1999), or Chakravartty's semi-realism (Chakravartty, 2007), or Worrall's epistemic structural realism (Worrall, 1989). Each of these views tries to limn what deserves to be believed and what should still be doubted using a criterion meant to preserve things from the nineteenth century onward, in some sciences at least, and not excluding fundamental physics. And each endorses IBE in some form, rather than looking for what simply can no longer be reasonably doubted.

What the reader may not see clearly, at the moment, is how TSR differs from van Fraassen's Constructive Empiricism (CE) (van Fraassen, 1980). Both decline to engage in IBE, which some take to be the hallmark, if not the *sine qua non*, of scientific realism. And when it comes to our fundamental physics theories, which is where SR debates often focus, there simply is no difference at the moment. CE and TSR recommend withholding belief, and at most granting empirical adequacy and usefulness.

But that is where the similarity ends. CE tries to offer a clear dividing line between what is to be believed and what is to be doubted in terms of what is *observable*, and (in van Fraassen's 1980 presentation, at least), gives a fairly restrictive explication of what counts as an observation. This dividing line will make its way across most of the sciences dealing with the very small (or the very ethereal, e.g., cosmology with its dark energy), cordoning off large portions of microbiology, chemistry, physics, material science, and electronics as still-to-be-doubted. TSR by contrast enjoins us only to doubt that which is *genuinely dubitable*, and my claim is that this puts very much more of the just-named sciences into the realm of that which we should take as true than CE does. Moreover, TSR is compatible with a future development of science in which our fundamental physics finally does deserve to be accepted as true. The fact that physics will, it seems, always have to deal with unobservable entities, does not entail that we can never have sufficient warrant for believing in the truth of a theory (or theories) of the fundamental level. We just need to get to the point where we have a theory that has no internal contradictions, tells a clear story about what exists and how it behaves, we have multiple epistemic handles on its entities and on the truth of its main claims, and have strong reason to think that no UD scenario can arise without the invocation of radical skeptical maneuvers. A tall order, but some of us still have faith that it is possible.

These differences, I submit, adequately distinguish TSR from CE or ordinary instrumentalism about science, and show that the 'realism' label is appropriate.

References

Baker, D. J. (2009). Against field interpretations of quantum field theory. *The British Journal for the Philosophy of Science* 60(3), 585–609.

Bostrom, N. (2003). Are we living in a computer simulation? *The Philosophical Quarterly* 53(211), 243–55.

Chakravartty, A. (2007). *A Metaphysics for Scientific Realism: Knowing the Unobservable.* Cambridge: Cambridge University Press.

Chang, H. (2012). *Is Water H2O? Evidence, Realism and Pluralism.* Dordrecht: Springer.

Doppelt, G. (2014). Best theory scientific realism. *European Journal for Philosophy of Science* 4(2), 271–91.

Fraser, D. (2008). The fate of 'particles' in quantum field theories with interactions. *Studies in History and Philosophy of Science Part B: Studies in History and Philosophy of Modern Physics* 39(4), 841–59.

Healey, R. (2015). How quantum theory helps us explain. *British Journal for the Philosophy of Science* 66(1), 1–43.

Laudan, L. (1981). A confutation of convergent realism. *Philosophy of Science* 48(1), 19–49.

McMullin, E. (1984). A case for scientific realism. In J. Leplin (Ed.), *Scientific Realism*, pp. 8–40. Los Angeles: University of California Press.

Miller, R. W. (1987). *Fact and Method: Explanation, Confirmation and Reality in the Natural and the Social Sciences.* Princeton: Princeton University Press.

Pearle, P. (2015). Relativistic dynamical collapse model. *Physical Review D* 91(10), 105012.

Psillos, S. (1999). *Scientific Realism: How Science Tracks Truth.* London: Routledge.

Putnam, H. (1981). *Reason, Truth and History.* Cambridge: Cambridge University Press.

Roush, S. (2010). Optimism about the pessimistic induction. In P. D. Magnus and J. Busch (Eds.), *New Waves in Philosophy of Science.* Basingstoke: Palgrave Macmillan.

Ruetsche, L. (2011). *Interpreting Quantum Theories.* Oxford: Oxford University Press.

Saatsi, J. (2019). Scientific realism meets metaphysics of quantum mechanics. In A. Cordero (Ed.), *Philosophers Look at Quantum Mechanics*, pp. 141–162. Dordrecht: Springer.

Stanford, K. P. (2006). *Exceeding Our Grasp: Science, History, and the Problem of Unconceived Alternatives.* Oxford: Oxford University Press.

Struyve, W. (2012). On the zig-zag pilot-wave approach for fermions. *Journal of Physics A: Mathematical and Theoretical* 45(19), 195307.

Van Fraassen, B. C. (1980). *The Scientific Image.* Oxford: Oxford University Press.

Worrall, J. (1989). Structural realism: the best of both worlds? *Dialectica* 43(1–2), 99–124.

3

Truth vs. Progress Realism about Spin

Juha Saatsi

3.1 Introduction

Spin, a theoretical concept at the heart of quantum theory, is scientifically as firmly established as any. Spin underlies numerous explanations and predictions in physics and chemistry, as well as a rapidly growing number of technological feats. Its central and multifaceted theoretical role strongly motivates a scientific realist attitude: we should be 'realists about spin'—as much as we are realists about anything in physics. But what does this realism amount to? I will answer this question by distinguishing in very general terms two conceptions of scientific realism, problematizing one of them, and articulating and defending the other. One of the conceptions I call *truth-content realism*. It incorporates both 'standard' realism defended by, e.g., Hilary Putnam, Richard Boyd, Stathis Psillos, as well as epistemic structural realism, construed as a principled qualification of theories' truth content by reference to theories' or the world's structure. The other conception of realism—the one I favour—focuses its commitments on theoretical progress in a way that renounces typical realist claims of (approximate) truth and knowledge about the unobservable world. I will argue that this more minimal *progress realism* offers a defensible positive epistemic attitude towards a theory such as quantum mechanics, while truth-content realism problematically involves 'deep' metaphysics not supported by the overall empirical evidence.

Traditionally much of the epistemological debate surrounding realism (e.g. relating to the historical record of science and underdetermination of theories by evidence) has been conducted in highly general and abstract terms (as has been customary in general philosophy of science), concerning all of 'mature' science in one fell swoop.[1] Recent years have seen growing interest in anchoring the debate much more firmly to local, case-specific details of particular theories and their evidential support.

[1] These epistemological issues should be distinguished from the metaphysical question of what is real according to a given theory (taken as approximately true). The latter question can only be answered in the context of specific theories, of course, and quantum theory has been an area of intense research in this regard. See Richard Healey (Chapter 7 in this volume) for further distinctions between various realism issues.

Juha Saatsi, *Truth vs. Progress Realism about Spin* In: *Scientific Realism and the Quantum*.
Edited by: Steven French and Juha Saatsi, Oxford University Press (2020).
DOI: 10.1093/oso/9780198814979.003.0003

This support can arguably substantially vary, upon philosophical analysis, from one theory or area of science to another, as it is prima facie plausible that the balance between the evidence for and against realism gets settled differently for different areas of science. Given the vast differences in theories' explanatory presuppositions and subject matters, there is no reason to think that theory-mediated epistemic access to them is equally (un)problematic across different areas of successful, well-established science as diverse as, e.g., cosmology, genetics, geology, palaeontology, and quantum physics.[2] It thus makes sense to have a good look at the details of specific areas of science to see how the case for or against realism plays out in different scientific domains. Quantum physics is particularly interesting in this regard, as it exhibits an increasingly broadly recognized and scientifically serious case of underdetermination by evidence that seems to breathe new life to a venerable challenge to scientific realism. This underdetermination has been discussed from different angles by various authors, including many represented in this volume. In this chapter I aim to highlight some subtleties of this challenge, arguing that its implications are best appreciated in the context of the broader dialectic of the realism debate and through more general reflections on what scientific realism amounts to.

It will pay off to further zoom in on spin to bring into focus the nature of the problematic underdetermination. For standard realism 'spin' is a central theoretical term of exceedingly well-confirmed science, taken to refer to an objective feature of mind-independent world. Traditionally realist commitments to our best theories' (approximate) truth are bound up with such terms' referential status, furnishing a sense in which we can gain knowledge from these theories as representations of reality, while also expecting future revisions in these theories. Prima facie, realism about spin is extremely well motivated by the lights of usual realist arguments, given spin's various theoretical roles (cursorily reviewed in Section 3.2). Briefly put: spin is at the heart of various scientific *explanations* furnished by quantum physics, exemplifying the kind of explanatory reasoning upon which standard abductive arguments for realism capitalize. Spin is also *unifying* in ways that matter for many realists. Finally, the increasingly robust theoretical handle on spin has led to myriad technological developments over the past couple of decades, rendering spin increasingly *manipulable* in ways that many realists have regarded as critical to justifying realism. In the spirit of the realist's classic 'no miracles' argument, we can sum this up by saying that the multifaceted success of quantum physics would be an astounding miracle if spin wasn't, well, real! Regardless of whether we endorse the letter of this argument, the prima facie case for realism about spin is forceful.

So runs a homely realist line of thought. But, as is well known, spin is also at the heart of interpretational conundrums of quantum physics, so it is immediately unclear what 'realism about spin' is *about*. I will argue that underdetermination of different candidate theories of spin—correctly understood—raises a serious challenge to the received way of understanding what realism amounts to. Although this challenge arises specifically in the context of quantum physics (which is notoriously 'weird' from

[2] For illustrations, see the various contributions to Saatsi (2018, Part IV).

our classical point of view), it also threatens the realist outlook on science much more broadly by demonstrating the unreliability of standard criteria for realist commitment towards indispensable theoretical notions more generally, since these criteria apply to spin just as well as they apply to any theoretical feature of science. This is one reason why it is difficult to delimit the implications of this challenge in a principled way to quantum physics, so as to give up on realism selectively in relation to the latter, while simultaneously withholding realism in relation to other areas of theoretical science.

One such strategy for limiting realist commitments has been proposed by Carl Hoefer (Chapter 2 in this volume), who adopts a selectively instrumentalist attitude towards all current *fundamental* theories of physics (including quantum physics), while retaining a realist attitude towards science for which we "cannot conceive of any way that they could be seriously mistaken, other than by appealing to scenarios involving radical skepticism". Hoefer offers various examples of epistemologically kosher objects of realist commitment from *non-fundamental* physics, including, e.g., the key properties of atoms and molecules employed in much of chemistry and statistical mechanics, and the key properties of electrons employed in much of electronics. He argues that realism about atoms and electrons (thus construed) can be underwritten by the fact that science without them is in a strong sense *inconceivable*, due to the ways in which the relevant scientific "theories and technologies are intertwined and entangled with each other". We cannot really conceive of an alternative science to ours that does without a notion of 'atom', for example, whose types and key properties are codified in the periodic table of elements.

> We [would] need to rewrite all of chemistry from the ground up, no small feat. But we also would need to rewrite much of astrophysics (what is the process that happens in stars to produce energy?), and statistical mechanics (with its theory of heat as molecular motion). We would need to revise all of microbiology as well (stories about how living cells get nutrients and oxygen, for example), and our understanding of what goes on in electron-tunneling microscopes.
>
> (Hoefer, this volume, p. 25)

Thus, numerous higher-level special sciences are thoroughly intertwined with the physics of atoms and electrons—both in their theories and core technologies—rendering the potential elimination of these theoretical posits inconceivable. By contrast, alternatives to our currently accepted 'fundamental' theories of physics are clearly not similarly inconceivable. On the contrary, they are actively considered by scientists, different quantum theories (e.g. Bohmian mechanics and dynamical collapse theories) being an obvious case in point.

Hoefer has identified an interesting, even if somewhat knotty contrast. It does not, however, serve to bracket off realism about quantum physics, some central notions of which are sure to be found in the good company of atoms and electrons, on the side of Hoefer's realist commitments. Spin presents a prime example of this, given its thorough entrenchment in a vast array of theories and accompanying technologies, as indicated below. Science as empirically successful (explanatory, unified, and technologically powerful) as ours without 'spin' is just as inconceivable, as I will next discuss.

3.2 Spin

Spin, the quantum property that Wolfgang Pauli in 1925 famously described (for the electron) very minimally as a "two-valued quantum degree of freedom", has become commonly known as a particle's intrinsic angular momentum, due to the way in which it contributes to the particle's magnetic moment in analogy and in combination with its orbital angular momentum. After initial attempts (by Goudsmit and Uhlenbeck, and others) to construe electron spin in quite literal mechanical terms as a charged spinning object—an idea quickly jettisoned as incompatible with relativity—the physics community rather quickly learned to accept spin as a real quantum feature that might indeed be irreducible or 'intrinsic': a property for which there is no further *dynamical* story to be told.[3] Furthermore, it is generally accepted that there is no classical counterpart to spin: unlike its energy or orbital angular momentum, say, there is no property in classical physics that corresponds to this property at the classical limit (as Planck's constant tends to zero, or whatever more sophisticated analysis of the limit one prefers).[4] In spatial terms, the issue is that if we think of spin as a particle's property, then it in some sense has direction but there is no classical vectorial quantity whose components in all possible directions belong to the same discrete set.[5]

Physicists' rapid acceptance of spin as such an irreducibly quantum property was driven by the indispensability of the new quantum degree of freedom in capturing and accounting for various phenomena, notably the anomalous Zeeman effect—observed deviations from the 'normal' Zeeman effect splitting of spectral lines for some elements in a strong magnetic field—which troubled the 'old quantum theory' of Bohr and Sommerfeld. Since these early scientific success stories the physics of spin has only gone from strength to strength, with hugely important theoretical implications in numerous subfields of physics, from elementary particle physics to condensed matter physics, optics, and physics of atoms and molecules (see Raimond and Rivasseau, 2009, for a selective review). In physics without spin, a huge amount of generally accepted theory would have to be rewritten from the ground up.

Spin also matters a great deal to chemistry and molecular biology, both disciplines of which now share a substantial interface with quantum physics. For example, spin is pivotal to understanding a broad range of phenomena involving electron transfer (from one chemical species to another) and spin exchange (changing of the orientation of oppositely polarized spins), both ubiquitous in important chemical and biological processes involving isolated molecules, ions and excess electrons in solution, electrochemical systems, and so on (see Likhtenshtein, 2016, for a review). Modern biological understanding of phenomena such as photosynthesis thus acknowledges the critical role of weak spin-related interactions in the relevant chemical and bio-

[3] In comparison, the spin of a baseball or other macroscopic body can be dynamically reduced to the orbital angular momentum of its constituent particles in a mechanical way. Whether or not *some* kind of dynamical interpretation for electron spin is possible is an open question. See, e.g. Ohanian (1986), Giulini (2008), and Sebens (2019).

[4] See, however, Keppeler (2003) and Giulini (2008).

[5] The quantum state of a spin-$\frac{1}{2}$ particle (a fermion) is represented by a two-valued spinor wavefunction, and the operators corresponding to orthogonal spin components are given by the Pauli spin matrices.

logical processes. At a more basic level, the periodicity of chemical elements and the notion of molecular bond—the two cornerstones of chemistry—profoundly rely on the Pauli exclusion principle, which 'forbids' electrons (as fermions with half-integer spin) from occupying the same quantum state, accounting in large part for the electron shell structure of the elements.[6]

Hoefer's observation that it is impossible to conceive of chemistry or molecular biology without atoms and molecules thus also applies to spin: many achievements of these disciplines, sans spin, are equally hard to conceive of. In terms of its integration to the special sciences, spin is not relevantly different from atoms or molecules.

Spin matters greatly to vast areas of chemistry not only as a theoretical notion, but through the technologies it has led to. The spin degree of freedom of both electrons and nuclei is critical in many areas of spectroscopy, which study atoms' or molecules' interaction with radiative energy. Since its inception spin has become absolutely indispensable to the quantum physical representation of possible particle states, their energy levels, and the 'selection rules' that govern possible state transitions. These involve, for example, details of the spin-orbit coupling—typically understood as the interaction between the magnetic moments due to spin and orbital angular momentum—as well as possible transitions between nuclear and electron spin states. The two-valued character of electron spin and its coupling with the orbital angular momentum famously explains the splitting of the main lines of atoms' emission spectra into two or more components—its fine structure—and the spin of the nucleus and its interaction with the magnetic field produced by the orbiting electron explains the further hyperfine structure.

Many modern spectroscopic techniques have important applications outside physics laboratories. Nuclear magnetic resonance (NMR) spectroscopy presents an important application that is vital, e.g., to modern organic chemistry in offering a pre-eminent technique for identifying organic compounds and determining their structure and functional groups. Molecules' electronic structure and individual functional groups (viz. groups of atoms or bonds responsible for the molecules' behaviour in chemical reactions) are determined by probing the intramolecular magnetic field, which is unique (or at least highly characteristic) to individual compounds. What theoretically underlies such probing is a quantum mechanical understanding of the magnetic moment of nuclear spin, and the way it precesses in an external magnetic field. The rate of precession of the bulk-magnetization due to the spins of particular kind of nuclei (e.g. ^{1}H or ^{13}C) in a given magnetic field is directly proportional to the nuclei's gyromagnetic ratio, which is a fundamental nuclear constant. A given type of nuclei in a sample can absorb energy from an electromagnetic field of appropriate (radio-length) frequency, which is dependent on the type of nuclei in question, as well as their chemical environment.[7]

[6] At the theoretical level the exclusion principle can be connected to spin via the spin-statistics theorem, according to which integral spin particles obey Bose–Einstein statistics, while half-integral spin particles obey Fermi–Dirac statistics and the exclusion Principle. First derived by Pauli, this theorem has subsequently been proven from numerous different starting points, typically involving non-elementary presuppositions in the context of quantum field theory. See Duck and Sudarshan (1997) for a classic review.

[7] This dependence can be determined on the basis of quantum mechanical understanding of nuclear spin and intra-molecular spin-spin interactions, and it offers the key to interpreting in chemical terms

Given the intimate connection between spin and particles' quantized magnetic moment, it is unsurprising that the notion of spin is central to understanding the magnetic properties of matter. Electrons' intrinsic angular momentum, in particular, is the key to ferromagnetism (of, e.g. iron fridge magnets), which is a purely kinematic consequence of spin and the Pauli exclusion principle (see e.g. Blundell, 2003). In general, magnetism at the macroscopic level is an example of collective phenomena studied in condensed matter physics, which provides understanding also of numerous more esoteric phenomena, such as spin waves and spin glasses, exhibited by ordered magnetic materials (Blundell, 2003). Theoretical advances regarding such phenomena bear promise of technological applications in quantum computing, for example, following the history of hugely important implications of spin and magnetism to electronics.[8] The steady growth of these technological achievements seems quite inconceivable without spin.

The increasing theoretical grasp on GMR and other spin-dependent magnetoresistance phenomena, such as tunnelling magnetoresistance, constitute the basis for *spintronics*. The development of increasingly varied spintronics devices evidences how the current theoretical handle on spin enables the control and manipulation of electric currents on the basis of electrons' spin (in addition to their electric charge). These devices utilize a 'spin valve' for controlling electrical resistance on the basis of GMR by manipulating a spin-polarized electric current by changing the direction of magnetic fields, as well as other spin-related phenomena (e.g. spin Hall effect and spin-torque transfer; see Blundell, 2003).[9] Such ability to employ particles' spin in electronic devices renders electric currents *as quantum theoretic objects* effectively manipulable in the sense of Ian Hacking's (1983) entity realism, sloganized: "if you can spray them, they are real". Hacking's realist criterion is one way to understand the distinction that Hoefer draws between (i) electrons as quantum mechanical posits, and (ii) electrons of electronics, the manipulability of the latter being based on a broadly causal grasp of electrons' charge, which is largely independent of quantum mechanics. With the advancement of spintronics we can now see spin also fitting Hacking's realist mould, despite it being a thoroughly quantum mechanical feature of the world.

the spectra of magnetic 'resonance' (viz. the Larmor frequency at which nuclear spins absorb energy). The key theoretical notions involved in all this—e.g. spin precession and nuclear magnetic resonance—ground many important technological applications, and the so-called Bloch equations provide a quantum theoretic description that unifies the physics of NMR with that of electron spin resonance spectroscopy, as well as Magnetic Resonance Imaging. From its Nobel Prize-winning inception (Felix Bloch and Edward Purcell in 1952) the NMR theory of nuclear spins has grown to yield an incredibly powerful technology for investigating all kinds of matter, ranging from brains, bones, cells, ceramics, liquid crystals, laser-polarized gases, proteins, superconductors, zeolites, blood, polymers, colloids, and so on (Keeler, 2005).

[8] 'Giant' magnetoresistance (GMR) is one momentous example, underlying modern computer hard drives, the magnetic field sensors of which are one device amongst many based on this phenomenon, accounted for in terms of electrons' spin-dependent scattering from magnetized layers (see, e.g., Blügel et al., 2009).

[9] This is even more striking in relation to recent successes to control and manipulate coherent single spin states of 'quantum dots', nanometer scale semiconductor quasi-particles, which can be suitably localized and isolated from environmental disturbances to ensure sufficient coherence timescales (Hanson and Awschalom, 2008).

At a more theoretical level, spin is central to high-energy elementary particle physics, and efforts to understand the nature of spin itself involve deep connections between quantum theory and special relativity. The wave equation for fermions in relativistic quantum field theory—the Dirac equation—automatically incorporates spin as a kinematic property of a spinor-valued field which transforms as an irreducible representation of the Lorentz group. A group-theoretic analysis of relativistic quantum theory yields the highest level of theoretical analysis of the origin of spin in terms of symmetries that unify it with other kinematic properties of quantum states.[10]

I have surveyed aspects of spin-related physics in order to evidence the claim made above: spin presents a prime example of a notion *so thoroughly in entrenched in different theories and technologies that we cannot really conceive of science as successful as ours without it*. The amount of research that has consolidated the 'science of spin' is colossal, and we have only selectively scratched its surface.[11] In relation to key realist criteria, spin surely ticks all the boxes, by virtue of being deeply explanatory, unifying, and even effectively manipulable. Hence, we should be realists about spin, as much as we are realists about any theoretical notion.

But what does this realism amount to?

3.3 Two Conceptions of Theory Realism

We get into a better position to answer the question above by first reflecting in general terms what scientific realism about theories like those of quantum physics amounts to. To this end I will first distinguish between two conceptions of realism, truth-content realism and progress realism, before problematizing the former in relation to spin in Section 3.4.

Taken at face value, realism about spin concerns the world: what exists, or is real. (Namely, spin.) This assertion makes recourse to quantum physics, of course: spin is a theoretical notion, and its cognitive content and epistemic access to it is through the theories in which spin figures. Realism about spin thus confirms or declares a positive epistemic attitude towards what our best theories say about spin-related unobservable features of the world. Presumably our current theories should not be regarded as providing a literally true description of spin, however, as we must leave room for future theoretical advances and revisions. Since today's physics should not be taken as the final word, the realist's epistemic confidence is better captured in terms of our current theories' 'approximate truth', or by reference to some other kind of representational fidelity that falls short of truth *simpliciter*. Realism about spin (in this way of thinking about it) is hence an assertion about the world that is made in terms of thus circumscribed confidence regarding our current best theories.

[10] See, e.g. Zee (2016, 256), whose "mathematical pragmatist" responds to the lack of "physical understanding" of spin (cf. Morrison, 2007) by saying that the electron is simply "a particle whose quantum state transforms like the 2-dimensional representation of the covering group of rotations carries spin 1, period". Ontological structural realism capitalizes on this kind of understanding of spin (French, 2014).

[11] At the time of writing this the physics preprint archive arXiv.org contains nearly 38,000 articles with 'spin' in their title (see https://arxiv.org/).

Notoriously, the notion of 'approximate truth' is as critical as it is difficult to spell out (particularly at the level of generality at which realist claims are typically expressed). For many realists a commitment to the referential status of central theoretical terms (e.g. 'electron', 'atom', or 'spin') has been a key to maintaining a sufficiently robust epistemic commitment towards the respective worldly features in the face of uncertainty regarding future theoretical developments, or (in other words) regarding the exact sense in which the current theoretical descriptions of those features are 'approximately true'. The future theories, whatever they exactly say about spin, for example, will at least be concerned with those very same entities (e.g. 'electron') and their properties (e.g. 'spin') to which our theories now refer. On this basis the realist can purportedly maintain that we have scientific knowledge about what entities exist and what these entities are like at the unobservable level: e.g. electrons exist and they have (inter alia) spin, which accounts for various observable phenomena.

So far, so familiar, at least to those acquainted with the classic writings of Hilary Putnam (1982), Richard Boyd (1984), Stathis Psillos (1999), and others. TRUTH-CONTENT REALISM is an appropriate label for variations of this traditional realist theme, the development of which is motivated by an ancient quest for knowledge about the reality behind the veil of directly observable phenomena, whose explanatory understanding science enables by reference to their unobservable causes. There is, of course, much more to be said about the two key notions at play here—reference and approximate truth—of which realists have developed various detailed accounts. Some have offered general formal accounts of approximate truth or 'verisimilitude' (e.g. Niiniluoto, 1999), while others have developed these notions in more informal and case-dependent terms (e.g. Psillos, 1999). The relationship between approximate truth and reference can also be understood in different ways: some saddle truth-content realism with a substantial causal reading of reference, for instance, while others associate it with much more minimal commitments (see e.g. McLeish, 2006).

Yet others regard reference altogether unnecessary for characterizing realist commitments. In particular, one natural reading of structural realism, according to which theories only provide us knowledge of structural features of reality, is that our current theories' truth content is limited to their correctly representing the world's 'structure' (suitably construed).[12] According to many structural realists, such now-rejected theories as Newtonian gravity or Fresnel's ether theory of light can contain appropriate truth content regardless of the referential status of 'gravitational force' or 'ether' (Frigg and Votsis, 2011).[13] Structural realism—at least in this epistemic reading—thus continues the quest of defending scientific knowledge about the unobservable reality; its novelty is in incorporating a principled limitation as to what can be known: only 'the structure of reality' is knowable. Structural realism (thus construed) is hence another variant of truth-content realism.

[12] This way of circumscribing theories' truth content may seem like a radical departure from the standard realist's idea that theories are 'approximately true'. But it actually only represents a natural continuation of the selective ('divide et impera') strategy already initiated by the traditional realists—especially Philip Kitcher (1993) and Stathis Psillos (1999)—for whom reference of central theoretical terms is a key component of realism.

[13] Some accounts of structural realism can be connected to the reference of theoretical terms, however. See Schurz (2009).

So much for truth-content realism. There is an alternative, much less ambitious way to conceive of the realist project. One can forgo many of the central tenets of truth-content realism concerning approximate truth, reference, and even scientific knowledge of the unobservable, while upholding its *most critical tenet*: the idea that the spectacular empirical success of our best scientific theories is due to their faithfulness as representations of reality. According to this realist tenet, theories of mature science, such as quantum theories, latch onto unobservable reality in ways that are responsible for their empirical successes—both predictive and explanatory—as recognized by scientists. PROGRESS REALISM is perhaps a good label for attempts to defend this realist tenet.

Progress realism has emerged in the debate revolving around the historical evidence that historicist critics of realism have presented against the realist credo that the empirical success of science would be a 'miracle' if the relevant theories weren't approximately true. Various realists have responded to this criticism by showing, case by case, how it is possible to maintain that the empirical successes of past scientific theories (concerning, e.g., luminiferous ether, phlogiston, gravitational force) can be accounted for in terms of their representational relationship to unobservable reality, as we see it *from our current vantage point*. The issue at stake in these historical analyses concerns first and foremost our current theories' representational relationship to the world: do we have reason to believe that these theories' empirical success is grounded in their representational faithfulness? According to progress realism we do, in the same sense as we can from our current vantage point regard the past theories' empirical successes being due to an appropriate representational success.

We can note, as a purely conceptual point first of all, that in defending this tenet a progress realist is not making an assertion about the world, or about what we can claim to know. Indeed, one may wish to defend this tenet (as I do) without defending claims about approximate truth or knowledge of the sort that truth-content realists hanker after. Indeed, it is possible that one is only able to argue that an appropriate representational relationship *holds*, without being able to tell exactly what that relationship *is like* (Saatsi, 2016). Progress realism maintains that the empirical successes of theoretical science are by and large due to theories latching onto reality in ways that ground those successes, and that there is genuine theoretical progress in science in how well theories' represent reality (Saatsi, 2016).

These two conceptions of realism are hence conceptually distinct, but they are rarely distinguished. This is unsurprising, since truth-content realism implies theoretical progress of the sort defended by progress realism.[14] Hence, from the point of view of truth-content realism the two conceptions are just different sides of the same realist coin. The core commitment of progress realism in and of itself is much weaker, however. In particular, it is possible that scientific theories' empirical success is due to their representational relationships to reality, and that science progresses with respect to how its theories represent reality, without us being in a position to

[14] As realists have tended to put it: as science progresses, its theories approximate the truth better and better (or, the degree of theories' verisimilitude increases), and their success is due to their approximate truth (or verisimilitude). Structural realists would capture this progress differently, e.g. in terms of theories' limiting relationships.

reliably pin down or precisify our current theories' truth-content in the absence of direct epistemic access to reality.

Whether it makes sense to capitalize on this conceptual distinction and limit one's epistemic ambitions to progress realism depends on one's perspective on the realism debate at large. In particular, one can be a lumper in relation to realist analyses of foregone scientific theories, hoping to extract from these analyses a unified story of how the empirical success of theoretical science is grounded in truths about, e.g., unobservable entities' causal powers (Chakravartty, 2007), or the world's structure in some broader sense (Ladyman and Ross 2007; French 2014). Or one can be a splitter, emphasizing instead the disunity and open-ended variety of different kinds of (possible) realist explanations of empirical success. If one is a splitter (as I am), it makes perfect sense to articulate and defend progress realism while forgoing the ambitions of traditional truth-content realism (Saatsi, 2015). If there is no projectable unified account to be given of the representational relationships responsible for theories' empirical success, we cannot possibly hope to extract from a handful of diverse case studies any kind of 'recipe' for identifying realist commitments towards our current theories with respect to their truth content. If there is no such generalizable sense of 'approximate truth' to be had, we had better give up on truth-content realism.

Returning to the topic of spin, should we be satisfied with such a progress realist stance towards quantum theories? Well, in the rest of this chapter I will argue that it is the best we can have. Our overall evidence towards a quantum physical grasp of spin—incorporating both the empirical evidence for quantum theories, as well as the available evidence about the relationship between our best theories and the world more broadly—supports progress realism, but goes no further. Hence, a naturalistic philosopher who takes on board all admissable evidence should be a progress realist, but not a truth-content realist.[15] In the next section I will discuss the challenge faced by truth-content realism, before elaborating on progress realist's commitments regarding quantum physics in Section 3.5.

3.4 Truth-Content Realism Challenged

Truth-content realism faces a subtle problem of underdetermination. The gist of the problem, in relation to spin in particular, is that the colossal empirical evidence amassed for quantum theories does not suffice to determine what we can claim to know about, e.g. the functioning of a Stern–Gerlach apparatus, the causes of spin-orbit interaction effects in atoms, or the workings of NMR- and GMR-based devices. There is, of course, an extremely well-established body of theory of these and other spin-related phenomena that we can undeniably trust in many ways; the problem is not that quantum theories are not delivering what they generally claim to deliver. Truth-content realism demands, however, that this body of theory yields *knowledge* about aspects of the unobservable world, regarding spin, spin-orbit interaction, spin-valves, and so on. In order to meet this demand a realist must provide two things: (i) a true interpretation of the relevant parts of the mathematized theory, and (ii) an empirically

[15] Spelling out how to think about 'overall evidence' here has to be left to another occasion.

well-grounded justification for this interpretation. A problem of underdetermination impedes the realist from achieving (ii), and hence we cannot know whether we have achieved (i). As a result the required kind of scientific knowledge about the world lies beyond our epistemic reach.

That is the problem in outline. Let me now elaborate on it and then highlight some often ignored subtleties due to the broader dialectic of the contemporary realism debate. It should be incontestable that some interpretation of the mathematical representation of spin (in terms of, e.g. spinor-valued wavefunction and Pauli spin matrices) in quantum theory is required for realism. Such interpretation should, furthermore, make reference to what there is in the mind-independent world—what the world is really like. Physicists often regard this as an obsolete quest for the 'meaning' of quantum theory, insisting that knowledge about, e.g., spin requires no such thing. This attitude risks deflating the content of the alleged knowledge about the unobservable world, vacillating unstably between: (i) committing to quantum states of affairs and quantum properties of entities like electrons, on the one hand, and (ii) to just using the theory as a reliable mathematical apparatus for calculating non-quantum features of the world and for 'understanding' these features in some sense that is entirely ambiguous in terms of its realist commitments, on the other. In order to secure a stable realist commitment with clear cognitive content, truth-content realism about spin is unavoidably imbued with metaphysics, in the sense of being committed to quantum theory delivering us identifiable, objective truths about unobservable features of the world.

Mind you, a professed realist interpretation of the theory need not be given in macro-physical terms that are somehow readily imaginable, familiar, or visualizable to us. As Ernan McMullin (1984, 14) puts it, "imaginability must not be made the test for ontology" when it comes to microphysics: "The realist claim is that the scientist is discovering the structures of the world; it is not required in addition that these structures be imaginable in the categories of the macroworld." (McMullin, 1984, 14) What is required, nevertheless, is that a realist interpretation of the theory yields truths about the relevant "structures of the world". This follows from the fact that knowledge—scientific or otherwise—is *factive*: if something is known, it is true. (Hence, a structural realist interpretation of Pauli spin matrices and such needs to comply with this requirement as well, in as far as it defends knowledge about spin.)

The issue is not that there are no *candidate theories* that could contain the requisite truths. (I use the term 'candidate theory' to refer to what is often called a 'realist interpretation': a coherent formulation together with an interpretation, such as Bohmian quantum mechanics, or quantum theory as the Everettians interpret it.) The issue is that we have good reasons to think that no candidate theory as such is worthy of the realist's epistemic commitment, due to the way in which such candidates involve metaphysical assumptions that go beyond what realists should deem responsible for quantum theories' explanatory and predictive successes. I have discussed this in detail elsewhere (Saatsi, 2019), introducing the notion of 'deep metaphysics' to capture those theoretical assumptions that transcend what the realist should regard as accounting for the empirical success of quantum physics, given the actual scientific practice of using quantum theory to predict, manipulate, and explain things. The involvement of such 'deep' metaphysics in all current candidate theories is manifested in the availability of several such candidates, which are all empirically adequate with respect to various quantum systems, whilst making radically different claims regarding spin, for example. In particular, Everettian quantum mechanics, pilot-wave theories, and dynamical collapse theories represent well-known alternative candidate theories of non-relativistic quantum mechanics, each offering a radically different account of the physical systems falling within their remit.

For illustration, consider an archetypal application of quantum mechanics: the Stern–Gerlach experiment. Why does an inhomogeneous magnetic field, as in Stern and Gerlach's experiment in 1922, appear to deflect neutral silver atoms (with zero orbital angular momentum), some 'up' and some 'down'? Classic textbooks account for this roughly as follows (e.g. Townsend 2000, Sakurai 1995). The silver atom has two-valued intrinsic angular momentum that is (almost) entirely due to the spin-$\frac{1}{2}$ of the 'lone' 47th electron in these atoms' 'outermost' electron shell. The quantum state of the atom evolves upon the magnetic moment's interaction with the magnetic field in a way that can be analysed as a (classical) force on the atom, deflecting its trajectory (or, rather, the support for the corresponding wavefunction) 'upwards' and 'downwards' (or into a superposition thereof), depending on the initial spin state.[16] Finally, measuring an atom's position at the end of the experiment 'collapses' the superposition exhibited by the atom's quantum state.

The more foundational discussions of quantum theory that aim to paint a realist picture of this kind of experiment invariably give up on the notion of collapse-upon-measurement as codified in the textbooks' notorious 'collapse' or 'projection' postulate. It is agreed that such a postulate cannot as such be part of a serious realist candidate theory of quantum mechanics, as it leads to the measurement problem and theoretical incoherence. Instead, a candidate theory suitable for a realist interpretation must somehow either do without such collapses, or it must change the fundamental quantum dynamics itself so as to make it empirically consistent with the (apparent) collapses of superpositions in measurement-like situations with Born-rule statistics. The latter option leads to dynamical collapse theories (such as that of Ghirardi, Rimini, and Weber, 1986, or Pearle, 1989), while the former option leads to either a hidden-variable formulation (de Broglie–Bohm), or to the Everettian many-worlds theory (which aims to interpret and make sense of standard quantum theory sans the problematic collapse postulate).

These different candidate theories incorporate radically different understanding of spin. Take dynamical collapse theories first. It is far from straightforward to spell out what silver atoms and their spin physically amount to in such theories, and there is a good deal of debate about their ontology (see, e.g. Lewis 2018; Myrvold 2018; Tumulka 2018). A dynamical collapse theory can be read as one that fundamentally just describes a wavefunction living in a very high-dimensional configuration space, so structured as to give rise to effectively three-dimensional reality, in which particle-like phenomena emerge from the fundamental wavefunction under suitable circumstances, with spin being just a feature of the (spinor-valued) wavefunction. An analysis of the Stern–Gerlach experiment looks at the magnetic field's effect on

[16] Typical 'textbook analyses' have been criticized in Hannout et al. (1998).

the wavefunction, which evolves according to a fundamentally stochastic dynamical law (instead of the deterministic Schrödinger equation), such that it is practically guaranteed to give rise to a determinate, randomly localized particle-like result upon its interaction with a macroscopic location-measurement device. One can try to interpret parts of the wavefunction before its stochastic collapse as effectively realizing the (non-fundamental) ontology of a spin-$\frac{1}{2}$ particle occupying a superposition state. Or one can introduce local beables—e.g. spatiotemporal matter density or 'flashes'—as collapse theories' further, 'primitive' ontology.[17] In the primitive ontology approaches spin is not a property of the theory's beables at all, but rather a nomological aspect of the world that just codifies facts about the beables' spatiotemporal positions and evolution.

Spelling out the workings of a Stern–Gerlach machine along these lines leads to specific realist accounts of what it *means* to attribute spin-$\frac{1}{2}$ to an electron or a silver atom, so as to explain the Stern–Gerlach experiment. These accounts diverge radically from a face-value textbook reading of quantum mechanics, according to which electrons have an *intrinsic (non-classical) property* of spin-$\frac{1}{2}$, which silver atoms also have due to the way in which electron spins quantum physically combine to yield the atom's total spin, which affords the atom the property of intrinsic magnetic moment that interacts with the magnetic field to yield the observed outcome (after a measurement 'collapse'). Dynamical collapse theories can similarly diverge from an Everettian account of the Stern–Gerlach set-up. The latter follows the face-value reading of the 'standard' quantum theory all the way up to the employment of the collapse postulate, which the Everettians jettison in favour of a deeper quantum physical account of the measurement process in terms of quantum decoherence theory (leading to the emergent branches of the multiverse; see Wallace, 2012). In particular, for the Everettians spin is not relegated to a nomological feature of the world any more than spatial location is, in contrast to the primitive ontology approaches to dynamical collapse theories.

De Broglie–Bohm theory is similarly revisionary regarding spin's status as particles' property, and the Bohmians naturally regard spin as a feature of the wavefunction in way that makes attributions of spin to particles *contextual*, and dependent on the experimental arrangement (see, e.g. Daumer et al., 1996; Norsen, 2014). Thus, there is no sense in which a 'measurement' of spin reveals some pre-existing, intrinsic property of a particle. In this respect the de Broglie–Bohm theory is similar to the primitive ontology approaches to dynamical collapse theories. The two differ radically, however, in their analysis of quantum physical probabilities. In dynamical collapse theories the outcomes of the Stern–Gerlach experiment obey the Born rule due to the fundamental dynamics of quantum reality being stochastic in a suitable way. In the Bohmian theory, by contrast, this is due to statistical aspects of deterministic dynamics and suitable initial conditions (Norsen, 2018).

This glimpse into some of the foundational analyses of quantum theory suffices to highlight the fact that different candidate theories can tell a very different story of what's going on with a Stern–Gerlach apparatus. The challenge to truth-content

[17] See, e.g. Lewis (2006) for an opinionated review of these options.

realism is that it seems forced to buy into 'deeply' metaphysical assumptions—assumptions that are epistemologically unwarranted by the realist lights—in trying to spell out what we can claim to know about, e.g. silver atoms in a Stern–Gerlach machine. On the one hand, without a specific candidate theory in mind the realist finds it difficult to specify substantial truths that ground her knowledge claims and support a realist explanation of the empirical success of quantum theory. On the other hand, it is hard to see how the realist can adopt any particular candidate theory on empirical grounds, given how each of them seems to be wedded to ('deep') metaphysical assumptions exactly of the sort that realists are at pains to avoid, having taken to heart the historical lessons regarding such assumptions' untrustworthiness as part of empirical science (see Saatsi 2019).

A natural reaction to this challenge is to look for a common theoretical ground at a more abstract, structural level. What substantial shared structural commitments can be found between the different candidate theories, however? We should not be content to just talk about abstract mathematical features of the quantum state attribution, since truth-content realism is concerned with the ontology *represented* by the mathematics. But without an ontological specification of what the relevant quantum states represent, it is hard to see how truth-content realism can provide a serious account of quantum theory's empirical success or the claimed knowledge of the unobservable reality. Realism about the structure of quantum reality in the spirit of truth-content realism requires that 'structure' can be explicated in a way that is compatible with all the candidate theories. It seems unlikely that this can be done by reference to these theories' dynamics or state-spaces, for example, as Ruetsche (2018, 300) notes:

> [I]t is not at all clear [these candidate theories] have in common any structure *of interest for realism*. Contender interpretations attribute QM different *types* of state spaces (for Everett, it's a Hilbert space; for Bohm, a space of particle configurations) and different *types* of dynamics (deterministic Schrödinger evolution, stochastic collapse, deterministic guidance equation).

A more metaphysical, modal characterization of 'structure' can be developed (French, 2014), but this risks opting for just another line of 'deep' metaphysics, beyond the bounds of the realist's epistemological humility (cf. Saatsi, 2019).

The problem of underdetermination thus is that none of the candidate theories seems worthy of the realist's epistemic commitment, given their involvement of metaphysical assumptions that go beyond what realists should deem responsible for quantum theories' explanatory and predictive successes. Calling these assumptions 'deeply' metaphysical is not derogatory, but just highlights the fact that practising physicists who successfully deal with, e.g. spin and magnetic fields, by and large do not (and seemingly need not) care about these candidates for making predictions, building instruments, or even explaining various phenomena. The theoretical details that seem unavoidable for spelling out the commitments of truth-content realism only play a role at the foundational and interpretational level of theorizing, which so far has not led to any empirical successes of the sort that realists (by their own, demanding lights) should regard as eliciting a realist commitment. In as far as realism is driven by a desire to account for the established empirical and explanatory successes of science, the realist should focus her commitments on those theoretical aspects of quantum physics that can be regarded as responsible for those successes.[18] And these aspects are strikingly independent from the foundational-cum-interpretational research on quantum theory, as Healey (2017; this volume, Chapter 7), for example, has forcefully emphasized.

There are a couple of subtleties about this underdetermination challenge that are worth emphasizing. First, note that I have framed it in terms of the metaphysical nature of the existing candidate theories of quantum physics (the problem being that their metaphysical 'depth' transcends the scientific realists' commitments). The alternative theories of non-relativistic quantum mechanics *manifest* the underdetermination, but the challenge to realism does not depend on the (historically contingent) fact that such alternatives have actually been developed. Even if we only had the de Broglie–Bohm theory on the table, say—not having conceived of the dynamical collapse or Everettian alternatives—a realist should want to be able to recognize upon rational reflection this theory's metaphysical 'depth' in relation to the theoretical commitments that are actually required for achieving the empirical successes of quantum mechanics.[19] The challenge rather depends on there being evidence of such metaphysical 'depth' in the current candidate theories, such that no substantial truth-realist commitments towards spin remain after bracketing the 'deeply' metaphysical assumptions. I have argued that we have such evidence in the relative independence of the empirical and explanatory successes of actual quantum physics from the theoretical assumptions that specify each candidate theory's ontological content.

Secondly, note that I have not claimed that the current candidate theories are evidentially on a par with respect to their: (i) prima facie metaphysical plausibility; (ii) metaphysical plausibility in relation to non-quantum physics; (iii) potential involvement of ad hoc assumptions; or (iv) prospects for providing a unified interpretation of all empirically successful quantum physics.[20] Such a further claim would be a red herring regarding the challenge at stake. The challenge is *not* that we have developed alternative theories each of which the realist would happily regard as delivering truths about unobservable aspects of reality (according to their preferred realist 'recipe'), were it not for the availability of an evidentially comparable competing

[18] One should also appreciate the considerable degree of humility of contemporary realist commitments in the broader dialectic of the realism debate (cf. Saatsi, 2019).

[19] Assume, for the sake of the argument, that in the scenario envisaged here most physicists are Bohmians about quantum mechanics, and that relativistic and field-theoretic extensions of Bohmian mechanics are the hottest area of research around. Such broad allegiance to Bohmianism should not convince the realist, since the commitments of truth-content realism are not read off from scientists' beliefs. This is comparable to how an enlightened realist should want to be able to recognize the undue metaphysical depth of the ether-laden construals of Maxwellian electrodynamics in the nineteenth and early twentieth centuries, given the role played by 'ether' in the actual physical theorizing by, e.g. the Cambridge Maxwellians around the time (cf. Gooday and Mitchell 2013).

[20] Expert opinions widely differ regarding (i) and (ii). Regarding (iii), the dynamical collapse theory risks being somewhat ad hoc. And regarding (iv), arguably Everettian quantum physics gets the upper hand here, for reasons laid out by Wallace (this volume, Chapter 6). Regarding spin, in particular, it is admittedly *not* the case that for any (or even for most) spin-related phenomena there are different candidate theories on the table, since many spin-related phenomena require relativistic treatment for which no adequate extension of, e.g. Bohmian mechanics, has been developed.

theory for which the realist's 'recipe' delivers different commitments. The challenge rather concerns the fact that each candidate theory in and of itself indispensably involves 'deep' metaphysical commitments that are epistemically unjustifiable by the realist's lights.

3.5 Progress Realism about Spin

We can avoid getting sucked into "deep" metaphysics by defending only progress realism about spin, while giving up the epistemic ambitions of truth-content realism. Progress realism about spin defends the claim that physics' astonishing empirical success with respect to various spin-related phenomena is due to its theories' and models' representational relationships to reality, and that this area of science has progressed and continues to progress with respect to how well its theories represent the unobservable reality. Progress realism maintains that we are warranted in believing this claim on the basis of the empirical evidence enjoyed by the relevant theories and models, and what we know about physics and its history at large.

Progress realism does not defend knowledge about what spin is like or what we can claim to know about the properties of electrons or some kind of worldly wavefunction (as in the currently fashionable 'wavefunction realism'; see Chen, 2019 for review). What it defends is the idea that the models and theories of quantum physics stand in a robust enough representational relationship to reality to ground its empirical successes: new predictions, increasing explanatory understanding, and manipulations of quantum systems. It is an attitude towards the success-yielding theoretical practices of quantum physics in its own terms: since these do not require foundational-cum-interpretational assumptions about what the quantum wavefunction or Pauli spin matrices represent, such 'deeply' metaphysical assumptions should not need to be part of the scientific realist account of the empirical successes of quantum physics either. Such assumptions belong to the *metaphysical* foundations of quantum theory, which is an extremely well-motivated and important endeavour, but one that the realist need not engage with in articulating her *epistemic* commitments.

Thus expressed, such commitments may immediately appear much too insubstantial and unsatisfying. And wholly unsatisfying they are, of course, *if* one is simply unwilling to give up on the idea that realism is a matter of defending knowledge of a certain sort and identifying theoretical truth-content underwriting that knowledge. As to the charge of lacking substance, what matters is how progress realism can be clearly differentiated from anti-realist or non-realist stances towards quantum physics. Progress realism aims at offering, along with physics itself, an account of the empirical successes of quantum physics, in terms of how our theories are latching better and better onto reality. Paradigmatic anti-realist positions see no role for such an account. For instance, van Fraassen's (1980) constructive empiricism maintains that a simple Darwinian story suffices, of science "red in tooth and claw" with theoretical representations' "fitness" measured entirely in terms of their degree of empirical adequacy. Likewise with instrumentalists of various stripes, such as Kyle Stanford's neo-instrumentalism, which acknowledges that there is probably some reason why 'foundational' theories (e.g. in quantum physics) are empirically successful, but maintains that nothing of substance can be said of these reasons.

Such anti-realist attitudes towards the realist's preoccupation with the success of science are in striking discord with physicists' own substantial accounts of their theories' and models' empirical success and ongoing attempts to further understand them. Consider spintronics, again, whose devices usually involve magnetic materials exhibiting various spin-related phenomena involving very large numbers of constituents. Magnetism and many relevant collective phenomena can be represented in classical or semi-classical terms (e.g., the direction of magnetism or spin-current is a classical vectorial quantity), which in models of spin-valves and such appear to represent interesting new physical properties (e.g., spin-currents and spin-waves), which have led to various novel predictions and experimental manipulations. (Indeed, some physicists are calling the ongoing 'quantum engineering' phase the second quantum revolution!) *How can these models be so incredibly successful?*

The basic realist idea is that their success is down to their representational faithfulness. For a naturalistic philosopher this idea can be motivated just by pointing out that the successful practice of spintronics is premised on this very idea. Regardless of what exactly it is that quantum physical models capture at the 'fundamental' level of individual electrons—a foundational-cum-interpretational issue on which spintronics research very rarely takes a stand—this area of research at large is premised on the idea that there is a detailed physical theory to be given of the relationship between particles' quantum physical spin (which has no classical analogue) and the collective 'classical' spin phenomena at the macro-level. Physicists' answer to the above question (in italics) can be thus summed up: because the world at the 'fundamental' level is quantum, involving quantum states and features that we effectively capture with e.g. the Pauli spin matrices, which are sufficiently well represented by quantum mechanics to enable the theoretical notions of spintronics, which capture collective quantum phenomena, to represent the relevant emergent features of semiconductors.

Anti-realists may be sceptical towards the notion that physics' theorizing itself could be genuinely explanatory of spintronics' success. Is the progress realist begging the question here? It is not clear what naturalistically respectable reasons there are for such scepticism, however, in the face of the established status of spintronics research, the central notions of which go back more than seventy years. By contrast, progress realism respects physicists' authority in thinking that an account of the empirical success of spintronics can be worked out by studying the relationship of quantum physical representations of single electrons and of macroscopically large collections thereof. Although this account is still in many ways in progress, substantial scientific understanding has already been established regardless of the foundational 'black boxes' that have to be drawn at the more fundamental level. As a part of this commitment a progress realist (along with the physicists) regards quantum notions like spin no less representational than any other notion in physics (while being quiet about what these notions actually represent).[21] This stands in contrast with the recent non-representationalist interpretations of quantum physics—such as quantum

[21] This is not in any way specific to quantum physics. The progress realist does not want to commit to representational truths about, say, the electromagnetic field and its polarization either, but just to the idea that that the relevant theoretical representations are latching onto reality in ways that are responsible for their theoretical success.

pragmatism of Richard Healey (2017; this volume, Chapter 7)—which aim to account for the empirical successes of quantum physics in terms that leave no representational role for any distinctly quantum notions.

In addition to there being a clear contrast between progress realism and established anti-realist views, progress realism also places substantial constraints on the kinds of developments in science that we can rationally expect, while being open to the possibility that future science will develop in radical, presently unforeseeable directions at the level of its (currently) foundational and 'deeply' metaphysical notions. In particular, it expects quantum physics to evolve in ways that make increasing sense of how current physics is latching onto the world in ways that make it successful, so as to complement and not profoundly revise our current understanding of the past theories' empirical success.[22] As to how physics can make increasingly good sense of its own empirical success, progress realism points to much studied exemplars of our best foundational studies of domain-specific inter-theoretic relations, between, e.g. classical and quantum physics, ray optics and wave optics, Newtonian and relativistic theories of mechanics and gravity, old quantum mechanics and contemporary quantum theory, and so on. These well-established exemplars of such realist understanding of how science theoretically progresses are provided from our current vantage point—where else?!—but they are not expected to be overthrown in the fullness of time. This places considerable constraints on our rational expectations regarding future scientific developments, yielding one kind of knowledge of how science progressively 'latches onto' reality. This, however, is quite different from truth-content realism.

Acknowledgements

A version of this chapter was presented at the 13th International Ontology Congress in San Sebastian, and I would like to thank the audience. I also benefited from discussions and feedback from various people, including Alberto Cordero, Douglas Earl, Matthias Egg, Steven French, Tetsuji Iseda, Simon Newey, Kohei Morita, and David Wallace. The research was funded by the Arts and Humanities Research Council.

References

Blugel, S., D. Burgler, M. Morgenstern, C. M. Schneider, and R. Waser (2009). *Spintronics – From GMR to Quantum Information*. Forschungszentrum, Zentralbibliothek.

Blundell, S. J. (2003). *Magnetism in Condensed Matter*. Oxford: Oxford University Press.

Boyd, R. (1984). The current status of the realism debate. In J. Leplin (Ed.), *Scientific Realism*, pp. 41–82. Los Angeles: University of California Press.

Chen, E. (2019). Realism about the wave function. *Philosophy Compass 14* (7), e12611.

Daumer, M., D. Dürr, S. Goldstein, and N. Zanghì (1996). Naive realism about operators. *Erkenntnis* 45(2–3), 379–97.

[22] This is closely related to structural realists' emphasis on inter-theoretic correspondence relations as explanatory of past theories successes. But progress realism recognizes nothing distinctly 'structural' in the plurality of such correspondence relations, and it denies that there is any kind of structuralist 'recipe' for extracting trustworthy truth-content from our current theories.

Duck, I. and E. Sudarshan (1997). *Pauli and the Spin-Statistics Theorem*. Singapore: World Scientific.
French, S. (2014). *The Structure of the World: Metaphysics and Representation*. Oxford: Oxford University Press.
Frigg, R. and I. Votsis (2011). Everything you always wanted to know about structural realism but were afraid to ask. *European Journal of Philosophy of Science 1*(2), 227–76.
Ghirardi, G. C., A. Rimini, and T. Weber (1986). Unified dynamics for microscopic and macroscopic systems. *Physical Review D 34*(2), 470.
Giulini, D. (2008). Electron spin or 'classically non-describable two-valuedness'. *Studies in History and Philosophy of Modern Physics 39*(3), 557–78.
Gooday, G. and D. Mitchell (2013). Rethinking 'classical physics'. In J. Buchwald and R. Fox (Eds), *Oxford Handbook of the History of Physics*, pp. 721–65. Oxford: Oxford University Press.
Hacking, I. (1983). *Representing and Intervening: Introductory Topics in the Philosophy of Natural Science*. Cambridge: Cambridge University Press.
Hannout, M., S. Hoyt, A. Kryowonos, and A. Widom (1998). Quantum measurement theory and the Stern–Gerlach experiment. *American Journal of Physics 66*(5), 377–9.
Hanson, R. and D. Awschalom (2008). Coherent manipulation of single spins in semiconductors. *Nature 453*(7198), 1043–9.
Healey, R. (2017). *The Quantum Revolution in Philosophy*. Oxford: Oxford University Press.
Keeler, J. (2005). *Understanding NMR spectroscopy*. Chichester: John Wiley.
Keppeler, S. (2003). *Spinning Particles - Semiclassics and Spectral Statistics*. Berlin: Springer-Verlag.
Kitcher, P. (1993). *The Advancement of Science: Science without Legend, Objectivity without Illusion*. New York: Oxford University Press.
Ladyman, J. and D. Ross (2007). *Everything Must Go: Metaphysics Naturalized*. Oxford: Oxford University Press.
Lewis, P. (2006). GRW: a case study in quantum ontology. *Philosophy Compass 1/2*, 224–44.
Lewis, P. J. (2018). On the status of primitive ontology. In S. Gao (Ed.), *Collapse of the Wave Function*, pp. 154–67. Cambridge: Cambridge University Press.
Likhtenshtein, G. (2016). *Electron Spin Interactions in Chemistry and Biology*. Dordrecht: Springer.
McLeish, C. (2006). Realism bit by bit: Part II. Disjunctive partial reference. *Studies in History and Philosophy of Science Part A 37*(2), 171–90.
McMullin, E. (1984). A case for scientific realism. In J. Leplin (Ed.), *Scientific Realism*, pp. 8–40. Los Angeles: University of California Press.
Morrison, M. (2007). Spin: all is not what it seems. *Studies in History and Philosophy of Modern Physics 38*(3), 529–57.
Myrvold, W. (2018). Ontology for collapse theories. In S. Gao (Ed.), *Collapse of the Wave Function*, pp. 97–123. Cambridge: Cambridge University Press.
Niiniluoto, I. (1999). *Critical Scientific Realism*. Oxford: Oxford University Press.
Norsen, T. (2014). The pilot-wave perspective on spin. *American Journal of Physics 82*, 337–48.
Norsen, T. (2018). On the explanation of born-rule statistics in the de Broglie-Bohm pilot-wave theory. *Entropy 20*(6), 422.
Ohanian, H. C. (1986). What is spin? *American Journal of Physics 54*(6), 500–5.
Pauli, W. (1925). Über den Einfluß der Geschwindigkeitsabhängigkeit der Elektronenmasse auf den Zeemaneffekt, *Z. Phys. 31*, 373–85.
Pearle, P. (1989). Combining stochastic dynamical state-vector reduction with spontaneous localization. *Physical Review A 39*(5), 2277.
Psillos, S. (1999). *Scientific Realism: How Science Tracks Truth*. London: Routledge.

Putnam, H. (1982). Three kinds of scientific realism. *The Philosophical Quarterly* 32(128, Special Issue: Scientific Realism), 195–200.
Raimond, J.-M. and V. Rivasseau (2009). *The Spin*. Basel: Birkhauser.
Ruetsche, L. (2018). Getting real about quantum mechanics. In J. Saatsi (Ed.), *Routledge Handbook of Scientific Realism*, pp. 291–304. London: Routledge.
Saatsi, J. (2015). Replacing recipe realism. *Synthese 194* (9), 3233–44.
Saatsi, J. (2016). What is theoretical progress of science. *Synthese 196* (2), 611–31.
Saatsi, J. (Ed.) (2018). *The Routledge Handbook of Scientific Realism*. London: Routledge.
Saatsi, J. (2019). Scientific realism meets metaphysics of quantum mechanics. In A. Cordero (Ed.), *Philosophers Look at Quantum Mechanics*, pp. 141–62. Dordrecht: Springer.
Sakurai, J. J. (1995). *Modern Quantum Mechanics*. Boston: Addison-Wesley.
Schurz, G. (2009). When empirical success implies theoretical reference: a structural correspondence theorem. *The British Journal for the Philosophy of Science* 60(1), 101–33.
Sebens, C. (2019). How electrons spin. *Studies in History and Philosophy of Modern Physics*. Online, https://doi.org/10.1016/j.shpsb.2019.04.007.
Townsend, J. S. (2000). *A Modern Approach to Quantum Mechanics*. Sausalito, CA: University Science Books.
Tumulka, R. (2018). Paradoxes and primitive ontology in collapse theories of quantum mechanics. In S. Gao (Ed.), *Collapse of the Wave Function*, pp. 134–54. Cambridge: Cambridge University Press.
Van Fraassen, B. C. (1980). *The Scientific Image*. Oxford: Clarendon Press.
Wallace, D. (2012). *The Emergent Multiverse: Quantum Theory according to the Everett Interpretation*. Oxford: Oxford University Press.
Zee, A. (2016). *Group Theory in a Nutshell for Physicists*. Princeton: Princeton University Press.

PART II

Underdetermination and Interpretation

4

Can We Quarantine the Quantum Blight?

Craig Callender

4.1 Introduction

The science fiction novel *Quarantine* portrays a world wherein interaction with human observers is necessary to collapse quantum wavefunctions. The author, Greg Egan, amusingly puts the emphasis on the observers being *human*—aliens can't do it. Aliens are therefore at a tremendous disadvantage. As we gaze at the night sky, we are constantly collapsing alien worlds, depriving them of their branch diversity. Whole civilizations are being snuffed out by our observations! Understandably the aliens grow tired of this. In response they erect an impenetrable shield around the solar system, one that blinds us to the outside universe. This shield protects the rest of the universe from harmful human observation, locking humanity into a starless Bubble.

When confronting scientific realism with quantum mechanics, many philosophers advocate the theoretical counterpart of this fictional strategy. Quantum mechanics is beset with notoriously difficult interpretational challenges. Different interpretations of the theory are compatible with present data. Only the most unreconstructed positivist thinks these different interpretations (different theories, really) are notational variants, i.e., different representations of the same facts. Scientific realism holds that most of the statements of our mature scientific theories are approximately true; but this claim is threatened by persistent underdetermination of theory by evidence, for one theory isn't better confirmed than its rivals. Faced with this threat, some try to lock the quantum interpretation problem into a theoretical Bubble, cordoning off the interpretational blight and leaving the rest of the world safe for scientific realism.

My goal in this chapter is to pop this Bubble. No shield can really protect the poor aliens in Egan's story, nor can any theoretical membrane protect scientific realism from dealing with the quantum measurement problem. One may be able to erect barriers around the observable or classical, preserving a realism about tables, chairs, and the like, but there is no safety zone within the quantum realm, the domain of our best physical theory. The upshot is not necessarily that scientific realism is in trouble. That conclusion demands further arguments. The lesson instead may be that scientific realists ought to stake their case on particular interpretations of quantum theory.

Craig Callender, *Can We Quarantine the Quantum Blight?* In: *Scientific Realism and the Quantum*. Edited by: Steven French and Juha Saatsi, Oxford University Press (2020). © Craig Callender.
DOI: 10.1093/oso/9780198814979.003.0004

In any case, the realist can't ignore the interpretational issues plaguing quantum mechanics.

4.2 The Quantum Blight

Quantum theory is one of the most successful sciences we have ever developed. It is a rigorous formalism attached to rich experimental practices. Together, the formalism and experimental practices allow us to make bold and novel predictions that have been confirmed time and again for over ninety years. Unfortunately, it's not clear what world is being described by this theory. We need to know the 'word-world' connections. What do the terms in the formalism represent in the outside world, if they represent anything? For example, does the wavefunction represent our knowledge, a real field evolving in a high dimensional space, a complicated field on a low-dimensional space, an aspect of a law of nature, or what? The question isn't solely directed at the quantum state. It applies to everything in the theory—the q's, p's, σ's, and more.

Word-world questions arise with every theory. Does classical particle mechanics portray a world with forces, with three equal types of mass (active, passive, inertial) or one type? Options exist. The main difference with the situation in quantum theory is that quantum theory, unlike classical mechanics, suffers from the infamous measurement problem. The measurement problem in effect shows that the word-world connections offered by the standard 'Copenhagen' interpretation are inconsistent—or at the very least, lack theoretical virtues that we normally expect of a theory (see, e.g., Bell, 1987). It's a huge flag calling attention to the need for clear and consistent representational connections for quantum mechanics.

Answers exist. Too many. There are Bohmians, followers of Bohm (1952), who hold that quantum mechanics is incomplete and supplement it with additional ontology. There are advocates of Collapse, like Ghirardi, Rimini, and Weber (1986), who propose modifications to the linear wave evolution. There are Everettians who posit a kind of multiverse (Everett, 1957). Hybrids of all three theories exist. For instance, one can divide Collapse interpretations into 'Everettian Collapse' and 'Bohmian Collapse' theories, depending on whether the theory posits beables in addition to the wavefunction (Allori et al., 2008). A similar claim can be made for Everett, as one can interpret Everett as positing a matter density distribution like GRW (Allori et al., 2011) or even create a kind of Bohmian multiverse (Sebens, 2015b). Answers with a more 'pragmatic' or 'instrumentalist' flavor exist, including Healey's recent pragmatic view and Fuchs' Quantum Bayesianism (see Healey, 2017 and this volume, Chapter 7, for discussion and references). The diversity of worlds possibly described by quantum mechanics is shocking. One might stubbornly insist that the difference with classical mechanics is one of degree, that both theories have unresolved word-world questions. Fine, but the number of degrees is huge. Nothing compares classically, for example, to the contrast between the sparse ontology of GRW (with 'flash' ontology) and the generous ontology of Everett.

The interpretations describe dramatically different ontologies, but more than that, they typically offer different laws of nature and different core theoretical edifices. Collapse theorists modify the linear dynamical evolution of the wavefunction. Bohmians offer a guidance equation for their particles or fields. It's hard to find a theoretical core that they all have in common such that we can regard them all as different interpretations of that core. Even the operator algebra that is taught in every quantum text is contested: for instance, for Bohmians, Hilbert space and the operator algebra are an emergent measurement formalism having no place in the fundamental description of nature, whereas for Quantum Bayesianism, that formalism is the core. For these reasons the many 'interpretations' are clearly different theories.

A clearer picture is painted by conceiving these 'interpretations' as different Lakatosian research programs (Lakatos, 1978). A research program is a series of theories sharing a 'hard core' of temporarily unimpeachable theoretical posits. Quantum mechanics is a live theory, one being extended to new forms and realms. The non-relativistic theory of 1925 is applied to new systems daily, from ever more sophisticated treatments of helium to the recent discovery of non-equilibrium time crystals. The theoretical structure was also transformed into QED, QCD, and the standard model, and we hope to integrate cosmology and gravity with the quantum. The interpretations typically have something to say about all these developments. Often how they respond changes the laws and ontology posited by the theory, e.g., as we'll see, Bohmian quantum field theory may posit a different ontology than non-relativistic Bohm theory. With so much different, in what sense can we speak of *an* interpretation or theory? The answer is that each 'interpretation' is really a research program.

Lakatos' 'negative heuristic' is that which is unrevisable in each program and defines its 'hard core': Everettians all hold that macroscopic superposition indicates multiplicity; Bohmians all postulate ontology guided by a new equation hooked up to a wave equation; advocates of Collapse all modify the wave equation to produce a unified story of the macro and micro realms; Quantum Bayesians are committed to the idea that wavefunctions represent the amount of information one has about a system. These hard cores are carried along when each 'interpretation' is applied to some new domain or theory. Lakatos' 'positive heuristic,' by contrast, is that which is revisable within each program. Bohmians can propose new choices of ontology to be guided by the wavefunction if they are better suited to the pheneomena, Collapse theorists can tinker with the size, timing, and triggers of collapse. Thus understood, even dramatic departures, such as Collapse approaches to semi-classical quantum gravity (Okon and Sudarsky, 2015) and Bohmian approaches to superstring theory (Weingard, 1996), are easily recognized as descendants of the original families.

Carve answers to the measurement problem into four broad research programs. Little hangs on this division, and I'm happy to acknowledge that different partitions and hybrid theories exist. Using one reasonable partition, we find four active research programs: Bohmian, Everettian, Collapse, and Pragmatist/Bayesian—each very broadly construed. The last of these programs doesn't aspire to characterize or represent a complete physical reality. These views are often accused of being instrumentalist interpretations. Whether this accusation is fair or not, this last set of programs will not be relevant to realism and the present worry of underdetermination because it doesn't offer us a representation of physical reality. We therefore have three live research programs (Bohm, Collapse, and Everett), each portraying dramatically different realities (for accessible discussions of each, see Maudlin, 2019).

Confined to nonrelativistic quantum mechanics, experimentally there is no way to confirm one over the other. Bohm and Everett use precisely the same algorithm for extracting predictions (the Born rule). Collapse typically uses a slightly modified one that differs negligibly in the macroscopic realm from what Bohm and Everett use. Philosophers sometimes raise the distinction between pairs of theory being 'in principle' undermined by data and other pairs merely being 'in practice' so underdetermined. Collapse and Bohm/Everett, they might say, are then in principle empirically distinguishable, unlike Bohm and Everett. But as we know from Laudan and Leplin (1991), it's not clear that the 'in principle versus in practice' distinction is itself an in-principle one. And when comparing research programs as opposed to artificially frozen theories and fixed empirical domains, it's not clear that this distinction is so useful. There are plenty of experiments we can imagine that would provide a crucial test of Collapse against other theories. Collapse posits two new constants of nature, the collapse width and the collapse rate. Some choices of these parameter pairs have already been 'falsified' by experiments involving spontaneous x-ray emission. Had the original GRW theory chosen such a pair, the theory would now be demonstrably false. However, with room remaining in the 'unfalsified' parameter space, an advocate of Collapse in that scenario could simply shift to a new parameter pair, saying that he or she had learned better. The Collapse research program can survive falsification of some particular parameter pairs. For this kind of reason, no Popperian crucial test between *research programs* is likely in the foreseeable near future.[1]

4.3 Dialing Up Underdetermination

Underdetermination of theory by data is a phenomenon that can happen, as the phrase suggests, when the empirical data do not narrow down the space of acceptable theories to one. As a logical matter, left at this, this situation is guaranteed to always obtain. We know from the curve-fitting problem that a finite number of data points can be connected via an infinity of curves. If we treat each curve as a theory, then we always face massive underdetermination of theory by data. In the philosophy of science, however, we curiously restrict the available theories to properly 'scientific' ones. I say this is curious because philosophers of science, of all people, know that the label 'scientific' is notoriously vague. What is meant?

Imagine a dial (Figure 4.1) that we can set to more or less 'scientific.' Admittedly oversimplistic, we might think of the settings as measuring increasing theoretical virtues. The lowest setting might be mere logical consistency. When the dial is set there all sorts of wild and intuitively 'unscientific' theories count. Skeptical nightmares

[1] Detractors of each program may object to what I've said here. Critics of Everett will insist that Everettians cannot recover the Born rule predictions due to the theory's well-known problem with probabilities, so we don't know if it's empirically adequate. Critics of Collapse will raise worries about the tails problem and related threats to the available parameter space (Sebens, 2015a). Critics of Bohm will point out that it hasn't been fully extended to quantum field theory and is therefore not empirically adequate in the relativistic realm. Each worry is very serious. I explicitly tackle the last in Section 4.6. However, for present purposes, because I'm discussing programs as opposed to static theories, I'm inclined to be generous and hope that each can overcome their challenges, especially the ones they're actively working on.

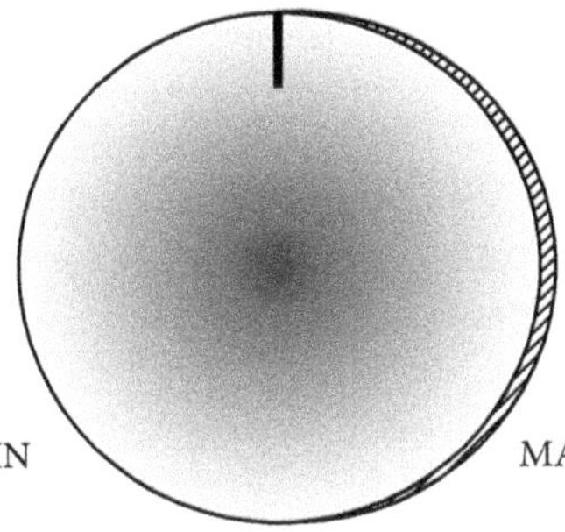

Figure 4.1 The Scientific Dial

like Descartes' demon theory and Putnam's brain-in-a-vat theory count as scientific because they are consistent. In terms of curve-fitting, this setting allows even the most 'wiggly' of curves. The result of this theory is massive underdetermination. Turning the dial up winnows down the number of contending research programs. Suppose we insist not only on consistency but also on the theory being predictively useful to human beings, unified, consilient, simple, etc. Then when we turn the knob Descartes' demon theory drops out because it isn't predictively useful to human beings. In principle, being super strict with what we mean by 'scientific' could winnow the acceptable theories down to one. Let's not go that far. The philosopher of science dealing with underdetermination instead sets the dial at a sense of scientific that a consensus would agree upon as genuinely scientific, a sense that would separate good scientific theories from pseudoscientific claims (e.g., evolution from creationism). What happens?

Arguably, at a sufficiently coarse-grained level, such a setting in biology restricts the available theories down to one, namely, the modern synthesis of molecular biology and Darwinian selection. No serious scientific rivals exist, although of course at a finer scale all manner of controversy erupts. By contrast, if we focus on quantum phenomena, we are left with our four programs, three of which describe in detail very different worlds. Although many of the terms used are a bit vague, still the contrast between the situation in quantum mechanics and modern biology is striking.

The three quantum research programs pose a prima facie threat to scientific realism. The realist holds that most claims about observable and unobservable facts made by a mature scientific theory are true or approximately true. Agree for the moment that Collapse is mature and successful. One can't defend a belief in collapses if one at the same time admits that the evidence equally well supports a theory without them.

Are we really in this situation? Prima facie, yes. In a perceptive paper John Norton (2008) attacks the idea that underdetermination is guaranteed. The artificial playthings of philosophers that serve to justify a state of permanent general underdetermination fail, he thinks. I agree. There is no automatic proof of general underdetermination. It's not easy meeting the demands of the dial when it is set high. However, Norton also hints that most cases, general or not, are probably not threatening. Is that the case here? Consider two theories underdetermined by the data, T and T^*. Let's examine the three cases he envisions.

First case: T and T^* predict the same observational evidence E and they are inter-translatable about unobservable content. That is certainly not the case for our three programs. There are no translations of a Collapse swerve nor a Bohm particle into their rivals, respectively.

Second case: T^* is parasitic upon T, but T^* is epistemically inequivalent to the original. Suppose T implies E and that $T^* = T\&H$. Then, as Norton points out, good theories of confirmation do not automatically agree that if E confirms T it also confirms $T^* = T\&H$, where H is some arbitrary hypothesis. Again, none of our three research programs are parasitic upon one another in this sense. One encounters the claim that Bohm is Everett 'in denial.' The thought is that Bohm simply adds a hypothesis about which Everett 'branch' is occupied by particles. However, no one would claim that this is parasitism in this cheap '&*H*' sense. If it were, it would be trivial to generate successful field extensions of Bohm—but it's not. So if parasitic, it's not in this automatic sense. In any case, the 'in denial' charge is in my opinion wrong for many reasons and on some interpretations not remotely plausible (see Callender 2015 and references therein).

Third case: T and T^* are not inter-translatable but 'similar,' and Norton thinks, therefore likely to be theoretically identical. Norton doesn't say why similarity makes identity likely, but let's grant him that it does raise the suspicion. He suggests that Bohm and Copenhagen are in this relationship. It's hard to understand the reason why, as Copenhagen isn't consistent, or if it is, it seems to fundamentally cleave the world into classical and quantum regimes according to fuzzy rules—neither of which is the case in Bohm. Nor are the theories similar structurally, as the Hilbert space formalism that is central to Copenhagen isn't a crucial part of Bohmian mechanics. Unless similarity is understood as simply empirical equivalence, there is little reason to take the three research programs considered here to collapse into one.

With these research programs, it seems that we face the realist's nightmare. Many philosophers suspected that the threat of underdetermination is artificial, confined to excessive flights of imagination and not genuinely scientific theories. However, these three programs are neither philosophers' toys nor notational variants (on any remotely reasonable semantics) and are clearly 'scientific' in letter and spirit. Quantum underdetermination is the real deal.

4.4 Underdetermination within Underdetermination?

Quantum underdetermination may be worse than just characterized. I described three programs, but there is potentially a lot of underdetermination within each too. My hope, however, is that many of these *empirically equivalent* theories will turn out to be uncontroversially *epistemically inequivalent*. The normal process of scientific discovery will weed them out. One might say the same about the underdetermination we just confronted—i.e., hope that it goes away—but what I have in mind at present are relatively uncontroversial choices dictated by theory development.

Consider Bohmian mechanics. Central to the theory is the choice of a beable (e.g., particles) and a guidance equation describing the dynamics of that beable. There is some freedom in both choices. Let's take a look.

Bohmians provide a dynamics for their beables to 'surf' wavefunctions. Wavefunctions evolve according to linear wave equations in quantum mechanics, such as the famous Schrödinger equation. Bohm's theory relies on a crucial feature of such equations, namely, that they imply a continuity equation, a local form of conservation. What is conserved is the probability density through time. That density determines the chance of finding a particle at a location at a time. Bohmian mechanics is based on the simple insight of using this conservation and its associated conserved current to define the velocity guiding the beables, just as one does in fluid mechanics and elsewhere. Supplemented with the claim that the particles are initially randomly distributed, the theory is empirically adequate in the non-relativistic regime.

Many other choices of guidance equation also prove to be empirical adequate. Add any divergence-less vector field (divided by the probability density) to the original velocity. The continuity equation does not 'see' this addition. Hence this new modified velocity will also be empirically adequate. Yet this additional vector field is arbitrary, characterizing indefinitely large and potentially wild deviations from the original Bohm velocity.

This case may be a benign form of underdetermination. When discovering scientific theories, scientists use a variety of non-empirical issues as guides. Some, like simplicity, raise worries because simplicity may be in the eye of the beholder, but other considerations seem uncontroversially 'scientific' and are a poor basis for serious worries about underdetermination. Such considerations may constrain the form of the Bohm velocity. Dürr, Goldstein, and Zanghì (1992), for instance, claim that the original choice is the unique Galilean-invariant velocity (but see Skow, 2010). Peruzzi and Rimini (2000) claim that it is the unique choice that works also for the center-of-mass of the Bohmian configuration, a desirable feature for many reasons. So there is plenty of reason to expect the form of the velocity to be whittled down by ordinary scientific reasons encountered in discovery.

Turn to the choice of beable. Consider two cases, one, the choice in the original theory, and two, the choice when we move to field extensions. The usual choice in the non-relativistic particle theory is to choose particles with determinate positions as the basic ontology. There are good reasons for this choice, as other choices such as momentum don't solve the measurement problem. However, it's well-known that one can add additional beables to the theory, such as spin. In fact, there is a general recipe for adding new 'basic' properties to Bohm particles (see Holland, 1993). These additions do seem akin to Norton's parasites. The measurement outcomes are recorded in position ('up,' 'down,' and so on). The wavefunction and particle positions together entail the spin vector representing the spin beables. Absent an independent reason to exploit the spin vector, it seems that Occam's Razor will quickly remove the basic spins from the Bohmian particles—and with them this alternative formulation.

Turning to quantum field theory, matters really open up (see Struyve, 2011 for an excellent review). Bohmians face choices between adding particle or field beables. In his original paper, Bohm proposed a field beable for the electromagnetic field, an actual field configuration corresponding to the transverse part of the vector potential.

One can do something similar for other boson fields, but this approach is hard to extend to fermionic fields. A radical response to this trouble is just to get rid of fermions altogether. Measurement results will get recorded in bosons, so in some sense they are 'enough' for empirical adequacy. On this 'minimalist' approach, there are no fermions but the wavefunctional carries a label representing what would be their degrees of freedom—so the boson field behaves as if there were fermions around. One can dress up the bosonic fields via the method used for spin vectors mentioned above, providing a sense in which there are fermions, but Occam's Razor will slice these properties away as quickly as it would the above spin properties. The other way to go is with particles rather than fields. Ironically, the particle picture works well for fermions but less well for bosons. In Bell (1987) fermion particle number is defined but there is no configuration for bosons. We can also entertain a hybrid theory, one treating fermions as particles and bosons as fields, which is how we treat electrons and photons in classical electromagnetism. In sum, we have choices between particles or fields and even whether bosons or fermions exist! Then again, the overdetermination here may be overstated. Right now approaches are getting eliminated or favored for normal reasons of physics, e.g., no natural measure for Grassman fields, Euler angles not solving the measurement problem. Work is ongoing, and as theories are extended they meet more constraints. It would be premature to say that quantum field theory yields rampant underdetermination for Bohmians.

I've concentrated on the Bohmian case, but the other two research programs face similar issues. If Collapse posits non-wavefunction beables (e.g., matter density, flashes), one will face similar questions about what is the right beable. There are also additional choices: the hit rate and collapse width, the 'trigger' for collapse (particle number, mass, Weyl curvature, Riemannian curvature), and more. Everettian theories likewise need to choose whether to add a beable (e.g., mass density) or not. Even if not, questions remain that can possibly lead to numerous theories, such as determining the microscopic ontology of the Everettian world. Wallace and Timpson (2010) make one proposal (spacetime points with properties) but there are alternatives. As in the Bohm case, I suspect that most of these decisions will be decided by normal theory development and not cause widespread underdetermination. That said, given the uncertainty, we enjoy no guarantee that this will be the case.

4.5 Quarantine

If what I've argued is on the right track, then we have serious scientific underdetermination of theory by data striking right at the heart of our most basic successful scientific theory of the world. One natural reaction is to quarantine this underdetermination to some specific regime and free some theoretical claims from epistemic danger. The intuition behind the quarantine strategy is that the disagreement between these camps is isolated to esoteric bits of the theory. These esoteric bits are where physicists have little confidence; instead, what they are confident about are the core explanations of typical quantum phenomena, and on these, each camp agree. Where the three camps agree on some theoretical claim, that claim is not subject to underdetermination. On its face this position strikes me as tempting and plausible.

Let's spell the position out slightly more carefully. To be interesting, the claims that the quarantine strategy must protect are

1. theoretical
2. not merely mathematical
3. specifically quantum.

Let me explain. Demand 1 should be uncontroversial. We're focusing on attempts to rescue realism. We already know that all three programs get the observables right—that's what it is to solve the measurement problem. Demand 2 should also be uncontroversial. We're after scientific realism, not mathematical realism. All the research programs agree that $2 + 2 = 4$. Yet that doesn't tell us much about unobservable contingent physics. Demand 3 is imposed because we're interested in whether realism can reach into the quantum realm. Just as the three programs may agree on the observables associated with macroscopic objects, they may also agree on some unobservables for some systems as we move into the classical limit. Quantum decoherence is a process whereby interactions among constituents of a system and its environment lead to the suppression of quantum interference. This process occurs in all three programs. The issue is tricky, but *arguably* after decoherence all three programs will agree on much. But that's not so satisfying if we want to know whether we should be realists about coherent quantum systems. Decoherence may provide a defensible quarantine strategy that frees some claims about unobservables, but hopefully we can do better than secure realism only in the classical domain of quantum theories.

The rules are set. Are there substantive claims about the specifically quantum realm that are shared by all three of our research programs? We know there is massive agreement from above: they all agree on the observables, perhaps even the classical limits. They all disagree way down below: for instance, we won't find continuous Bohmian particle trajectories in Collapse or Everett. What about in between? Although the real world and real theory of it are way more complicated, Figure 4.2 provides a toy model of the setup.

4.5.1 'Textbook' quarantine zone

In a little discussed paper, Alberto Cordero (2001) offers what I think is one of the best ways of finding a quarantine zone. His idea, as above, is that the three camps overlap considerably. In his own words:

> the underdetermination at hand is clearly of limited scope ... all the mentioned competitors associate the quantum state with a peculiar physical field, all include the Schrödinger equation centrally in the dynamics, all endorse a strong form of ontic-structural nonseparability, and all agree on geometrical relations between subsystems (internal molecular shapes, atomic and quark structure, etc.). By contrast, divergence between the competing models is peculiarly confined to certain specific respects and degrees of precision, with clear significance limited to some fundamental questions ... So, although the case makes for an intense ontological debate, its corrosive power on belief seems confined to just some aspects of the full narrative. The encountered underdetermination does strike realist theorizing from a certain depth down, but then again only along certain lines of inquiry. (p. 307)

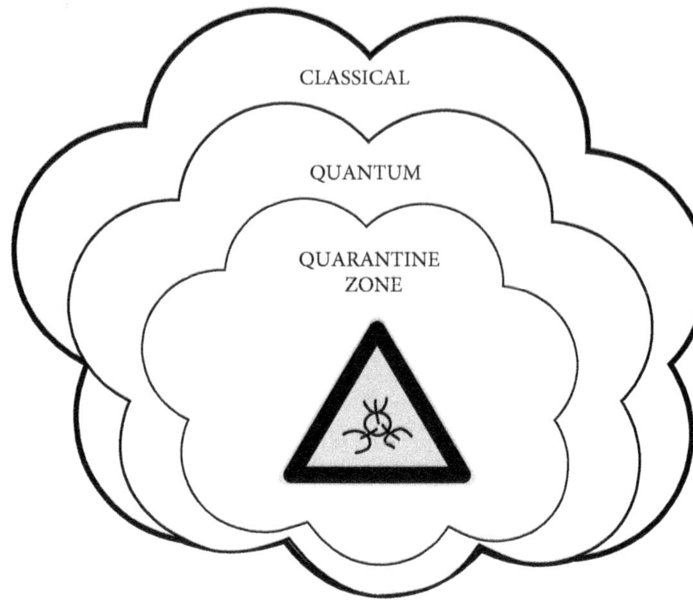

Figure 4.2 Quarantine Zone

Having provided examples of convergence amongst our three programs, Cordero concludes that many hypotheses about the quantum world are safe from the quantum blight.

Cordero mentions many areas where the programs converge, but I want to focus on what I'll dub *textbook quantum mechanics*. By this I mean the narratives found in quantum textbooks about what's going on in quantum systems. Here I'm thinking about Cordero's claim that the research programs agree on "geometrical relations between subsystems (internal molecular shapes, atomic and quark structure, etc.)" but also similar examples he gives throughout, e.g.,:

> Take, for example, the basic quantum mechanical model of the water molecule, with its atom of oxygen bonded to two atoms of hydrogen, the latter making with the former an angle of about 103* in "normal" thermodynamical conditions. Stuff like this is contained in approximate partial models shared by all the viewed theories. . . . [M]uch in the quantum mechanical story about water molecules and their interaction seems at least as credible as the most ordinary talk about, say, cats and common objects. (p. 309)

Cordero doesn't associate his position with textbooks, so he shouldn't be saddled with my interpretation. If preferable, we can say this idea is inspired by Cordero's examples and not his position. In any case, I want to focus on this claim because it strikes me as new and interesting. Textbook physics seems interpretation neutral. We feel that we can trust what they say about typical quantum systems (e.g., claims about orbits, molecular structure, the behavior of energy, tunneling phenomena, and so on) while bracketting the measurement problem. At a certain 'depth' trust runs out, e.g., whether the theory is deterministic or based on particles. Yet these are deep metaphysical questions that do not touch quantum textbook claims.

In fact, I think the conjecture about what I'm calling textbook quantum mechanics is more plausible than his initial claims quoted above. Tease apart what I'm calling the textbook claim from his initial claim that "all the mentioned competitors associate the quantum state with a peculiar physical field" and so on. That initial claim about the quantum state might strike readers as true—but at best it's only approximately true. The normalization one does in Collapse will slightly change the state used, and differences in decoherence might imply slightly different Bohmian effective wavefunctions from the wavefunctions associated with branches in Everett (the analogues of the quantum state assigned to sub-systems in each program). More importantly, the agreement may only exist at the mathematical level. What the quantum state represents can vary dramatically amongst research programs. Similar claims can be made about the dynamical equation, although here the differences between Collapse and the rest are starker. The most mathematically sophisticated versions of Collapse, continuous spontaneous localization theories, propose stochastic nonlinear wave equations; the stochastic modifications are crucial to the theory and have huge structural ramifications (e.g., regarding norm preservation). Structurally they are importantly distinct from the Schrödinger equation. For these reasons I want to reject Cordero's first set of claims.

Proponents of structural realism (McKenzie, 2017) may insist that all three programs share substantial core structure, namely, the structure of Hilbert space, the operator formalism, commutation relations, Born's rule, and more. Structural realists modify realism by retreating to the mathematical or structural relations in a theory. But the claim of a common structure here would be overblown. The operator formalism—for all three programs—is simply a measurement formalism, a tool added to the core theory, not the theory itself. For the Bohmian, for instance, experiments define maps from initial wavefunctions to distributions of particle positions. Bohmian commitments imply this map is bilinear. Bilinear maps are equivalent to positive operator valued measures and the traditional quantum operators are particularly simple expressions of these (Daumer, Dürr, and Zanghì 1996). Bohmians could in principle just speak of the particle distributions directly and skip all of this—at least as far as fundamental theory goes. The same goes for the Everettian but regarding the quantum state. The conventional 'word-world' connection used by Copenhagen is abandoned. That interpretation understood the operator formalism as being a guide to the 'properties' of a system (a system has a property associated with an operator iff its state lives in the subspace that the operator projects the state onto). But none of the programs we're looking at employ this connection. Stripped of its connection to the unobservable quantum world, what's left behind is a useful algorithm for predicting measurement outcomes and no more. True, at some emergent measurement level that algorithm is shared. Yet that level is observable and therefore not relevant to rescuing any kind of realism. The operator formalism is the wrong place to look for substantial physical overlap.

Back to quantum textbooks. Cordero's claim, disentangled from structural realism, is independently attractive. When we probe it, however, it faces trouble. Recall that for realism to be interesting in this context, theoretical claims must satisfy at least three demands. They must be non-mathematical, theoretical, and specifically quantum. Do Cordero's examples meet these criteria? Do other quantum textbook claims? I think it's pretty clear that these claims are not held in common amongst programs, not with each other, and not with what textbooks say. Divergent physical pictures emerge as soon as we peek into anything quantum. Extracting 'what textbooks say' can be a bit of an art, but I suspect my readings agree with conventional wisdom. I'll typically focus on the Bohm case, as it provides many worked out physical systems that diverge sharply from the quantum textbooks, and crucially, the other interpretations; but this focus is for convenience only.

4.5.2 *Water and bonds*

Let's begin with the assertion quoted above about water, its bonds, internal angles, and so on. These propositions are non-mathematical and about the unobservable level, but are they quantum? Not necessarily. Claims about the composition of water, derived from stoichiometry, go all the way back to Lavoisier! Crystallography and x-ray diffraction then added to our knowledge of water's structure, but this was based on theoretical work by G.N. Lewis and experimental work by van Laue—both safely pre-quantum. The currently used angle of 104.5 degrees is based on crystallography experiments. Cordero may respond that the experimental value is often considered a confirmation of quantum effects, as a simple textbook quantum treatment puts the value at the nearby 109 degrees. But we could still say what Cordero does about water had neither Schrödinger nor Heisenberg ever existed. Experiment plus some minimal non-quantum theory would have been enough.

The danger of confusing insights from experiments for quantum posits exists even with water's subcomponents. Peeking at hydrogen (more in a moment), note that the accepted bond strengths and bond lengths are based entirely on classical physics and experiment. Other claims about orbits, deriving from the famous Bohr model, are based upon semi-classical theories. Claims from such theories are not quantum. A fully quantum treatment of hydrogen will include specifically quantum effects. I grant that a vague boundary exists between what is quantum and what is not in chemistry. Linus Pauling's famous work on the chemical bond is probably the beginning of a fully quantum treatment, although I'm no historian and I'm happy to concede the boundary to Cordero.

In any case, our first lesson is that many plausibly 'safe' statements about bonds and angles are not truly quantum.

4.5.3 *Tunneling*

Tunneling is without doubt a purely quantum effect, one studied in every quantum textbook. It was used by Gamow and Gurney and Condon to explain the emission of alpha particles from unstable nuclei. Because the attractive potential of the nuclei is much larger than the kinetic energy of the alpha particles, such observed emission is impossible classically but possible quantum mechanically.

Textbooks typically explain how tunneling is possible by treating a system of particles of energy E incident from the left beamed at a one-dimensional potential step of height V, where $E < V$. An approximate plane wave solution to the Schrödinger equation is given, where the wavefunction to the left of the barrier is supposed to represent a superposition of a wave going to the right toward the barrier and a reflected wave going to the left, and the wavefunction to the right of the barrier represents a wave that is transmitted through the barrier. Textbooks calculate the probability of transmission by showing that it is a function of the incoming and reflected fluxes, demonstrating that for certain ratios this probability is non-zero—hence demonstrating the possibility of tunneling.

Given this wavefunction, one can work out the Bohmian trajectories that are then implied. Assuming it is a case of tunneling—and therefore that the incident wave is bigger than the reflected one—it follows that the probability current is positive. Because the probability density is positive, the velocity of the particles is therefore positive too. Hence there is nothing *reflected* at all. The alleged 'particles that reflect' actually all have positive velocity *toward the right*! So there is no reflection even when the reflection coefficient is non-zero and the transmission coefficient is not one. The reflection/transmission coefficients used in the textbooks don't have anything to do with the actual Bohmian motion. Here we have a massive departure from the physics promoted by the textbooks. You might reasonably have hoped that all the interpretations would agree with the minimal implication found in the textbooks on the most canonical system, namely, that *something is reflected to the left*. That is not the case.[2]

4.5.4 Hydrogen

This case is simple but instructive. The hydrogen atom was the first system treated quantum mechanically and is a staple of every textbook. The electron is said to orbit the nucleus—or sometimes something vague about a 'probability cloud' extending a certain distance from the nucleus's center is mentioned. But in Bohmian mechanics, as is well known, stationary states such as $\psi = |1s>$ and $\psi = |2s>$ have constant phase and therefore the electron is *at rest* with respect to the nucleus. (In the quantum potential approach to Bohm, what happens is that the so-called 'quantum potential' Q balances the classical potential V, holding the electron a fixed distance from the nucleus. Q thus provides the quantum 'pressure' keeping the electron from crashing into the nucleus as predicted classically.) Here we have *no orbits* at all, contrary to the textbook picture!

This situation can happen even when the orbital angular momentum quantum number (l) is greater than zero. One can have (a kind of) momentum without motion. The lesson is that "[q]uantum numbers do not directly represent dynamical properties" (Holland, 1993, 156). This is an important point, as the textbooks mostly assume that quantum numbers do reflect properties of a system. An advocate of the quarantine strategy might hope for that much. If we briefly turn to collapse theories, note that on a 'flash' ontology picture a lone hydrogen atom is most likely literally

[2] I admit that this example is an artifact of an inadequacy of the textbook treatment of tunneling. The normal textbook treatment via plane waves is flawed: it's not clear how such stationary states justify talk of entities moving from the left and so on; worse, these waves are not renormalizable, so they aren't physical. A better treatment is possible (see Norsen, 2013). Nonetheless, this case is a nice one to use here because this (albeit flawed) textbook treatment of tunneling is so common and the straightforward Bohmian consequence is so at odds with it.

nothing. The world is populated with flashes only when collapses occur, but we would in all likelihood have to wait thousands of years for a hydrogen atom to collapse. On a mass density interpretation, by contrast, the mass or matter will be smeared out across all components of the superposed atom.

Bohmians and textbooks don't agree about hydrogen until the observable level. Agreement exists because the observables require an experiment, and hence a physical interaction, one that changes the state of the hydrogen atom (and in particular, gets the electrons to where they need to be). Below the surface of the observable, Bohmians also disagree with Collapse and Everett about the behavior of hydrogen atoms.

4.5.5 *Two-path experiments*

Consider the sort of typical two-path interference experiment commonly found in textbooks. A spin-1/2 particle enters a Stern–Gerlach device oriented so as to separate x-up from x-down particles. Depending upon its initial position, the Bohm particle either follows the upward wavepacket or the downward one. Deflectors are added to the setup that deflect the upward wavepacket downward and the downward one upward. The wavepackets meet at location I, symmetric between the up and down paths, and then continue on their way, the initially upward wavepacket heading downward to A' and the initially downward wavepacket heading upward to B'. Measurements can be made at A' and B'.

Textbooks describing such a case of course do not assume that there are particles traveling definite trajectories. Often we're told that because the particle is in a superposition it simultaneously travels both paths. Whatever is going on, it is assumed that *if the particle is measured at A' then it came from A* and *if it is measured at B' then it came from B*. Something is traveling from A to A' or B to B' or both.

What happens in Bohm's theory? Because the Bohm velocity equation is first order and deterministic, trajectories cannot cross in configuration space. That fact, coupled to the additional fact that spin is a feature of wavefunctions and not particles, forces Bohmian trajectories sometimes to behave in highly non-classical and surprising ways. In the experiment at hand, due to the symmetry of the setup, trajectories would have to cross at I for a particle from A to go to A' or from B to go to B'. Hence the probability of finding a Bohm particle at the exact line of symmetry intersecting what would be the intersection point is zero. In terms of the quantum potential, what happens is that it grows infinitely large at that point in configuration space, pushing all particles away. The result: Bohm trajectories bounce at I! Loosely put, a particle from A will 'ride' that wavepacket until location I, but there, where the x-up and x-down wavepackets overlap, the particle will jump ship and hitch a ride with the originally downward wavepacket, creating the bounce. Particles found at B' originate at A and those found at A' originate at B, just the opposite of what is normally assumed (Figure 4.3).

You might reply that the standard textbook doesn't clearly commit to some ontology traveling from A to A' or B to B'. I agree that, ontologically speaking, the standard interpretation says little that is clear. Yet conventional quantum wisdom here is not ambiguous. In fact, the understanding that what is found at A' came from A is so strong that it was a central premise in an attempt by Englert et al. (1992) to falsify Bohm's theory. In the so-called 'surrealistic trajectories' debate, beautifully diagnosed

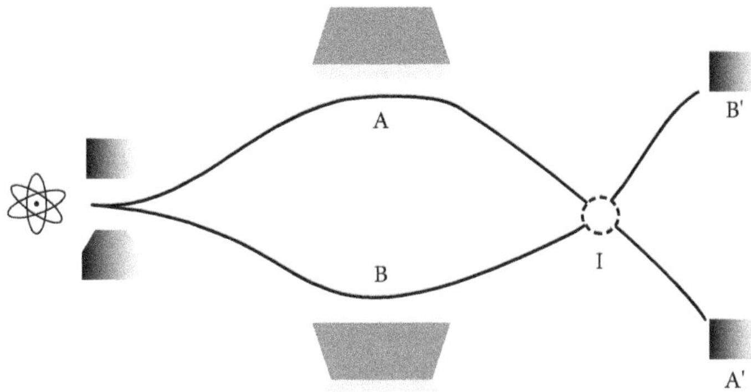

Figure 4.3 Two-Path Setup

by Barrett (2000), the main issue was that Englert et al. assumed that they knew, from quantum theory, that the 'real' trajectories didn't bounce.

4.5.6 Bohr orbits

Perhaps I'm missing Cordero's point. Cordero suggests that there are layers of models that exist in quantum physics. The idea might be that many of these models are shared in common amongst quantum programs. Although not purely quantum, there is, for example, the Bohr model of the atom, and the BCS model of superconductors. There are also whole theories and interpretations of this 'middle layer,' such as semiclassical mechanics and the semiclassical interpretation. Again I find myself drawn to Cordero's position. His picture of physics as complex sets of models covering different regimes is much more realistic than that of many philosophers. One wouldn't be surprised if Everett, Collapse and Bohm, ever the enemies, end up holding hands in peace at the semiclassical level. Alas, I don't think that is so either. Let's agree to relax what we mean by quantum. Now we just want significant claims about the unobservable that aren't purely in the classical domain. That might not be enough to justify a realism about the quantum, but it would be a start.

Start with Bohr orbits. Consider hydrogen again and now its orbits. In the Bohm theory, when the electron is not at rest it orbits the z-axis with constant speed and radius and is independent of mass. In the Bohr model, by contrast, the electron traces out orbits in the equatorial plane and the radius is a function of mass. The orbits are around different axes and one is a function of mass yet the other isn't. (See Fortin, Lombardi, and González, 2017 for this case and more.) They're different and it's hard to see how one approximates the other in any way. Again, at the measurement level, the Bohmian will be able to explain why the Bohr model worked as well as it did; but the reason isn't that Bohm particles travel the same orbits.

4.5.7 Semiclassical particle in a box

A field known as semiclassical physics (associated with the physicist Gutzwiller) develops and examines connections between classical orbits and quantum fluctuations. In this area, a system is understood to be 'semiclassical' if the classical action is large compared to quantum $\hbar$. One finds semiclassical trajectories in this approach as one deals with chaos and other topics in dynamical systems theory. A tempting thought is that we'll find that as we approach the observable level, there will be agreement below the surface of the observable on the semiclassical trajectories. However, Matzin and Nurock (2008) show that semiclassical orbits differ dramatically from Bohmian orbits.

A simple example of the difference is the particle in a box. Using the typical wavefunction for such cases and assuming one dimension and fixed energy, the Bohm particle will just sit still (unless you pull a wall away rapidly (Callender and Weingard, 1998)). The semiclassical orbits, by contrast, give two classical orbits at each position in the box, one going in each direction, i.e., particles bouncing back and forth. Matzin and Nurock display other examples of divergence too, including (not surprisingly) hydrogen. Bohm's theory suggests a physical picture demonstrably at odds with the trajectories used in a semiclassical treatment.

4.5.8 *Summary*

Most of what we say about the quantum realm is 'interpretation' dependent. The research programs described here portray radically different worlds from top to bottom, agreeing on little more than what is observable. I provided some examples but could multiply them easily, e.g., Bohmian Fermi–Dirac particles are not always repelled nor Bose–Einstein particles always attracted (Holland, 1993, 310). I could also have used more examples from Collapse or Everett, so the response that these cases just show that Bohm is weird or unusual isn't sustainable. I cannot prove Cordero wrong. I have not gone through all of quantum mechanics and shown that there is nothing safe from the blight. Some models and systems may be safe. But these would be more like small disconnected islands of reprieve, not anything like a full quarantine zone.

4.6 Reaction: There Can Only Be One?

If I am right, there are at least three major research programs that each portray different worlds but that are compatible with the current empirical evidence. This situation poses a threat to the epistemic ambitions of the realist, someone who believes that mature successful theories are well confirmed and approximately true. We can't say one program is well confirmed and approximately true if we know that there are two others, equally well confirmed, that contradict its hypotheses.

Before considering realist reactions, note that the situation isn't horribly dire. We hardly have *guaranteed* underdetermination by an *indefinite* number of theories. Three is a small number. We could apportion our degrees of belief over these three programs and not be at a complete loss when it comes to claims about the unobservable quantum world.[3]

[3] Could we repartition our programs and narrow down to two? What I have in mind is focusing on ontology and tipping theories into wavefunction-only and wavefunction-plus camps. Some versions of Everett and Collapse will then join Bohm in the wavefunction-plus camp, as they add a beable to the wavefunction in their basic ontology. I'm not a big fan of this repartitioning as it obscures major differences.

Quantum underdetermination isn't, therefore, a disaster for realism. It is still disappointing. Can we do better? Ultimately there are two options, fight or flight. Realism could retreat by restricting its ambitions to claims that are interpretation neutral, as Hoefer (this volume, Chapter 2) does. Quarantine works if the wall is placed around the observable level. Arguably, it also works in the potentially unobservable classical domain sector of quantum theory—although this is a tricky question and I have my doubts. Alternatively, we can fight this judgment by turning up the 'scientific' dial. That is, we can use traditional realist features such as simplicity, unification, explanatoriness, and so on to decide which research program is best. We agreed that there are at least three options when the dial is set at 'scientific.' That might be too low a standard. None of the programs are cheap philosophical playthings, but that doesn't mean they are all equally well confirmed.

Let's briefly explore the more aggressive option of turning up the dial. I began the essay with a metaphor based on Greg Egan's clever and sophisticated book, *Quarantine*. I end with one based on a terribly acted and weakly plotted fantasy film, *The Highlander* (1986). In the film a group of (nearly) immortal warriors battle through history, dying only through decapitation. The last remaining will win the Prize. Warriors get stronger each time they kill one of their own. They know that, in the end, *there can only be one*. While I don't expect proponents of the different research programs to go away any time soon—and I certainly hope that they don't resort to Highlander-like tactics—it may be that when we turn up the dial, only one remains.

But which one?

Wallace (this volume, Chapter 5) asserts that there is no underdetermination in quantum mechanics, that there is only Everett. His argument is that Everett and only Everett has been successfully applied to *all* of current physics. Bohm and Collapse lag behind, slogging their way through the history of quantum theory. Specifically, those research programs must develop relativistic and field-theoretic versions of quantum theory, whereas it takes no time at all to make these versions Everettian. Put in terms of research programs, the idea is that the Everettian program is ahead of its rivals. When the dial is set to include empirical reach or size of domain, there is no underdetermination serious enough to cause alarm.[4]

Some proponents of structural realism also claim that there is no underdetermination; indeed, some motivate this type of realism via its ability to overcome

For instance, Wallace (2014) shows that the branches corresponding to wavefunction-only GRW tails are qualitatively different from their Everettian counterparts, so the wavefunction-only camp includes very different worlds. More importantly, scientific realism is not only about the ontology but also the laws of the theory, and this partitioning ignores that fact.

[4] To be clear, the issue is more subtle than just described. There are indeed plenty of Bohmian field theories. There are also extensions of the theory to quantum gravity, quantum cosmology, superstring theory, quantum chemistry, and more. If Bohmian or Collapse answers to problems in new realms can't be reproduced by Everett then it's not clear who is more progressive. The Bohmian answer to the problem of time in canonical quantum gravity, for example, is not replicable in an Everettian framework, nor is the Collapse approach to the information loss paradox. The reason I put these points aside for the moment is that, overall, I agree that quantum field theory as a target dwarfs these examples in importance, and it's also fair to say that Bohmian or Collapse versions of the standard model of particle physics are a ways off.

quantum underdetermination. Since the mathematical structure of Collapse, Bohm, and Everett differ so markedly, how could such a position ever get off the ground? In at least one case, the answer is that such realists deny that Collapse and Bohm merit consideration! Ladyman and Ross (2007), in their polemic against unmoored speculative metaphysics, claim that any response to the measurement problem that takes standard quantum mechanics to be an incomplete description of reality is an example of extravagant epistemically irresponsible metaphysics rather than good metaphysics or science (p. 181). We can understand this as turning the dial to a very specific (and odd) setting.[5]

For a quite different judgment, consider the position of Jean Bricmont (2016) when confronting quantum underdetermination. He argues that "there is no existing alternative to de Broglie–Bohm that reaches the level of clarity and explanatory power of the latter" (p. 228). Bricmont sees the threat of quantum underdetermination and adopts the Highlander move of eliminating alternatives. Unlike Wallace or Ladyman and Ross, he opts for Bohm as the last one standing, a judgment based on emphasizing explanatory virtues. We might understand Bricmont as employing confirmation understood as inference to the best explanation (Lipton, 1991). According to such theories, for T to be confirmed by E, T must not only imply E but T must be the best explanation of E. Bohmians feel that the nuts-and-bolts accounts the theory provides of the stability of matter, uncertainty relations, interference, apparent wavefunction collapse, tunneling phenomena, and more, is a major reason to adopt the theory. In a note added to the paperback edition of his Bohmian masterpiece, Holland stresses that the primary virtue of the Bohm program is its "quality of explanation" (1993, xix), much like that he finds in Darwinian reasoning. If this is right—and it certainly fits with my thoughts on the matter—one can imagine a Lipton-style argument to the effect that Bohm is better confirmed than its rivals. We turned up the dial once and eliminated the Cartesian demon. We turn the knob again and the same kinds of explanatory virtues leave only Bohm.

The debate between Wallace and Bricmont isn't likely to be settled anytime soon. The reason is that the *very features that allow the Everettian interpretation its easy extension to new physics* are *precisely the same features that invite its problems and whose solutions by other programs lead to their explanatory virtues.* Collapse swerves and Bohmian beables make probabilities relatively straightforward in these theories, for instance, whereas understanding probability is massively problematic in Everett. Moreover, it's these beables and swerves that allow the nuts-and-bolts explanatory

[5] I cannot fully respond to this charge here due to the editors' demand for polite, professional language. I'll just note that the following people have developed 'completions' of quantum mechanics: Louise de Broglie, John Slater, Erwin Madelung, Albert Einstein, Nathan Rosen, Jean-Pierre Vigier, David Bohm, Hans Freistadt, and John Bell. This list includes some of the top physicists who have ever lived. Scores of mathematicians and physicists the world over continue this work, e.g., Peter Holland, Sheldon Goldstein, Detleff Dürr, B.J. Hiley, Nino Zanghì, and Roderich Tumulka, and publish rigorous advances in the best physics journals, e.g., *Physical Review*. Bohm's original paper has over 5000 citations. Appeal to authority is an improper argument form, I agree, but we can use it as a shortcut to a longer case I could make and ask: given the extrinsic markers of epistemic quality just listed, which is the more likely to be extravagant metaphysics, or a progressive research program—the work of Einstein, Bell, and others in our best physics journals, or ontic structural realism?

narratives that attract Bohmians (and presumably, Collapse theorists). Yet it's these very swerves and beables that demand new physics to be developed when quantum theory is applied to new regimes, making extensions hardly automatic. The Everettian program is a 'minimalist' one, little more than the quantum formalism itself coupled to a new rule for reinterpreting our definite empirical outcomes. So it is little wonder that it can be 'successfully applied' to new physics. To someone engaged in a competing research program, trumpeting Everett's easy application to new physics sounds like a thief bragging about how little they had to work for their reward. After all, the Copenhagen interpretation—and Cartesian demon theory, for that matter—can make the same boast, yet many would agree that Copenhagen—and Cartesian demon theory—are so theoretically deficient as to take themselves out of the running. They earn their empirical progressiveness at the cost of providing a decent theory. Naturally, the Everettian sees matters precisely the other way around. Why are these 'nuts-and-bolts' explanations, whatever they are, so great, and in particular, relevant to confirmation? Absent a compelling reply, and failing to see Everett as theoretically unsatisfactory, the choice for the empirically more progressive program is to them a no-brainer.

My own sympathies lie with the more 'nuts-and-bolts' approaches. The point I want to make, however, is that we're pretty close to philosophical bedrock at this point. The Everettian and Bohmian described above aren't merely disagreeing on the correct dial setting, but they are disagreeing on the nature of the dial. Put somewhat simplistically, the Everettian uses a dial that represents size of empirical domain whereas the Bohmian uses a dial that represents explanatory virtues. The choice of research program therefore hangs on deep, hotly contested and familiar matters in philosophy of science—in particular, the relationship, if any, between explanatory virtues and confirmation.

In sum, we have serious scientific underdetermination. The nightmare of scientific realists is real. We're unlikely to secure any quarantine zone that retains much one can trust in the quantum realm. Because the differences between programs are so stark, it's tempting to go into Highlander mode and declare that there can be only one. But that debate—who remains standing—won't go away soon because its resolution hangs on philosophical matters that are gridlocked. Welcome, scientific realists, to the quantum foundations.

Acknowledgments

Many thanks to Adam Chin, John Dougherty, Carl Hoefer, Anncy Thresher, and especially Kerry McKenzie, Chip Sebens, and the referees for incredibly helpful feedback on a draft of this paper.

References

Allori, V., S. Goldstein, R. Tumulka, and N. Zanghì (2008). On the common structure of Bohmian mechanics and the Ghirardi–Rimini–Weber theory. *The British Journal for the Philosophy of Science* 59(3), 353–89.

Allori, V., S. Goldstein, R. Tumulka, and N. Zanghì (2011). Many-worlds and Schrödinger's first quantum theory. *The British Journal for the Philosophy of Science* 62(1), 1–27.

Barrett, J. A. (2000). The persistence of memory: surreal trajectories in Bohm's theory. *Philosophy of Science* 67(4), 680–703.
Bell, J. S. (1987). *Speakable and Unspeakable in Quantum Mechanics*. Cambridge: Cambridge University Press.
Bohm, D. (1952). A suggested interpretation of the quantum theory in terms of "hidden" variables, parts I and II. *Physical Review* 85, 166–93.
Bricmont, J. (2016). *Making Sense of Quantum Mechanics*. Dordrecht: Springer.
Callender, C. (2015). One world, one beable. *Synthese* 192, 3153.
Callender, Craig and R. Weingard (1998). Nonlocality in the expanding infinite well. *Foundations of Physics Letters* 11, 495–8.
Cordero, A. (2001). Realism and underdetermination: some clues from the practices-up. *Philosophy of Science* 3, S301–S312.
Daumer, M., D. G. S. Dürr, and N. Zanghì (1996). Naive realism about operators. *Erkenntnis* 45, 379–97.
Dürr, D., S. Goldstein, and N. Zanghì (1992). Quantum equilibrium and the origin of absolute uncertainty. *Journal of Statistical Physics* 67, 843–907.
Englert, B. G., M. Scully, G. Süssman, and H. Walther (1992). Surrealistic Bohm trajectories. *Zeitschrift für Naturforschung A* 47, 1175–86.
Everett III, H. (1957). "Relative state" formulation of quantum mechanics. *Reviews of Modern Physics* 29, 454–62.
Fortin, S., O. Lombardi, and J. C. M. González (2017). The relationship between chemistry and physics from the perspective of Bohmian mechanics. *Foundations of Chemistry* 19, 43–59.
Ghirardi, G. C., A. Rimini, and T. Weber (1986). Unified dynamics for microscopic and macroscopic systems. *Physical Review D* 34, 470–91.
Healey, R. (2017). Quantum-Bayesian and pragmatist views of quantum theory. *Stanford Encyclopedia of Philosophy*: https://plato.stanford.edu/entries/quantum-bayesian/.
Holland, P. (1993). *The Quantum Theory of Motion: An Account of the de Broglie–Bohm Causal Interpretation of Quantum Mechanics*. Cambridge: Cambridge University Press.
Ladyman, J. and D. Ross (2007). *Every Thing Must Go: Metaphysics Naturalized*. Oxford: Oxford University Press.
Lakatos, I. (1978). *The Methodology of Scientific Research Programmes: Philosophical Papers*, Volume 1. Cambridge: Cambridge University Press.
Laudan, Larry and J. Leplin (1991). Empirical equivalence and underdetermination. *Journal of Philosophy* 88, 449–72.
Lipton, P. (1991). *Inference to the Best Explanation*. London: Routledge Kegan & Paul.
McKenzie, K. (2017). Ontic structural realism. *Philosophy Compass* 12(4): https://doi.org/10.1111/phc3.12399.
Matzin, A. and V. Nurock (2008). Classical and Bohmian trajectories in semiclassical systems: mismatch in dynamics, mismatch in reality? *Studies in History and Philosophy of Modern Physics* 39, 17–40.
Maudlin, T. (2019). *Philosophy of Physics: Quantum Theory*. Princeton: Princeton University Press.
Norsen, T. (2013). The pilot-wave perspective on quantum scattering and tunneling. *American Journal of Physics* 81, 258–66.
Norton, J. (2008). Must evidence underdetermine theory? In D. H. M. Carrier and J. Kourany (Eds), *The Challenge of the Social and the Pressure of Practice: Science and Values Revisited*, pp. 17–44. Pittsburgh: Pittsburgh University Press.
Okon, Elias and D. Sudarsky (2015). The black hole information paradox and the collapse of the wave function. *Foundations of Physics* 45, 461–70.

Peruzzi, G. and A. Rimini (2000). Compoundation invariance and Bohmian mechanics. *Foundations of Physics 30*, 1445–72.

Sebens, C. (2015a). Killer collapse: empirically probing the philosophically unsatisfactory region of GRW. *Synthese 8*, 2599–615.

Sebens, C. (2015b). Quantum mechanics as classical physics. *Philosophy of Science 82*, 266–91.

Skow, B. (2010). On a symmetry argument for the guidance equation in Bohmian mechanics. *International Studies in the Philosophy of Science 24*, 393–410.

Struyve, W. (2011). Pilot-wave approaches to quantum field theory. *Journal of Physics: Conference Series 306*, 012047.

Wallace, D. Life and death in the tails of the GRW wave function https://arxiv.org/pdf/1407.4746.pdf [quant-ph].

Wallace, D. and C. G. Timpson (2010). Quantum mechanics on spacetime i: spacetime state realism. *The British Journal for the Philosophy of Science 61* (4), 697–727.

Weingard, R. (1996). Exotic (quixotic?) applications of Bohm theory. In R. Clifton (Ed.), *Perspectives on Quantum Reality*, The University of Western Ontario Series in Philosophy of Science. Dordrecht: Springer.

5

On the Plurality of Quantum Theories: Quantum Theory as a Framework, and its Implications for the Quantum Measurement Problem

David Wallace

Philosophy is what you do when you don't yet know what the right questions are to ask.

Daniel Dennett[1]

5.1 Introduction

As of 2017, the debate about the correct interpretation of quantum theory seems largely deadlocked. There is a reasonable consensus within philosophy of physics on what the main viable interpretative strategies are, which goes something like this (see subsequent discussion for references):

1. 'Realist' strategies, dominated by:
   (a) Hidden-variable theories, most notably the de Broglie–Bohm theory (aka Bohmian mechanics);
   (b) Dynamical collapse theories, most notably the Ghirardi–Rimini–Weber theory and Pearle's continuous-state-localization theory;
   (c) The Everett interpretation, in its modern (decoherence-based) form.
2. 'Non-realist' strategies, notably (but not exhaustively):
   (a) The (various forms of the) Copenhagen interpretation;
   (b) Physics-is-information approaches, most notably the 'Quantum Bayesianism' or 'QBism' of Chris Fuchs and co-workers;
   (c) Healey's quantum pragmatism.

[1] Blackmore (2005), 91.

David Wallace, *On the Plurality of Quantum Theories: Quantum Theory as a Framework, and its Implications for the Quantum Measurement Problem* In: *Scientific Realism and the Quantum*. Edited by: Steven French and Juha Saatsi, Oxford University Press (2020). © David Wallace.
DOI: 10.1093/oso/9780198814979.003.0005

The debate over realist-vs-non-realist strategies has been fairly cursory in recent discussions and has largely turned on general disagreements about the legitimate aims of science: Maudlin (1995), for instance, simply takes as read that quantum theory ought to have a representational role, while according to Fuchs and Peres (2000a; 2000b) or Healey (2012; 2017b), a central lesson of quantum mechanics is that it does not have such a role and needs to be understood more as some kind of predictive or calculational tool. In the bulk of philosophy of physics (in particular in its more metaphysically inclined corners), indeed, the non-realist strategies are set aside almost without comment.[2]

Meanwhile, comparative assessment of the realist strategies has tended to turn on relatively detailed, and fairly metaphysical, concerns with those strategies. Can the Everett interpretation explain probability? Is Bohmian mechanics just Everett in denial? How do dynamical-collapse theories resolve the problem of tails?[3] The form of the discussion normally takes as read that these various approaches would succeed in 'solving the measurement problem' if only these metaphysical problems could be resolved. Indeed, it is common (see, e.g. Cordero, 2001; Egg, 2014; Lewis, 2015; Lyre, 2010) to describe the choice between realist interpretations as a classic case of *underdetermination of theory by data*, with Bohmian mechanics, the GRW theory, and (sometimes) the Everett interpretation all on a par empirically and to be distinguished only by assessments of 'extra-empirical virtues', again of a largely metaphysical nature: preservation of determinism, avoidance of ontological extravagance, conformity with intuition, conformity with the spirit of relativity. If so, there seems to be little realistic likelihood of consensus, or even significant progress, on the interpretation of quantum mechanics any time soon. But perhaps we can console ourselves that even if the final answers to the questions of quantum interpretation will continue to elude us, at least we have a clear understanding of what the questions are and what the space of possible answers looks like.

But in philosophy, most of the work, and most of the controversy, lies precisely in stating and understanding the questions, and so apparent consensus on how to frame a problem is often a sign of hidden assumptions and communication failure.[4] And so it is in quantum theory (I shall argue). Advocates of different interpretative strategies differ not just on technical details of those strategies, or on assessment of whether and how those strategies overcome their own specific problems, but on basic questions of philosophy of science: notably, on how theories confront experiment and on how they represent physical facts. And (relatedly) they differ on what the theory is that we

[2] Productive debate is not helped by the fact that most critics *and* defenders of what I here call 'non-realist' strategies regard 'realist' as a virtue term: witness in particular Fuchs' (2017) plea to critics *not* to call QBism anti-realist, but also Healey's description of his position as a 'third way' between realism and non-realism. I don't myself think the label is particularly helpful (I will later argue that 'non-representationalist' is a more neutral description); my summation here is intended as descriptive of the sociology of the field.

[3] See Wallace (2008) for references in each case.

[4] For examples elsewhere in philosophy, consider free will, or the mind-body problem. In the former case, careful formal proofs to the effect that we do or do not have free will tend to pack most of the philosophical work into apparently uncontentious premises (Dennett, 1984, 3); in the latter case, even deciding to frame the issues in terms of an 'easy' and 'hard' problem (Chalmers, 1995) concedes most of the ground on which hardline functionalists will want to fight (Dennett, 2005, ch. 6).

are supposed to interpret: to a first approximation, advocates of different strategies are trying to interpret very different theories in the first place. Put succinctly, if the central question in philosophy of quantum theory is 'what is the correct interpretation of quantum mechanics', then as well as open disagreement about the assessment of 'correct', there is hidden disagreement about the meaning of 'interpretation' and, even more so, of 'quantum mechanics'.

To expand slightly, the case I will make is that:

1. Quantum theory is a framework theory, under which many specific quantum theories stand. These theories are related by different (and only partially understood) instantiation relations (where the instantiation relation holds between two specific quantum theories, describing the same system at different levels of detail) but a satisfactory interpretation of quantum mechanics must be a strategy for interpreting each of these theories in its own terms. No satisfactory interpretation can be an interpretation of the 'fundamental' quantum theory, partly because we do not have any such theory to interpret at present but mostly because the way in which quantum theory makes contact with experiment is through this plurality of different theories and cannot be cleanly described in the vocabulary of any one quantum theory, however fundamental.
2. Neither the so-called 'non-realist' interpretations (I will argue that 'non-representationalist' is a fairer name), nor the strategies that aim to modify or supplement quantum theory, adequately deal with the theory as it is actually used and tested in science. Rather, they attempt to interpret either the abstract framework (in a way which is not at present adequate to recover the empirical success of particular instances of that framework), or else they interpret a particular theory within that framework, under the fiction that it is by itself adequate to all quantum phenomena (and as such rely on a once-and-for-all account of how quantum theories make contact with empirical data which is again not adequate to recover the full empirical successes of quantum theory). Only the Everett interpretation attempts to provide a general recipe to interpret the various particular quantum theories in a way that is compatible with their interrelations.

These conclusions are not neutral: they give strong support for the Everett interpretation, of which I have been a longstanding advocate. That's not a coincidence: this chapter is a codification and development of what I have long thought to be the strongest reasons for that interpretation (building on preliminary versions of these arguments presented in Wallace (2008, 83–5) and Wallace (2012, 33–5)). But I hope even readers sceptical of this particular conclusion might be persuadable of the general point that the deadlock in discussions of interpretation can be broken—or at least, our understanding of the problem can be deepened—by focusing less on specifics of the particular interpretative strategies, and more on what we should reasonably require of a solution to the measurement problem in the first place.

The structure of the chapter is as follows: in Section 5.2 I develop the idea of quantum theory as a framework, and in Section 5.3 I consider how to think about inter-theoretic relations between concrete theories in that framework. In Sections 5.4–5.6 I consider, sequentially, Everettian quantum mechanics, so-called non-realist strategies, and strategies that try to supplement or modify the quantum formalism, and explore how each fits into the conception of quantum mechanics developed in Sections 5.2–5.3. Section 5.7 is the conclusion. The physics I discuss in this chapter is for the most part standard and well established, and I do not attempt to give original references.

5.2 Quantum Theory: Frameworks and Instances

Asked what 'quantum theory' is, two initially plausible answers might be:

Abstract quantum mechanics: A quantum theory is specified by:

1. A Hilbert space $\mathcal{H}$ (the rays of which represent possible states of the system);
2. A collection of self-adjoint operators on $\mathcal{H}$ which represent the *observables* of the system (this is a term of art; a more neutral term might be 'physical quantities').
3. A preferred observable $\widehat{H}$, the *Hamiltonian*, which generates the system's dynamics via the *Schrödinger equation*

$$i\hbar\frac{\partial}{\partial t}\left|\psi(t)\right\rangle = \widehat{H}\left|\psi(t)\right\rangle. \tag{5.1}$$

Physical content is extracted from the theory by the *Born rule*, which states that the expectation value, on measurement, of the observable corresponding to operator $\widehat{O}$, if the system's quantum state is $|\psi\rangle$ is

$$\langle\widehat{O}\rangle_\psi = \langle\psi|\widehat{O}|\psi\rangle. \tag{5.2}$$

Quantum particle mechanics: The quantum theory of N particles of mass $m_1, \ldots m_n$ is specified by a function ψ from the $3N$-dimensional configuration space of those N particles to the complex numbers, satisfying the *Schrödinger equation*

$$i\hbar\frac{\partial\psi}{\partial t}(\mathbf{x}, t) = -\sum_{1\leq i\leq N}\frac{\hbar^2}{2m_i}\nabla_i^2\psi(\mathbf{x}, t) + \sum_{1\leq i<j\leq N} V_{ij}(|\mathbf{x}_i - \mathbf{x}_j|)\psi(\mathbf{x}, t) \tag{5.3}$$

where V_{ij} is the interaction energy between particles i and j, $\mathbf{x}$ represents, schematically, the N-tuple of coordinates $\mathbf{x}_1, \ldots \mathbf{x}_N$ on configuration space, and ∇_i is the gradient with respect to $\mathbf{x}_i$. Physical content is extracted from the theory by the *Born rule*, which states that the probability density, on measurement, of finding the particles in configuration $\mathbf{x}$ at time t is

$$\Pr(\mathbf{x}; t) = |\psi(\mathbf{x}, t)|^2. \tag{5.4}$$

As accounts of quantum mechanics as a scientific theory, both accounts are for different reasons very deficient (even before we address more philosophical concerns like the quantum measurement problem). To begin with 'abstract quantum mechanics': in an important sense this is not really a scientific theory at all. By itself it makes no predictions and explains no phenomena; by itself it cannot be tested or falsified. It is a *framework* within which concrete quantum theories can be stated: a given theory within the framework is specified (typically) by a specific choice of Hilbert space, a specific choice of dynamical variables (represented by self-adjoint operators), a specific dynamics (represented by some choice of Hamiltonian) and often a specific decomposition into subsystems. These *specific, concrete* theories can of course be tested, and those tests can indirectly confirm or falsify the viability of the framework, but without being supplemented by a particular concrete realisation of the framework, it can do only very little scientific work.

When we recognize this, we can also see what is deficient about 'N-particle quantum mechanics': it is only one example of a quantum theory, and one of quite restricted applicability. Indeed, strictly speaking there is to my knowledge only one small class of applications of wave mechanics in this form: to the hydrogen atom, and to other ions with only one electron. If the theory is modified to include identical particles, background potentials, and spin, its applicability becomes much wider: most of atomic physics, and most of quantum chemistry, and a large part of solid state physics falls within its remit, taking the 'particles' to be electrons and the nuclei of atoms. But even there, the theory rarely suffices to describe *all* the relevant phenomena of a system: for instance, atomic energy levels are calculated within this theory, but the theory cannot incorporate the de-excitation of those energy levels by photon emission which is the main route by which we test the theory's energy-level predictions.[5]

A really satisfactory description of 'quantum mechanics' would have to characterize it as a large *collection* of theories, each fitting (more or less) within the abstract framework, and each applicable to different systems and at different energy- and length-scales. One of those theories would be N-particle quantum mechanics. Or more accurately: *many* of those theories would be different forms of N-particle quantum mechanics, with different numbers and statistics of particles, different background potentials, and different interpretations of the 'particles'. (Sometimes they are electrons, sometimes ions, sometimes the centers of mass of larger bodies—even moons or planets (Zurek and Paz, 1995).) Others would be quantum field theories of various kinds, applicable to various situations involving light, collective excitations of solid bodies, or relativistic effects. Others still would be discrete theories with finite-dimensional Hilbert spaces, applicable to the internal degrees of freedom of certain systems whose spatial degrees of freedom can be idealized away.

It should not be surprising to find that 'quantum theory' needs to be characterized this way. After all, the same is true for 'classical mechanics'. There we could have tried to characterize the theory as the abstract form of Hamiltonian or Lagrangian mechanics, or as the theory of point particles interacting under long-range forces. There, too, the first characterizes a framework for dynamical theories rather than a concrete theory; the latter characterizes a theory which captures only a small part of classical mechanics (failing to capture, for instance, electromagnetic effects, or the physics of rigid bodies or fluids). And classical particle mechanics, like quantum particle mechanics, is best thought of not as a single theory but as a wide class of different theories, characterized by different particle masses, interactions, background potentials, and interpretations of the 'particles'. (Recall that in the most famous application of classical mechanics, the 'particles' are the centers of masses of the Sun and planets.) If we were to set out to 'interpret classical mechanics', our task would not

[5] The theory is also somewhat misrepresented in the form I have given it by its focus on configuration space and on the position representation, but that can more readily be fixed.

be completed *either* by trying to interpret the abstract form of Hamiltonian mechanics as a concrete theory (which would simply be a category error), *or* by trying to interpret classical particle mechanics (which has no univocal interpretation, and in any case is only a small part of classical mechanics).[6]

At least prima facie, the same would seem to apply to quantum theory. Before developing this point, though, I need to develop further the picture of quantum theory I have sketched by considering how different versions of quantum theory relate to one another.

5.3 Quantum Theory: Inter-Theoretic Relations

A tempting and popular picture of inter-theoretic relations is that of a tower of theories, each approximating the theory below it in the appropriate limit.[7] For physics, at the bottom of the tower would lie the Standard Model of particle physics (perhaps with its base shrouded in mist to leave room for the hoped-for theory of quantum gravity that *it* approximates). Above it, perhaps, would be quantum electrodynamics; above that, the quantum theory of photons and nonrelativistic atoms; above that, nonrelativistic quantum mechanics; above *that*, perhaps, classical particle mechanics, and then classical fluid mechanics.

But this picture is badly misleading. It implies there is some unified class of phenomena all simultaneously describable by the Standard Model, some subclass of that class simultaneously describable by quantum electrodynamics, and so forth. In reality in physics, modelling is local. A given system—say, the electrons of a metal, or the electrons and nuclei of an ion—might be describable by quantum particle mechanics, but it does not follow that in anything but the most formal sense there is a single system comprising both of these that is describable by it. Indeed, sometimes the same system is described by the same theory at different levels: for instance, the electrons in the metal might be treated as evolving against a classical background potential which, at a deeper level of description, is produced by a lattice of ions. Nor is there any consistent hierarchy of theories: in classical mechanics, for instance, a collection of lattices of masses connected by stiff springs could be describable at one level by point-particle mechanics (with the 'particles' being the lattice elements), at a higher level by rigid-body mechanics (with the 'bodies' being the lattices), and at a higher level still by point-particle mechanics again (with the 'particles' being the centres of masses of the lattices).

So a better picture is more like a patchwork than a tower: for any given system there are various levels of description at which various theories are applicable. And since the notion of 'system' is itself theory-laden, there is no theory-free starting language with which we can describe this picture. For instance, one 'system' is composed of the centres of masses of the bodies in the Solar system, but 'centre of mass' is itself theory-laden and theory-relative. (Again, see Wilson (2013) for further development of this point in the classical context.)

[6] Mark Wilson has developed (essentially) this view of classical mechanics in much more depth; see, in particular, Wilson (2013).

[7] The classic statement of this view is Oppenheim and Putnam (1958).

What do inter-theoretic relationships look like on *this* picture? If they exist, they would have to look like this: if theory X describes a system at some level of description, and theory Y describes that same system at a more detailed level, then the description from theory X can in some sense be derived from that of theory Y. And if several systems S_1 through S_n can be described separately by theory X at a higher level, and collectively by theory Y at a lower level, we can derive from theory Y both the applicability of X to the systems separately and the validity, at that level, of the decomposition into subsystems. Ultimately we might hope to find a sufficiently fine-grained level of description at which the whole Universe can be jointly regarded as a single system, and the applicabilities of all of the higher-level theories derived directly or indirectly from it.

It is not a priori obvious that things must work out this way: Nature could be inherently disunified, with many different descriptions on different lengthscales and no systematic relations between them. Cartwright (1983, 1999) has argued, based on careful study of case studies in physics and other areas of science, that indeed this is the case. But (although there is far more to say here than space permits) this kind of disunified picture—whatever its metaphysical coherence or incoherence—does not seem to give adequate weight to the great many cases in which physicists have actually succeeded in constructing local inter-level relations. Most of solid state physics, for instance, and most of non-equilibrium statistical mechanics, and much of particle physics and astrophysics, is concerned with deriving higher-level descriptions from lower-level ones, and there are a great many successes that have led to novel empirical predictions, albeit they often involve approximations and assumptions, and contain conceptual puzzles of their own. And the progress of physics does seem to be a progression towards greater unification: the many different solid-state systems are all instantiations of the same nonrelativistic quantum theory; the physics of stellar interiors and of stellar atmospheres seem to be derivable from the same underlying physics under different assumptions. We have reached the point where one theory, the Standard Model of particle physics (with the spacetime metric treated as one more quantum field) is at least a candidate to underlie all the various applications of high-level physics, and to provide the basis for explanation of all physical phenomena outside the extremes of the early universe and the singularities within black holes. So notwithstanding Cartwright's criticisms, I will continue for this chapter to assume—tentatively—that the various applications of physics are indeed interrelated by locally valid derivations, and ultimately that all can be underpinned by the Standard Model.

If we want to interpret quantum mechanics, then, should we simply interpret the Standard Model, and regard every other quantum theory as derivative upon it, or even just as useful calculational tools? There are two problems with so doing. Firstly, almost all our evidence for quantum mechanics is in the first instance evidence for some higher-level quantum theory, not for the Standard Model directly. In most concrete applications we do not have an explicit derivation of that higher-level theory from the Standard Model; even in the most favourable cases, we have a long and indirect chain of derivations. As such, the way in which the Standard Model confronts experiment is almost exclusively indirectly, via tests of higher-level theories.

For instance, quantum theory allows us to calculate the heat capacity of a crystalline insulating solid at low temperatures: it scales with the fourth power of the temperature. That prediction cannot even be stated, much less derived, in the Standard Model, where 'crystalline insulating solid' is not a well-defined term. Other quantum predictions concern the spectral lines of excited atoms, or the conductivity of metals, or the phase transitions of superfluids, or the neutrino emissions of stars; again, these cannot directly be stated, let alone tested, within the Standard Model. To understand quantum mechanics well enough to recover its empirical success, we need to understand the various different quantum theories and their theoretical content. That understanding might in principle be derivative on some understanding of the Standard Model, but it cannot simply be skipped—not if our goal is to understand quantum mechanics as it is in fact used and tested.

Secondly, the standard model—like almost all empirically relevant quantum field theories—is an *effective field theory*. What this means, in outline (see Wallace (2011; forthcoming) for a more detailed discussion) is that while the theory is *formally* a continuum field theory defined on arbitrarily short lengthscales, in fact it must be 'cut off' below some short lengthscale, which represents the point at which the theory breaks down and ceases to be accurate. (For the Standard Model, defined as I have to include the spacetime metric, that lengthscale is the Planck length.) Effective field theories are by their nature not candidates for a fundamental theory; indeed, they are not really theories at all in the philosopher's usual sense, but equivalence classes of theories, with different cutoffs and different implementations of the cutoff, but the same larger-scale structure. Insofar as the interpretation of 'quantum mechanics' ought to be concerned with a single well-defined, universal theory, that theory cannot be the Standard Model but would have to be the yet-unknown theory of quantum gravity which, we hope, underlies it. But since we do not have that theory, it would be premature to try to interpret it, and more premature still to conflate the task of understanding that theory with that of making sense of quantum theory.

5.4 Interpretative Recipes and the Everett Interpretation

Given this account of quantum theory—or rather, given this account of the *many* quantum theories and the system-local, not-fully-hierarchical relations between them—what should we expect from an 'interpretation' of quantum mechanics? Here is one natural answer: we should expect an *interpretative recipe*, a set of instructions which tells us, for any given quantum theory, how to understand *that* theory. Furthermore, the recipe must be compatible with inter-theoretic relations: if theory X instantiates theory Y in certain regimes, our understanding of theory X ought to tell us how to understand theory Y, and it should deliver the *same* understanding as we would get by applying the recipe to Y directly. (I use 'understanding' here as a term neutral between pure interpretations of the quantum formalism, and strategies which supplement it with additional variables or modify its equations.)

The interpretation of *classical* mechanics has exactly this 'recipe' form. The interpretative recipe says, simply: interpret the classical phase space as a space of possible states that a physical system is in, so that a system with dynamical history $x(t)$ has, at time t, the physical features represented by $x(t)$. At this level of abstraction, there is little more to be said: in particular, it would be a category error to ask for 'the' ontology of classical mechanics. For any *particular* classical theory, we can ask for the ontology of *that* theory (and at that level, there is space for controversy as to interpretative matters even in classical physics: the substantivalist/relationist dispute is such a controversy, for instance). But that answer will have to be fairly structuralist, because the ontology of a theory instantiated by another theory must be compatible with the ontology of the latter theory, and inter-theoretic relations in classical mechanics are cashed out in terms of structure and dynamics and don't fit into (for instance) any standard mereological form. If there were a classical *ur*-theory underlying all other classical theories, we could regard ontological questions about that particular theory as questions of 'fundamental classical ontology' and see the ontologies of all other classical theories as derivative from it; but there is no such theory, and so *all* questions of ontology in classical physics will have a somewhat local, scale-dependent nature. (For further development of these points see Wallace (2016a).)

Turning back to quantum mechanics, the Everett interpretation also has the recipe form. To be more specific, by 'the Everett interpretation' I have in mind here the interpretation in its modern form, with decoherence understood as providing an emergent branching structure for the macroscopic degrees of freedom, with higher-level ontology understood in structural terms, and with no modification of the quantum formalism, as developed by, e.g. Gell-Mann and Hartle (1989; 1993), Saunders (1993; 1995; 1997), Zeh (1973; 1993), Zurek (1991; 1998), and myself (Wallace, 2003; 2010; 2012). (The extent to which this conforms to Everett's own views is moot, though see Barrett (2011a; 2011b), Bevers (2011), and the commentary in Everett (2012) for some considerations.) Indeed, at this level of abstraction the Everett interpretation is pretty much the same as the interpretation of classical mechanics: the formalism is left unchanged, the state space of a system is interpreted as a space of possible physical states for that system, and the evolution of the quantum state under the Schrödinger equation is interpreted as describing the change over time of the system's physical properties. The 'emergence of the classical world from quantum mechanics', in any particular physical situation, is just a special case of inter-theoretic reduction. (And to ask about 'the' ontology of Everettian quantum mechanics is again to commit a category error.)

Of course, this level of abstraction hides subtlety and controversy. Applied to microscopic systems, the Everett interpretation delivers an ontology that is not readily described in the categories of ordinary experience. Applied to macroscopic systems, it describes the state as representing not one, but an indeterminately large multiplicity, of classical states (and it represents each of them only approximately and emergently). And to connect to experience in the latter case, we must interpret the branch-weights assigned to the classical states probabilistically. Each of these contains philosophical subtleties sufficient to allow a sceptic to reject the interpretation, even before any weight is given to the ontological extravagance of Everett-interpreted quantum theory.

But this is not the place to explore the viability of the Everett interpretation. (I defend it *in extenso* in Wallace, 2012.) My point here is more modest: because the Everett interpretation has the recipe form, it is at least a *candidate* for an interpretation of quantum mechanics suitable for physics as we find it. Among the commonly discussed alternatives to the Everett interpretation, none has the recipe form, and so—

I will argue—none is actually in a position to explain quantum theory as it is in fact used in physics. I begin with the so-called 'non-realist' strategies.

5.5 Non-Representational Approaches to Quantum Theory

As a starting point to understand the common core of 'pragmatist', 'Copenhagen' and 'information-based' approaches to quantum theory, consider the classic Schrödinger-cat state

$$\alpha\,|\text{live cat}\rangle + \beta\,|\text{dead cat}\rangle \tag{5.5}$$

which unitary quantum theory can straightforwardly produce. If the quantum state can be understood *representationally*—that is, if distinct quantum states correspond to distinct objective ways a physical system can be—and if the theory is unsupplemented by hidden variables, then it looks as if such a state must somehow represent a cat that is simultaneously alive and dead, or perhaps neither alive nor dead. Quite apart from the weirdness of such a thing, it seems in conflict with what quantum mechanics itself tells us, via the Born rule, that we should expect to observe given a system in such a state: namely, either a live cat, or a dead one, and with probabilities $|\alpha|^2$, $|\beta|^2$ respectively.

But there are other roles than representation for a piece of mathematical formalism to play. Consider the probability distributions of classical statistical mechanics, for instance: mathematically they are functions on phase space but different such functions correspond not to different states of the actual world, but to different probability distributions over the possible states of the world (however the notion of probability is to be understood here).[8] And in particular, distributions can have nontrivial support on regions of phase space corresponding to macroscopically distinct states of affairs: for instance, the probability distribution that statistical mechanics assigns to a ferromagnet cooled below its critical temperature is, in the first instance, an equally weighted sum of distributions describing the magnet with magnetization in all possible directions. But this does not represent a ferromagnet in some weirdly indefinite state of magnetization, but merely a magnet with equal probability of being magnetized in each possible direction. Non-representational strategies take the quantum state as playing a role at least somewhat akin to the statistical-mechanical distribution: as encoding the various probabilities associated to a physical system. Different strategies differ in the way they understand those probabilities: as encoding our information about a system, or our partial beliefs about it, for instance, or as providing a practical guide to how to use and interact the system. And just as it would be misleading to call classical statistical mechanics non-realist simply because the distribution function does not play a representational role, so would it be misleading to call these approaches to quantum theory non-realist *simply* because in those theories the quantum state does not play a representational role either.

[8] I have in mind here classical statistical mechanics as it was understood before the quantum revolution: that is, on the false view that classical physics was exact. I argue elsewhere (Wallace, 2016b) that even supposedly classical systems in the actual universe should not be understood in purely classical terms: rather, they should be understood as quantum-mechanical states (mixed or pure) in the decoherent limit.

However, we can reasonably ask: if the quantum state does not represent the physical state of a system, what—if anything—does? Classical mechanics provides a straightforward answer to its analogous question: the *points* of phase space represent physical states, and the classical observables—those functions that correspond to position, momentum and the like—represent the physical properties of those states. In doing so, classical statistical mechanics also offers a straightforward answer to the question of what the probabilities are probabilities *of*: they are the probabilities of the system being in one physical state or another. And this representational story will vary from one classical-mechanical system to another, and in each case the theory itself will provide the representational machinery. In this way, an 'interpretation' of classical statistical mechanics in terms of probability distributions straightforwardly is compatible with the idea of classical statistical mechanics as a whole being a framework theory: it provides, for each concrete instantiation of statistical mechanics, an interpretation of the distribution function in that instantiation. In the classical dilute gas it represents probabilities for the particles to have various positions and momenta; in a classical model of the ferromagnet it represents probabilities for the individual magnetic atoms to have one orientation or the other; and so forth through the indefinitely many physical systems to which classical statistical mechanics can be applied. (Note that this interpretation of classical statistical mechanics has the 'recipe form' which I discussed in Section 5.4.)

In the dawn of quantum theory it was possible to imagine essentially the same story playing out. Quantum systems are equipped with algebras of 'observables' that correspond to the dynamical variables that describe a physical system, and the quantum state provides—and, conversely, is completely characterized by its description as—a probability distribution over those variables. So it would be extremely natural to suppose that the physical state of any quantum system is given by some definite value of each of its observables and that the quantum state is a probability distribution over those physical states. Such an interpretation would seem to be a very natural generalization of classical statistical mechanics (and, again, to have the recipe form). A shame, then, that such an interpretation is impossible: von Neumann's original no-hidden-variable theorem (von Neumann, 1955) shows that it cannot be done straightforwardly, and the Kochen–Specker theorem (Kochen and Specker, 1967) rules it out pretty much completely. So any viable non-representational account of quantum mechanics will owe us another account of what the probabilities encoded in the quantum state are probabilities of, of what the *physical* features of systems are. Or, put in more pragmatic terms: what the non-quantum features of a system are such that the quantum state is a tool for answering questions about those features.

In the heyday of the Copenhagen interpretation, the standard answer (made explicit in, e.g. Landau and Lifshitz, 1977) was: the physical description of a system is a classical description; quantum mechanics cannot be understood except against a background of classical mechanics. This line is quite rarely defended in modern physics,[9] for familiar reasons: separating off 'classical' from 'quantum' physics is not at all trivial in an era where the quantum–classical transition is a major topic of theoretical

[9] A notable exception is Peres (1993).

and empirical study, where the classification of a system as 'classical' or 'quantum' has become a matter of degree, and where the language of experimental physics is rich with terms—'laser', 'superconductor', 'LCD', whose very definition is quantum-mechanical.

Richard Healey (2017a), in discussing his 'pragmatist' interpretation of quantum theory (developed in more detail in Healey, 2012; 2017b; this volume, Chapter 7), is keenly aware of this issue:

> A successful interpretation must explain how quantum mechanics may be formulated as a precise physical theory and unambiguously applied to real-life physical situations . . . by applying quantum mechanics we become able better to describe and represent those situations in non-quantum terms. I say 'non-quantum' rather than 'classical' to acknowledge that the progress of science naturally introduces novel language to describe or represent the world (Bose-Einstein condensate, Mott insulator, quark-gluon plasma). (Healey, 2017a)

But *how* does science 'naturally introduce' this novel language? Take 'quark-gluon plasma', for instance. I can give you a verbal gloss on what that means—it's a state of matter so hot that protons and neutrons—normally made up of three tightly bound quarks—break up into individual quarks. But that verbal gloss isn't the real physical description—it is a metaphor at best, a fiction at worst. Let's see what happens if we ask for a proper (albeit still a bit simplified) account of 'quark-gluon plasma'.

Q: What's the quark-gluon plasma?

A: It's the state of a quantum-chromodynamics (QCD) system above a certain temperature, at which a phase transition occurs to a state where the fermionic elementary excitations are associated to the quark field rather than to colour-neutral products of that field.

Q: Slow down. What's 'temperature' in QCD?

A: A quantum system, including a field-theoretic system, is at (canonical) thermal equilibrium when its quantum state is

$$\rho(\beta) \propto \exp(-\beta \widehat{H}) \tag{5.6}$$

where $\widehat{H}$ is the Hamiltonian and β is a real number. For a system at thermal equilibrium—or that is reasonably close to thermal equilibrium—its temperature T is given by $\beta = 1/k_B T$.

Q: And what's an 'elementary excitation'?

A: Generally in quantum field theory, we can analyse systems in states reasonably close to the thermal equilibrium state as gases of weakly interacting particles. Those weakly interacting particles are the elementary excitations.

Q: 'Particles' as in classical point particles?

A: Not really. 'Particles' as in subsystems whose Hilbert space bears an irreducible representation of the Poincaré group, at least in the interaction-free limit.

Q: So the quark-gluon plasma is associated with one sort of particle, colder systems with another. Shouldn't I be able to say what the 'particles' are once-and-for-all?

A: Not in quantum field theory: the optimal choice of particle depends on the state of the system. Hot systems are described most naturally in terms of quarks, colder systems, in terms of protons and neutrons.

Q: Can't I just think of a proton or neutron as an agglomeration of three quarks?

A: Only heuristically. The more precise way to explain the relation between the protons and quarks is at the field level: the proton is associated with a certain triple product of the quark field.

Q: How is a particle supposed to be associated with a field?

A: If a quantum system is in thermal-equilibrium state $\rho(\beta)$, the 'two-point function' of that system with respect to field $\widehat{\phi}(x)$ and that state is

$$G_2(x-y;\phi,\beta) = \mathrm{Tr}(\rho(\beta)\widehat{\phi}(x)\widehat{\phi}(y)). \tag{5.7}$$

If the Fourier transform of that state has a pole, there's a particle associated with it.

Q: That's a weird postulate.

A: It's not a postulate; it's something you derive, by looking at the dynamics of states obtained by excitations of the thermal-equilibrium state. Where there's a pole, there's a subspace of states which can be interpreted as superpositions of singly localized excitations and which is preserved under the dynamics.[10]

Q: What's a 'plasma', anyway?

A: In 'quark-gluon plasma' it's a bit metaphorical. This phase of QCD has a lot in common with ordinary electromagnetic plasmas, which are characterized by screening of the Coulomb force, or equivalently by the photon acquiring mass.

Q: I thought the photon was massless?

A: The mass of a particle depends on the quantum state. If the state is the vacuum or some small excitation of the vacuum, then yes, photons are massless. But in a state that's an excitation of a hot, high-density system of protons and electrons, the photon acquires mass.

Q: What do you mean, 'acquires mass'?

A: It's a dynamical statement about elementary excitations again, derived from looking at the plasma dynamics. There's again a formal statement in terms of poles of two-point functions.

It doesn't matter if you followed all of the details of the above; they're not the main point. (I could have carried out essentially the same exercise with 'Bose–Einstein condensate' or 'Mott insulator'.) The main point is that I have *not the faintest idea* how to make sense of any of this without taking the quantum state of the QCD system, and its dynamical evolution under the Schrödinger equation, as representational. (Even the claim that the system has temperature T is a claim about its state.) I don't know how to begin eliminating representational uses of the state from my account of the quark-gluon plasma—and, to the best of my knowledge, neither does Healey.

If Healey departs from Copenhagen-style use of classical physics in order to seek a broader conception of objectivity in physics, the 'quantum Bayesianism' (or QBism) of Chris Fuchs and co-workers moves radically in the opposite direction:[11]

[10] Experts will recognize this as (a corollary of) the Callan–Lehmann representation of the two-point function; see, e.g. Peskin and Schroeder (1995 ch. 7) for the details.

[11] For presentations, see, e.g. Caves, Fuchs, and Schack (2002), Fuchs (2002) or Fuchs, Mermin, and Schack (2014).

the primitive concept of experience is fundamental to an understanding of science. According to QBism, quantum mechanics is a tool anyone can use to evaluate, on the basis of one's past experience, one's probabilistic expectations for one's subsequent experience.

(Fuchs, Mermin, and Schack, 2014)

Taken at its most straightforward, quotes like this seem to suggest that the probabilities of quantum theory are probabilities of agents having certain experiences. It can be tempting for more conventionally inclined physicists or philosophers to argue against such a view on the grounds that science should give us something more objective, less apparently solipsistic—but the real place to object is more mundane. Namely: the formalism of any particular quantum theory doesn't say anything *about* experiences. In any concrete instantiation of quantum theory, the observables over which probability distributions are defined are particle positions, field strengths, collective spins, and the like. The only way to say anything non-circularly about an agent's experience in quantum mechanics is to characterize it externally, as an experience *of* something describable in a more physical language. And then the problems confronted by the Copenhagen strategy, and by Healey, reappear.

It's reasonable then to ask how it is that Fuchs et al. (and Healey, for that matter) actually manage to develop and apply a non-representational version of quantum mechanics. (And it would be unfair not to recognize that they do indeed demonstrate a significant number of applications.) The answer is that the explanations they provide of quantum phenomena are (I think without exception) explanations of features of quantum theory in the abstract—Bell inequality violations, EPR effects, quantum state tomography and the like. They treat measurement as primitive and make no use of details of any particular quantum system (sometimes a two-state system is described using spin variables but even then there is no particular connection to the physics of spin).

Non-representational strategies, we might say, provide an interpretation of the *abstract framework* of quantum theory. They do not appear to provide a method—at least at present—to interpret any particular instance of that framework. Explanations of, say, superconductivity, or the heat capacity of crystals, or the thermodynamic features of the quark-gluon plasma, or the colour of gold, or any of the thousands of concrete applications of quantum theory that form its real empirical base, seem out of reach for QBism, or for pragmatism, at least as they are currently stated. From this perspective it is unsurprising to find that advocacy of these approaches within physics is predominantly found in quantum information, a field whose power indeed comes from studying those features of quantum theory that are common to all instances of the theory, rather than those physical explanations which are irreducibly quantum but which require engagement with the features of one or other specific quantum theory.

Let me close this section with an observation drawn from more general philosophy of science. The non-representationalist strategy—interpreting some large part of the content of physics not as representing things in themselves but as an inferential or pragmatic tool used in describing some smaller class of physically-real phenomena—is not new, nor is it specific to quantum theory. It is, rather, the central idea in the logical-positivist and logical-empiricist pictures of science (see Creath, 2017 and references therein for historical details). There, we make a principled distinction between the 'observation language' in which our observations are described, and the 'theory language' in which the non-observational parts of our scientific theories are stated. The observation language is to be understood literally, but the theory language is not; rather, it is an inferential or pragmatic tool to help us derive truths stated in the observation language.

It is almost universally accepted today that these approaches are not viable. But the predominant reason, historically, that they fell from grace was not some awakening realization among philosophers of science that scientific theories 'should' give a more robust account of the unobservable.[12] It was the increasingly clear realization—notably (though by no means exclusively) by Kuhn (1962) and Quine (1951) and in the recognized failure of Carnap's project in the *Aufbau* (1928)—that observation is theory-laden, that there *is* no clean separation of the vocabulary of science into a part which represents our observations, or the macroworld, or 'medium-size dry goods', and a part which does not. And so no actual scientific theory can be analysed as the logical positivists propose.

Non-representationalist strategies at least seem to be committed to making the same division, whether the analogue of the 'observation language' is Copenhagen's use of classical mechanics, or pragmatism's 'non-quantum' description, or QBism's appeal to direct experience. The problem with these approaches, as with positivism, is not that making such a distinction is *unreasonable* or *illegitimate*, but that—at least at present—we do not know how to do it.

5.6 Modifications of Quantum Theory

Non-representational strategies continue to have a significant following among physicists, but philosophers have generally given them short shrift. Their general attitude is well captured by Tim Maudlin's paraphrase of Hume:

> [W]e can be clear on the questions that must be asked of an interpretation. Is it an additional variables interpretation whose dynamics guarantee solutions to the problem of statistics and the problem of effect? Is it a collapse theory that leads to appropriate outcome states with the right probabilities, and whose fundamental terms all have clear physical significance? If the answer in each case is "no", then commit it to the flames, for it can contain nothing but sophistry and illusion. (Maudlin, 1995)

To a substantial degree, philosophers of quantum mechanics have followed Maudlin's line, focusing on solutions to the measurement problem that either (a) augment quantum theory with additional 'hidden' variables whose task it is to represent the physical world, or (b) modify the Schrödinger equation to introduce 'dynamical collapse' and so remove macroscopic superpositions (there is, perhaps, increasing willingness to consider the Everett interpretation alongside these two strategies).

In principle, either strategy could fit the 'recipe model' I laid out in Section 5.4. A hidden-variable theory would be a fairly general recipe to assign additional variables to a quantum system, presumably picked out in some way in terms of the

[12] Which is not to say that arguments of this sort played no role in the fall of logical positivism: Putnam (1962), for one, gave arguments of this kind. (I am grateful to an anonymous referee for the observation.)

observables and the dynamics of that system; a dynamical collapse theory would be a recipe for modifying the Schrödinger equation given the same inputs. For the recipe to be consistent, it would have to behave properly in cases of inter-theoretic reduction: if $h(X)$ is the hidden-variable theory associated with quantum theory X, and some particular quantum theory $X1$ instantiates higher-level quantum theory $X2$ in some regime, then hidden-variable theory $h(X2)$ would have to be derivable from $h(X1)$ in the same regime (and similarly for dynamical-collapse theories). For the recipe to solve the measurement problem, it would have to handle the quantum–classical transition correctly: in those regimes where quantum theories reduce *formally* to classical theories, the hidden-variable or dynamical-collapse theory would need to reduce to the appropriate classical theory in a way which properly predicts unique classical outcomes.

There is even a concrete class of hidden-variable theories that actually aspire to realizing this picture (I don't know of any dynamical-collapse strategy which aims to do so). The *modal* interpretation[13] really does give a general recipe for hidden variables, applicable to any quantum system, and which at first sight does reproduce definite outcomes in the classical limit. The generally (not universally) accepted *failure* of that approach, conversely, can be attributed in large part to a mixture of (a) its failure to tell a fully consistent story at different levels of description (the hidden variables depend sensitively on the details of the system/subsystem decomposition: Arntzenius, 1990, 1998; Clifton, 1996) and (b) the failure of its account of the quantum-classical transition to handle the necessary approximations involved in that transition (Bacciagaluppi, 2000; Bacciagaluppi, Donald, and Vermaas, 1995; Donald, 1998). (See Wallace, 2008, s.2.6.3 for further discussion.)

But the bulk of discussion of dynamical-collapse and hidden-variable theories takes a very different form. It is almost entirely concerned with theories which supplement or modify one *specific* quantum theory: non-relativistic particle mechanics. Indeed, it is almost entirely concerned with two specific such theories: de Broglie's and Bohm's pilot-wave theory, aka Bohmian mechanics (Bell, 1987; Bohm, 1952; Bohm and Hiley, 1993) and the 'GRW' collapse theory (Ghirardi, Rimini, and Weber, 1986), sometimes with lip service paid to the need to modify the latter along the lines of Pearle's 'CSL' theory (Pearle, 1989) in order to account adequately for identical particles.

Furthermore, the *way* Bohmian mechanics, and GRW theory, are normally discussed in philosophy of physics (especially in more metaphysical contexts) is sharply at odds with the relatively humble role nonrelativistic particle mechanics plays in real quantum theory (where, recall, it is a useful model to describe various particular systems of nonrelativistic particles in the absence of radiation, normally with some phenomenologically understood background fields and potentials). The only way I know to make sense of (most of) this literature is to interpret it as discussing nonrelativistic quantum particle mechanics *under the fiction that it is a fundamental and universal theory*. The literature is too large to demonstrate this exhaustively, but

[13] Originally proposed by van Fraassen (1991) and developed by many authors; see Dieks and Vermaas (1998) and references therein.

here are some illustrative recent examples (all drawn from Ney and Albert's recent anthology, *The Wave Function*, 2013):

1. Albert (2013, 53) describes the dimensionality of configuration space as 'three times as large as the total number of elementary particles in the universe', despite the fact that most such particles are not describable in the configuration-space formalism in most regimes.
2. Lewis (2013, 111) follows Albert: 'quantum mechanics represents the state of the world via a $3N$-dimensional wave function, where N is the number of particles in the universe.'
3. Allori (2013), in her discussion of 'primitive ontology'—of which more later—is explicit (in her title, no less) that she is discussing 'the structure of *fundamental* physical theories', and then proceeds to apply her framework to nonrelativistic particle mechanics.
4. North (2013, 184–5) is refreshingly explicit: '[T]he fundamental structure of a world's space(time) may be more properly given by a relativistic theory. Still, it is plausible that the fundamental theory of our world will be quantum mechanical. So it is worthwhile to think about what the world's fundamental space would be if [nonrelativistic quantum particle mechanics] is its fundamental theory.'

But why work under this fiction? I think the most charitable assumption (which assumes authors realize that it *is* a fiction[14]) is that it is a warm-up exercise: 'nonrelativistic quantum particle mechanics' is a standin for the *real* fundamental quantum theory, and we hope that as many as possible of our conclusions carry over to hoped-for hidden-variables or dynamical-collapse theories formulated for the real fundamental theory.

In previous work, I have taken the referent of 'real fundamental quantum theory' to be something like the Standard Model, and have argued that this strategy is problematic both (i) technically (Wallace, 2008, 83–5; 2012, 33–5), because it is much harder than is generally recognized to construct a quantum-field-theory version of Bohmian mechanics or GRW theory and so confidence that such a theory even exists is premature, and (ii) conceptually (Wallace, 2016a), because most of the features of nonrelativistic quantum theory appealed to by metaphysicians of quantum mechanics are emergent approximations at best in QFT. But from this paper's perspective, these criticisms do not get at the heart of the problem: that any strategy that works only when applied to a universal and fundamental theory does not seem to have the resources to explain the success of quantum mechanics *in general*, which is mediated through a very large number of explicitly *non*-fundamental physical theories. Absent any strategy for systematically constructing hidden-variable or dynamical-collapse versions of higher-level theories from lower-level theories, any hypothetical Bohmian or dynamical-collapse version of the standard model has a gap when it comes to connecting quantum theory with phenomenology and empirical confirmation.

[14] Philosophers of physics mostly do so realize, I think; certainly, all the authors I cite above do. But the very fact that the fiction is rarely spelled out explicitly creates a real risk of confusion, especially in more mainstream metaphysics discussions which get their physics second-hand from the philosophy of physics literature.

Can the gap be filled? Doing so would seem to require us to find a way to provide information, *directly in the language of the fundamental*, about what can be observed. And the current thrust of work in the metaphysics of hidden-variable and dynamical-collapse theories aims to do exactly that. A particularly clear example is the 'primitive ontology' strategy of Allori (2013; this volume, Chapter 11) and Allori et al. (2008) (Maudlin's 2013 'primary ontology' is a close relative; Esfeld et al., 2017, adopt essentially the same framework). The idea is that to any purportedly fundamental theory must be associated a primitive ontology of spatially localized matter (point particles or extended continua), such that the macroscopic world is identified with composites of the primitive ontology. (A table, for instance, is by definition a collection of primitive-ontology elements arranged in a table shape—the mereological sum of that collection, in metaphysical parlance.) We are assumed to have direct empirical access to at least coarse grainings of the primitive ontology, and so the condition for empirical success of a fundamental theory is that it generates a distribution of primitive-ontology elements whose coarse grainings match our observations of macroscopic matter. Non-primitive ontology might also be admitted to our theory (the electromagnetic field is supposed to be a classical example of non-primitive ontology; the quantum state represents non-primitive quantum ontology) but it has no direct role in empirical confirmation: we learn of it only indirectly, through its dynamical effect on the primitive ontology. Primitive ontologists seldom discuss *non*-fundamental theories, but so far as I can see, in their framework these are of purely instrumental value, providing calculational techniques to extract information about the primitive ontology but having no ontological significance of their own.

Four observations follow naturally. Firstly, although advocates of primitive ontology often write as if theirs is an unproblematic account of the physics-world relation before quantum mechanics (usually cashed out in terms of a primitive ontology of classical point particles), they do not provide historical arguments for this and it looks most implausible. The 'particles' to which classical particle mechanics was applied historically were typically the centres of mass of large bodies (planets, moons, and comets in celestial mechanics, in particular). Other applications of classical mechanics were directly to extended objects like fluids and rigid bodies. Whatever the *hope* might have been that these large bodies could be understood as swarms of smaller bodies, that hope was never realized, and we now know that it *could not* be realized: classical microphysics does not support stable matter, and systems like comets or liquids (*pace* Allori, 2013) cannot be analysed as agglomerates of classical particles.

Secondly, if we want to look for a primitive ontology for extant physics, presumably we had better look in the Standard Model. But the task of finding a characterization of the macroscopic *within the basic vocabulary of the Standard Model* looks monumentally difficult. Most of the connections between the Standard Model and experiment are fantastically indirect, proceeding through layers upon layers of only-partially-understood emergence—so indirect, indeed, that philosophers like Cartwright can deny that the predictions of higher-level physics are grounded in the Standard Model at all with at least prima facie plausibility. Even those predictions which are 'direct' results of the Standard Model—the Higgs boson, say—are not stated in anything like a primitive ontology: the experimental signature of the Higgs boson is a certain resonance in the cross section for hadron–hadron scattering, and of course 'resonance', 'cross section', 'scattering', and 'hadron' are all highly theory-laden terms.

(Does the primitive-ontology strategy succeed in defining the observable in terms of the microscopic in non-relativistic particle mechanics, under the fiction that this theory is universal and fundamental? Never mind why the question is of any interest; it is not even well posed. For beings like us (who, inter alia, make most of our 'observations' using electromagnetic radiation) could not exist under the assumptions of that fiction. They who do not exist, do not observe.)

So the primitive ontologist is committed to forging a direct link between high-energy physics and observation, with little or no help from actual physical practice. And (my third observation) the task is made still harder by features of quantum field theories in particular. Recall that the Standard Model, like pretty much any empirically relevant quantum field theory, is an *effective* field theory, regularized by some short-distance cutoff. The physics at the cutoff lengthscale is far from negligible, but it is possible to absorb its effects via 'renormalization', whereby those degrees of freedom that describe the quantum field theory at large lengthscales are redefined in a complicated, cutoff-dependent way that absorbs most of the effects of the short-distance physics. (See Wallace, forthcoming, and references therein, for a more detailed discussion from a conceptual point of view and for further details.) This means that the relation between the 'fundamental' and the empirically relevant in quantum field theory is complicated, indirect, dynamically mediated and cutoff-dependent. It is very hard to see how this could be made compatible with the primitive-ontology approach, or indeed with any approach committed to a description of a theory's empirical content directly in its microphysical vocabulary. It is notable that none of the various extant suggestions for Bohmian quantum field theories—based on associating the hidden variables to fermion number density (Bell, 1984; Colin, 2003; Colin and Struyve, 2007), or particle number (Dürr et al., 2004), or field-configuration-strength (Struyve, 2007; Struyve and Westman, 2006; 2007)—have discussed renormalization, or given more than a qualitative verbal plausibility argument for how these theories recover the macroworld. The acid test of such a theory is to demand a full model of how some nontrivial quantum-field-theoretic prediction—say, the cross section for electron–electron scattering, calculated to loop order where renormalization matters—actually plays out as a physical process, but no such theory is currently close to passing that test.

The considerations of effective field theory are really just a reminder that even though the Standard Model is the closest we have to a fundamental physical theory, it is not such a theory, and could not be: its emergent, approximate status is built into its characterization as an effective field theory. And this brings me to the last observation: if the primitive-ontology strategy makes a direct connection between *fundamental* physics and observation, and higher-level, emergent physics is to be treated purely instrumentally, what—beyond idle curiosity—is the point of trying to develop such an approach until and unless we have a physical theory which we have reason to think *is* fundamental? (Not that I am sanguine about the prospects of developing a primitive ontology for that theory either: its 'observational' claims are likely to be even more indirect and theory-laden than the Standard Model's.) Even in its own terms, the primitive-ontology strategy seems to be a strategy for postponing the measurement problem until the dust of fundamental physics has otherwise settled.

There is something rather ironic about this situation. Advocates of hidden-variable or dynamical-collapse theories are normally ardently committed to some form of scientific realism; to compare them to the logical positivists would be a killing insult. But what are the advocates of primitive ontology looking for, if not something like the observation language that the logical positivists sought in vain?[15] And again, the problem with this strategy is not so much that the *metaphysical* distinction between primitive and non-primitive ontology is ill-defined or unmotivated; it is that we do not know how to make it, for realistic physics, in a way that achieves the task it is supposed to perform.

For expository clarity I have focused here on primitive-ontology strategies, but I believe the observations generalize. The mainstream modificatory approaches to the measurement problem are committed to (1) developing modifications specifically to 'fundamental' quantum theory with only an instrumentalist attitude to non-fundamental quantum theories; (2) finding a direct way to characterize the observational evidence for that fundamental theory in its microphysical vocabulary. These commitments disconnect modificatory approaches from physics as it is practised, and leave it opaque at best how they can hope to account for the empirical predictions of quantum theory writ large.

5.7 Conclusions

I have argued that when we recognize the real structure of quantum theory—an abstract framework realized by indefinitely many concrete theories that are realized by indefinitely many concrete systems and whose relations one with another are complicated and not really hierarchical—then most extant approaches to the quantum measurement problem should be recognized as inadequate to that real structure. Of currently extant approaches, only the Everett interpretation, in its modern decoherence-based form, provides the interpretative *recipe* to make sense of the multiplicity of quantum theories in a self-consistent way. Other approaches either try to make sense of the abstract structure of quantum theory and thus fail to give an adequate account of scientific representation in concrete applications, or else analyse one particular quantum theory as if it alone could exhaust the theoretical content of the subject, and as if all experimental predictions could be described directly in its vocabulary. If this is correct, there is no underdetermination in quantum mechanics: either the Everett interpretation is viable, in which case it alone provides an adequate interpretation of quantum theory; or else it must be rejected for some philosophical or technical reason, in which case there is at present *no* adequate interpretation of quantum theory.

[15] The parallels extend to attempts to treat the non-primary ontology as lawlike (Dürr, Goldstein, and Zanghì, 1997; Goldstein and Teufel, 2000; Goldstein and Zanghì, 2013), or to eliminate it entirely at the metaphysical level via a Humean account of laws (Esfeld et al., 2014; Miller, 2014). See Dewar (2017) for further discussion.

To the reader who resists this conclusion, there is a straightforward way to prove me wrong. Take any moderately complicated, moderately concrete application of quantum theory in a regime that is not fully covered by nonrelativistic particle mechanics: take the BCS model of superconductivity, for instance (with electromagnetic radiation present), or take the quark-gluon plasma, or take the Higgs mechanism, or take the colour of gold; come to that, take the photoelectric effect, or the (photon) two-slit experiment. Explain, in reasonable detail, how that application is to be described, understood, and tested inside your preferred interpretation of quantum mechanics. And check, in particular, that your preferred interpretation can be used to make the quantitative calculations whose match with experiment is the reason why those applications are deemed successful. The Everett interpretation can do this, at least if its account of probability and the structuralist notions of emergence that it relies on are deemed adequate, simply by working through the standard accounts of these applications in physics and interpreting each particular version of quantum mechanics appealed to in Everettian terms. If your preferred alternative to Everettian quantum mechanics cannot do it, you have not yet solved the quantum measurement problem.

This might seem like cheating. What I have called an 'Everettian' way of making sense of concrete quantum applications is, after all, really just the explanation of those applications found within mainstream physics. But this simply underlines the fact that Everettian quantum mechanics—ontological extravagance at the macro-level notwithstanding—is a modest, conservative project, aimed at *legitimating* and *making sense of* the ordinary practice of quantum theory. Other approaches to the measurement problem, by and large, are less conservative: the ordinary practice of quantum theory is fundamentally confused and in large part needs to be reformulated or replaced, not 'made sense of'. But the great virtue of modesty and conservatism, in this context, is that it minimizes the need to redo from scratch much of the last century of physics.

Let me finish, as did Maudlin (1995), by paraphrasing Hume: We can be clear on the questions that must be asked of an interpretation. Does it provide a way to legitimate and make sense of the actual practice of quantum physics, across the various interrelated domains to which quantum theory has been applied? Does it set out to reform the practice of quantum physics, and does it provide evidence that this is more than a bluff by actually doing the hard work in some non-trivial, concrete examples across multiple instantiations of the quantum framework (and not just in an abstract quantum-information setting, or in nonrelativistic particle mechanics)? If the answer in each case is 'no' . . . , well, maybe don't hastily commit it to the flames, as it may contain valuable insights and be the seed of a yet-to-be-completed research programme, but don't kid yourself that it is *at present* a viable interpretation of quantum mechanics, and maybe be a bit cautious exploring all its metaphysical implications until you've done some more work to see if it plausibly might be made viable.

Acknowledgments

I am grateful to Neil Dewar, Richard Healey, James Ladyman, Juha Saatsi, Simon Saunders, and Chris Timpson for useful conversations, and to two anonymous referees for helpful comments and corrections.

References

Albert, D. Z. (2013). Wave function realism. In A. Ney and D. Z. Albert (Eds), *The Wave Function: Essays on the Metaphysics of Quantum Mechanics*, pp. 52–7. Oxford: Oxford University Press.

Allori, V. (2013). Primitive ontology and the structure of fundamental physical theories. In *The Wave Function: Essays on the Metaphysics of Quantum Mechanics*, pp. 58–75. New York: Oxford University Press.

Allori, V., S. Goldstein, R. Tumulka, and N. Zanghì (2008). On the common structure of Bohmian mechanics and the Ghirardi–Rimini–Weber theory. *British Journal for the Philosophy of Science* 59, 353–89.

Arntzenius, F. (1990). Kochen's interpretation of quantum mechanics. *Proceedings of the Philosophy of Science Association* 1, 241–9.

Arntzenius, F. (1998). Curioser and curioser: a personal evaluation of modal interpretations. In D. Dieks and P. E. Vermaas (Eds), *The Modal Interpretation of Quantum Mechanics*, pp. 337–77. Dordrecht: Kluwer Academic Publishers

Bacciagaluppi, G. (2000). Delocalized properties in the modal interpretation of quantum mechanics. *Foundations of Physics* 30, 1431–44.

Bacciagaluppi, G., M. J. Donald, and P. E. Vermaas (1995). Continuity and discontinuity of definite properties in the modal interpretation. *Helvetica Physica Acta* 68, 679–704.

Barrett, J. A. (2011a). Everett's pure wave mechanics and the notion of worlds. *European Journal for Philosophy of Science* 1, 277–302.

Barrett, J. A. (2011b). On the faithful interpretation of pure wave mechanics. *British Journal for the Philosophy of Science* 62, 693–709.

Bell, J. S. (1984). Beables for quantum field theory. CERN preprint CERN-TH 4035/84. Reprinted in J. S. Bell (1987), *Speakable and Unspeakable in Quantum Mechanics*, pp. 173–80. Cambridge: Cambridge University Press.

Bell, J. S. (1987). *Speakable and Unspeakable in Quantum Mechanics*. Cambridge: Cambridge University Press.

Bevers, B. M. (2011). Everett's "many-worlds" proposal. *Studies in the History and Philosophy of Modern Physics* 42, 3–12.

Blackmore, S. (2005). *Conversations on Consciousness*. Oxford: Oxford University Press.

Bohm, D. (1952). A suggested interpretation of quantum theory in terms of "hidden" variables. *Physical Review* 85, 166–93.

Bohm, D. and B. J. Hiley (1993). *The Undivided Universe: An Ontological Interpretation of Quantum Theory*. London: Routledge and Kegan Paul.

Carnap, R. (1928). *Der Logische Aufbau der Welt*. Leipzig: Felix Meiner Verlag.

Cartwright, N. (1983). *How the Laws of Physics Lie*. Oxford: Oxford University Press.

Cartwright, N. (1999). *The Dappled World: A Study of the Boundaries of Science*. Cambridge: Cambridge University Press.

Caves, C. M., C. A. Fuchs, and R. Schack (2002). Quantum probabilities as Bayesian probabilities. *Physical Review A* 65, 022305.

Chalmers, D. J. (1995). Facing up to the problem of consciousness. *Journal of consciousness studies* 2, 200–19.

Clifton, R. (1996). The properties of modal interpretations of quantum mechanics. *British Journal for the Philosophy of Science* 47, 371–98.

Colin, S. (2003). Beables for quantum electrodynamics. Available online at http://arxiv.org/abs/quant-ph/0310056.

Colin, S. and W. Struyve (2007). A Dirac sea pilot-wave model for quantum field theory. *Journal of Physics A* 40, 7309–42.

Cordero, A. (2001). Realism and underdetermination: some clues from the practices-up. *Philosophy of Science 68*, S301–S312.

Creath, R. (2017). Logical empiricism. In *The Stanford Encyclopedia of Philosophy* (Fall 2017 Edition), Edward N. Zalta (ed.), online at https://plato.stanford.edu/archives/fall2017/entries/logical-empiricism/.

Dennett, D. C. (1984). *Elbow Room: The Varieties of Free Will Worth Wanting*. Oxford: Oxford University Press.

Dennett, D. C. (2005). *Sweet Dreams: Philosophical Objections to a Science of Consciousness*. Cambridge, MA: MIT Press.

Dewar, N. (2017). La bohume. http://philsci-archive.pitt.edu/13235/.

Dieks, D. and P. E. Vermaas (Eds) (1998). *The Modal Interpretation of Quantum Mechanics*. Dordrecht: Kluwer Academic Publishers.

Donald, M. J. (1998). Discontinuity and continuity of definite properties in the modal interpretation. In D. Dieks and P. E. Vermaas (Eds), *The Modal Interpretation of Quantum Mechanics*, pp. 213–22. Dordrecht: Kluwer Academic Publishers.

Dürr, D., S. Goldstein, R. Tumulka, and N. Zanghì (2004). Bohmian mechanics and quantum field theory. *Physical Review Letters 93*, 090402.

Dürr, D., S. Goldstein, and N. Zanghì (1997). Bohmian mechanics and the meaning of the wave function. In R. S. Cohen, M. Horne, and J. Stachel (Eds), *Potentiality, Entanglement and Passion-at-a-Distance—Quantum Mechanical Studies in Honor of Abner Shimony*. Dordrecht: Kluwer. Available online at http://arxiv.org/abs/quant-ph/9512031.

Egg, M. (2014). *Scientific Realism in Particle Physics: A Causal Approach*. Berlin: de Gruyter.

Esfeld, M., D. Lazarovici, M. Hubert, and D. Dürr (2014). The ontology of Bohmian mechanics. *British Journal for the Philosophy of Science 65*, 733–96.

Esfeld, M., D. Lazarovici, V. Lam, and M. Hubert (2017). The physics and metaphysics of primitive stuff. *British Journal for the Philosophy of Science 68*, 133–61.

Everett, H. (2012). *The Everett Interpretation of Quantum Mechanics: Collected Works 1955–1980 with Commentary* ed. Jeffrey A. Barrett and Peter Byrne. Princeton: Princeton University Press.

Fuchs, C. (2002). Quantum mechanics as quantum information (and only a little more). Available online at at http://arXiv.org/abs/quant-ph/0205039.

Fuchs, C. (2017). On participatory realism. In I. T. Durham and D. Rickles (Eds), *Information and Interaction: Eddington, Wheeler, and the Limits of Knowledge*, pp. 113–34. Dordrecht: Springer. Online version at https://arxiv.org/abs/1601.04360.

Fuchs, C. and A. Peres (2000a). Quantum theory needs no "interpretation". *Physics Today 53*(3), 70–1.

Fuchs, C. and A. Peres (2000b). Fuchs and Peres reply. *Physics Today 53*, 14.

Fuchs, C. A., N. D. Mermin, and R. Schack (2014). An introduction to QBism with an application to the locality of quantum mechanics. *American Journal of Physics 82*, 749–54.

Gell-Mann, M. and J. B. Hartle (1989). Quantum mechanics in the light of quantum cosmology. In W. H. Zurek (Ed.), *Complexity, Entropy and the Physics of Information*, pp. 425–59. Redwood City, CA: Westview Press.

Gell-Mann, M. and J. B. Hartle (1993). Classical equations for quantum systems. *Physical Review D 47*, 3345–82.

Ghirardi, G., A. Rimini, and T. Weber (1986). Unified dynamics for micro and macro systems. *Physical Review D 34*, 470–91.

Goldstein, S. and S. Teufel (2000). Quantum spacetime without observer: ontological clarity and the conceptual foundations of quantum gravity. In C. Callender and N. Huggett (Eds), *Physics meets Philosophy at the Planck Scale*, pp. 275–89. Cambridge: Cambridge University Press.

Goldstein, S. and N. Zanghì (2013). Reality and the role of the wave function in quantum theory. In A. Ney and D. Z. Albert (Eds), *The Wave Function: Essays on the Metaphysics of Quantum Mechanics*, pp. 91–109. Oxford: Oxford University Press.

Healey, R. (2012). Quantum theory: a pragmatist approach. *British Journal for the Philosophy of Science* 63, 729–71.

Healey, R. (2017a). How pragmatism reconciles quantum mechanics with relativity etc. (interview by Richard Marshall). http://www.3ammagazine.com/3am/how-pragmatism-reconciles-quantum-mechanics-with-relativity-etc/.

Healey, R. (2017b). *The Quantum Revolution in Philosophy*. Oxford: Oxford University Press.

Kochen, S. and E. Specker (1967). The problem of hidden variables in quantum mechanics. *Journal of Mathematics and Mechanics* 17, 59–87.

Kuhn, T. (1962). *The Structure of Scientific Revolutions*. Chicago: University of Chicago Press.

Landau, L. and E. Lifshitz (1977). *Quantum Mechanics (Non-Relativistic Theory)* (3rd edn). Oxford: Butterworth Heinemann.

Lewis, P. (2013). Dimension and illusion. In A. Ney and D. Albert (Eds), *The Wave Function: Essays on the Metaphysics of Quantum Mechanics*, pp. 110–25. Oxford: Oxford University Press.

Lewis, P. J. (2015). Internet encyclopedia of philosophy. http://www.iep.utm.edu/int-qm/.

Lyre, H. (2010). Why quantum theory is possibly wrong. *Foundations of Physics* 40, 1429–38.

Maudlin, T. (1995). Three measurement problems. *Topoi* 14, 7–15.

Maudlin, T. (2013). The nature of the quantum state. In A. Ney and D. Albert (Eds), *The Wave Function: Essays on the Metaphysics of Quantum Mechanics*, pp. 126–53. Oxford: Oxford University Press.

Miller, E. (2014). Quantum entanglement, Bohmian mechanics, and Humean supervenience. *Australasian Journal of Philosophy* 92, 567–83.

Ney, A. and D. Albert (Eds) (2013). *The Wave Function: Essays on the Metaphysics of Quantum Mechanics*. Oxford: Oxford University Press.

North, J. (2013). The structure of a quantum world. In A. Ney and D. Albert (Eds), *The Wave Function: Essays on the Metaphysics of Quantum Mechanics*, pp. 184–202. Oxford: Oxford University Press.

Oppenheim, P. and H. Putnam (1958). The unity of science as a working hypothesis. *Minnesota Studies in the Philosophy of Science* 2.

Pearle, P. (1989). Combining stochastic dynamical state-vector reduction with spontaneous localization. *Physical Review A* 39(5), 2277–89.

Peres, A. (1993). *Quantum Theory: Concepts and Methods*. Dordrecht: Kluwer Academic Publishers.

Peskin, M. E. and D. V. Schroeder (1995). *An introduction to Quantum Field Theory*. Reading, MA: Addison-Wesley.

Putnam, H. (1962). What theories are not. In E. Nagel, P. Suppes, and A. Tarski (Eds), *Logic, Methodology, and Philosophy of Science*, pp. 240–51. Stanford: Stanford University Press.

Quine, W. V. O. (1951). Two dogmas of empiricism. *Philosophical Review* 60, 20–43.

Saunders, S. (1993). Decoherence, relative states, and evolutionary adaptation. *Foundations of Physics* 23, 1553–85.

Saunders, S. (1995). Time, decoherence and quantum mechanics. *Synthese* 102, 235–66.

Saunders, S. (1997). Naturalizing metaphysics. *The Monist* 80(1), 44–69.

Struyve, W. (2007). Field beables for quantum field theory. Available online at http://arxiv.org/abs/0707.3685.

Struyve, W. and H. Westman (2006). A new pilot-wave model for quantum field theory. *AIP CONFERENCE PROCEEDINGS* 844, 321.

Struyve, W. and H. Westman (2007). A minimalist pilot-wave model for quantum electrodynamics. *Proceedings of the Royal Society of London A* 463, 3115–29.

Van Fraassen, B. C. (1991). *Quantum Mechanics*. Oxford: Oxford University Press.

Von Neumann, J. (1955). *Mathematical Foundations of Quantum Mechanics*. Princeton: Princeton University Press.

Wallace, D. (2003). Everett and structure. *Studies in the History and Philosophy of Modern Physics* 34, 87–105.

Wallace, D. (2008). The interpretation of quantum mechanics. In D. Rickles (Ed.), *The Ashgate Companion to Contemporary Philosophy of Physics*, pp. 16–97. Burlington, VT: Ashgate. Available online at http://arxiv.org/ under the title 'The Quantum Measurement Problem: State of Play'.

Wallace, D. (2010). Decoherence and ontology: or: how I learned to stop worrying and love FAPP. In S. Saunders, J. Barrett, A. Kent, and D. Wallace (Eds), *Many Worlds? Everett, Quantum Theory, and Reality*, pp. 53–72. Oxford: Oxford University Press.

Wallace, D. (2011). Taking particle physics seriously: a critique of the algebraic approach to quantum field theory. *Studies in the History and Philosophy of Modern Physics* 42, 116–25.

Wallace, D. (2012). *The Emergent Multiverse: Quantum Theory According to the Everett Interpretation*. Oxford: Oxford University Press.

Wallace, D. (2016a). Lessons from realistic physics for the metaphysics of quantum theory. https://doi.org/10.1007/s11229-018-1706-y.

Wallace, D. (2016b). Probability and irreversibility in modern statistical mechanics: classical and quantum. https://dornsife.usc.edu/assets/sites/1045/docs/oxfordstatmech.pdf.

Wallace, D. (forthcoming). Quantum field theory. Forthcoming in Knox and Wilson (Eds), *Handbook of the Philosophy of Physics* (Routledge).

Wilson, M. (2013). What is "classical mechanics" anyway? In R. Batterman (Ed.), *The Oxford Handbook of Philosophy of Physics*, pp. 43–106. Oxford: Oxford University Press.

Zeh, H. (1973). Toward a quantum theory of observation. *Foundations of Physics* 3, 109–16.

Zeh, H. D. (1993). There are no quantum jumps, nor are there particles! *Physics Letters* A172, 189.

Zurek, W. and J. P. Paz (1995). Quantum chaos: a decoherent definition. *Physica D* 83, 300–8.

Zurek, W. H. (1991). Decoherence and the transition from quantum to classical. *Physics Today* 43, 36–44. Revised version available online at http://arxiv.org/abs/quant-ph/0306072.

Zurek, W. H. (1998). Decoherence, einselection, and the quantum origins of the classical: the rough guide. *Philosophical Transactions of the Royal Society of London* A356, 1793–820.

6

Naturalism and the Interpretation of Quantum Mechanics

J. E. Wolff

6.1 Introduction

Quantum mechanics is a theory notoriously 'in need of an interpretation'. Providing and assessing such interpretations of quantum mechanics and other physical theories has become a central task of philosophy of physics. Being able to contribute to such interpretive endeavours would accordingly seem to be of great importance to philosophical naturalism, which places science at the heart of inquiry and views philosophy as continuous with science. Interpreting physics is commonly regarded as good strategy for arriving at a *naturalistic* ontology (Ladyman and Ross, 2007). It is hence surprising that Bas van Fraassen has recently made the case that naturalists, unlike empiricists, are unable to make sense of interpretive tasks (2015). Indeed, he claims that resistance to the project of interpreting science is a *characteristic feature* of philosophical naturalism. To make matters worse, van Fraassen further suggests that this refusal poses a problem for philosophical naturalism as a view of science, because the interpretation of scientific theories makes a key contribution to our understanding of these theories. If van Fraassen is right, then naturalism would seem to be a rather unattractive philosophical position vis-à-vis the sciences.

The question I will pursue in this chapter is whether there are interpretive projects a naturalist might be able to engage with, and if so, whether these are of a sort that would permit naturalists to respond to van Fraassen's challenge. I will argue that naturalists can improve their understanding of scientific theories just like empiricists can, but that they will not do so by engaging in an explicitly extra-scientific or philosophical project of interpretation. The key difference between van Fraassen's empiricist interpreter and more naturalistic approaches is that naturalists take understanding, including explanatory understanding, to be an epistemic aim of science and scientific practice, whereas van Fraassen treats it as an aim of some further, philosophical activity.

For van Fraassen, to interpret a scientific theory is to ask, "under what circumstances is this theory true? What does it say the world is like?" (van Fraassen, 1991, 242) The model of interpretation he puts forward is that of interpreting a text and its relation to us and to the world; we improve our understanding of the text by

J. E.Wolff, *Naturalism and the Interpretation of Quantum Mechanics* In: *Scientific Realism and the Quantum*. Edited by: Steven French and Juha Saatsi, Oxford University Press (2020). © J. E.Wolff.
DOI: 10.1093/oso/9780198814979.003.0006

offering a range of competing interpretations. While van Fraassen is right to point out that his kind of interpretive project is hard to make sense of from a naturalistic viewpoint, I will argue that naturalists can achieve understanding, including explanatory understanding, by means other than interpretation in the empiricist's sense. Naturalists and empiricists disagree over what it takes to understand a theory. Van Fraassen's argument hence succeeds as a diagnosis of the difference between empiricist and naturalist stances, but it is less compelling as a critique of naturalism, as it is unlikely to sway committed naturalists, who reject the idea that we need to engage in interpretation to achieve understanding.

The chapter is divided into three main parts. First, I present van Fraassen's challenge to naturalism (Section 6.2). Then I look closely at the case of quantum mechanics as a paradigmatic physical theory subject to diverse interpretive projects (Section 6.3). Finally I show how the difference over interpretation between naturalists and empiricists results from a disagreement over what it takes to understand a scientific theory (Section 6.4).

6.2 Van Fraassen's Challenge

Some terminological housekeeping is in order to set the stage for van Fraassen's challenge. Both naturalism and empiricism will be treated as *stances*, not as theses. A stance, unlike a philosophical thesis or dogma, involves both belief-like elements, and non-belief elements.[1] Treating broad philosophical positions as stances is a way of acknowledging that there is no entirely neutral ground from which to evaluate philosophical positions and that there is an element of voluntarism in adopting or rejecting a particular such stance. Stances may be criticized, but such criticism will typically have to show how a stance fails by its own lights, not that it is incompatible with, or unattractive from, the point of view of a competing stance.

Naturalism is further characterized by epistemic anti-foundationalism and fallibilism—any belief or theory is potentially subject to revision, and no source of beliefs or method of inquiry is regarded as incorrigible. Science plays a central role in this epistemic stance: our best current science provides the starting point for all inquiry and is typically regarded as the only conceivable model for inquiry. This form of naturalism has been developed especially clearly in Penelope Maddy's *Second Philosophy* (2007), where second philosophy is the activity of an idealized inquirer, the *Second Philosopher*. Unlike 'first philosophers', the second philosopher is deeply immersed in the practices of current science, and she approaches all questions, including 'philosophical' questions, entirely 'from within' scientific inquiry. The second philosopher, as Maddy describes her, is "from birth" a busy sailor on Neurath's boat (Maddy, 2007, 85). As a result of her position embedded within scientific practice, she rejects certain philosophical questions and projects as unintelligible or as of interest only from a perspective outside of science, which is a perspective the second philosopher does not adopt.

[1] For more discussion of the notion of a stance, see van Fraassen (2002) and Rowbottom and Bueno (2011); for naturalism as a stance, see Wolff (2015).

Empiricism is also to be understood as a stance, as anti-foundationalist, and as a view that takes science to be central to inquiry. Given these similarities, it is important for empiricists to distinguish empiricism from naturalism, and to show that empiricism is a more viable stance. Since both views are treated as stances, however, it is particularly important that naturalism must be shown to be unsuccessful by its own lights; it won't do to show that *from an empiricist point of view* naturalism does not look attractive.

Van Fraassen sets up his challenge to Maddy's naturalism by drawing a distinction between the 'naturalistic native', who is Maddy's ideal inquirer, and the naturalistic philosopher, who, like Maddy, offers the naturalistic native as a paradigmatic participant of contemporary science. This distinction is important for van Fraassen, since he suggests that Maddy's strategy for rejecting certain types of questions relies crucially on sliding back and forth between the position of the naturalistic native and that of the naturalistic philosopher (van Fraassen, 2015, 69). To treat the naturalistic native as the paradigmatic participant in the scientific enterprise is to offer an interpretation of what science is. A problematic feature of naturalism is that this interpretive stance (often) remains unacknowledged as such; in particular, the possibility of alternative interpretations of what it means to be a paradigmatic participant in science is denied, or engagement with these alternatives is refused. Just as the naturalistic philosopher puts forward her naturalistic native as the paradigmatic practitioner in scientific practice, an empiricist or pragmatist might put forward different model inquirers (van Fraassen, 2015, 78). Who one takes to be a model participant will reflect what one takes to be the aims and methods of scientific inquiry. The conception of an ideal inquirer is not neutral between different philosophical stances one might take towards science. At the very least this non-neutrality needs to be acknowledged by the naturalists, even if one grants them the right to take such a stance without first offering a stance-independent defence for it, as van Fraassen does.[2]

Instead, Maddy and other naturalistic philosophers seem to refrain from defending their take on science against possible alternatives. Maddy's criticism of 'first philosophy' (which addresses a wide range of viewpoints, including van Fraassen's empiricism) typically takes the form of rejecting the philosophical concerns that motivate these alternative views of science from the perspective of the 'second philosopher'. The second philosopher, Maddy argues, sees no need for a philosophical stepping back from scientific practice, and she struggles to understand what might possibly motivate others to take such a step (Maddy, 2007, e.g. 308–9). From an empiricist perspective, however, this refusal to engage with philosophical questions already betrays a particular outlook on science.

> [T]he fundamental Naturalistic impulse is not so much to take science for granted as to take for granted a particular but unacknowledged philosophical view of what science is. Precisely if that is the case, the stance is not taken consciously and explicitly, but seen as unavoidable or inherent in what it is to be scientific at all. (van Fraassen, 2015, 80)

Failure to acknowledge one's own position as a stance is to fail to acknowledge that a choice was involved in adopting it, and that other choices are possible.

[2] Granting this right is part of van Fraassen's stance voluntarism (van Fraassen, 2002).

Could the naturalist simply respond by acknowledging that she has taken a stance? The cost of doing so would be to admit van Fraassen's distinction between the naturalistic native as the ideal scientific practitioner, and the naturalistic philosopher as the philosopher who puts forward the naturalistic native as an ideal inquirer. The second philosopher, as conceived of by Maddy, is both a paradigmatic practitioner in scientific practice, *and* a plausible candidate for a successful philosopher. But while the paradigmatic practitioner in scientific practice may indeed be oblivious to certain kinds of philosophical questions, it is implausible for a philosopher to fail to understand such questions, precisely because she needs to acknowledge that her view is one stance among several others. In failing to acknowledge her stance as a stance, the naturalist philosopher is failing as a philosopher, even though the naturalistic native could perhaps carry on with her practice without taking any philosophical stance at all.

But is the naturalistic native even a plausible candidate for being an ideal inquirer in scientific contexts? The naturalistic native, van Fraassen argues, is not only just one among several candidates for the paradigmatic participant in science; as portrayed by the naturalists, she is not even a particularly plausible one. The naturalistic native is unable and unwilling to take an interpretative stance towards her own scientific practice. Yet the history of science suggests that even within science there are situations in which scientists need to be able to 'step back' from the practice at hand and reflect on the aims, methods, and content of the science they usually engage in. So the naturalistic native, as a paradigmatic participant in science, cannot be as oblivious to interpretation as the naturalistic philosopher makes her out to be, or else she will fail in such situations.

The situations van Fraassen seems to have in mind are cases of Kuhnian 'crisis science'—situations in which established methods and theories fail, and practitioners are forced to step back from the practice at hand to reflect on their presuppositions, aims, and methods. A paradigmatic case of such a situation is the development of quantum mechanics, where the need to interpret the newly developing theory became especially clear. If the naturalistic native is indeed unable to engage in such reflective and interpretive endeavours, she cannot be regarded as a paradigmatic participant, as she will be unable to handle situations of crisis. To handle a crisis, van Fraassen suggests, she would have to be able to engage in a stepping back of just the sort the second philosopher earlier claimed was so alien to the naturalistic native that the questions and concerns raised by typical philosophers were unintelligible. Van Fraassen hence poses a dilemma for naturalists: "(a) to be in a position to share a point of view from which the sciences can be discussed, bracketing our beliefs about to what extent they are true, or (b) to be in no position even to survive as a Naturalistic Native in times of trouble" (van Fraassen, 2015, 85).

The dilemma hinges crucially on the claim that there are situations in which practising scientists must be able to step back from their work and reflect critically on it *in the same manner* that would also force the naturalistic philosopher to reflect critically on her stance towards science and on the naturalistic native as a paradigmatic inquirer. I will come back to this distinction below when discussing understanding, but first let's turn to the case of quantum mechanics.

6.3 The Case of Quantum Mechanics

Quantum mechanics is a paradigmatic case of science itself demanding an interpretation, both because when it was first developed, physics was 'in trouble' in the sense of a Kuhnian 'crisis', but more importantly because the theory as developed seemed difficult to understand. Unsurprisingly, van Fraassen chooses quantum mechanics as his example when he insists that theories are open to interpretation, and that a failure to engage in such interpretation is therefore a failure not just of the naturalistic philosopher, but of the naturalistic native as a paradigmatic participant of science.

> Theories are formulated, their formulation is investigated in the context of the alternatives that are open: for example, quantum mechanics is understood better now that we have seen Bohmian mechanics and the GRW theory [as alternatives to 'orthodox quantum mechanics']. We could see all three, and compare them, discuss agreements and possible disparities in the empirical predictions, try to imagine at least thought experiments in which their differences would become manifest... We could much more clearly, because of the displayed contrasts, address the question what the world could possibly be like if it were as quantum mechanics says it is. What could I call this except a 'level of analysis above that of science'?
>
> (van Fraassen, 2015, 83)

Van Fraassen here connects three ideas: (i) that greater understanding of quantum mechanics was achieved through the development and investigation of alternatives, (ii) that this investigation allowed us to address the interpretive question of 'what the world could possibly be like if it were as quantum mechanics says it is', and (iii) that interpreting theories in this sense involves stepping outside of science proper. The reason for this last step is that van Fraassen claims that to investigate alternatives requires treating our best science as open to re-interpretation; it requires bracketing what we think we know (van Fraassen, 2015, 84). In calling this interpretive endeavour a 'level of analysis above that of science', van Fraassen poses a challenge to the naturalist: if her naturalistic native is unable to engage in interpretation, she has to give up on understanding (or on improving her understanding) of the scientific theories she takes for granted; but if she can and does engage in such interpretations, even the naturalistic native is engaged in stepping back from science, and there is hence no reason for the naturalistic philosopher to refuse to do so as well. We are right at the dilemma set out above.

A naturalist might wish to respond to this challenge by questioning, on the one hand, whether the interpretive activities described require a level of analysis above that of science, and on the other hand the move from offering interpretations of particular scientific theories to taking an interpretative stance towards science as a whole. Both points make interpretation out to be distinctively *philosophical* in character, whereas the naturalist might want to suggest that there are forms of interpretation and understanding that are not beyond the native's reach. Indeed, there seem to be three different moments of interpretation arising in connection with quantum mechanics: interpretive questions in the development of quantum mechanics, the development of alternative viewpoints in competition with 'orthodox quantum mechanics',[3] and

[3] The term 'orthodox quantum mechanics' is itself problematic; for discussion, see Wallace (2019).

finally the project of articulating what the world is like if quantum mechanics is true. When van Fraassen speaks of interpretation, he primarily has in mind this last project. It will nonetheless be worthwhile to talk about all three forms of interpretation, since what counts as an interpretation and what counts as understanding gained through the project of interpretation are not neutral ground between empiricists and naturalists. I now turn to these three interpretive projects, beginning with the initial development of quantum mechanics.

6.3.1 The development of quantum mechanics

The development of quantum mechanics is notorious as a tumultuous and exciting period in the history of science (Beller, 1999), and it is clearly a case where questions of interpretation arise from within science. On the one hand, a new theory was needed to make sense of a range of experimental results and effects, and on the other hand, newly developed approaches needed to be compared, assessed, and understood. Can naturalists make sense of these debates? Responding to empirical data by developing new bits of theory, comparing and evaluating potentially competing theoretical proposals, and deriving new empirical predictions from the proposals under discussion are all part and parcel of what it means to engage in science. Even in the especially excited debates over quantum mechanics, many of the activities in question simply look like (theoretical) physics and the debates are just what one might expect at the cutting edge of research.

It is nonetheless fair to describe these debates as involving interpretive questions. To acknowledge the mismatch between extant theory and experimental phenomena, and more specifically, to assess this mismatch as so significant as to warrant the development of an entirely new theory, requires entertaining the possibility that the theory that served as a starting point is not entirely true or possibly false altogether. That is to say, acknowledging that a scientific paradigm is in crisis requires stepping back from it to some extent. Moreover, once a new candidate paradigm has been proposed, it too seems to require interpretation: its concepts and principles are unfamiliar, and its relation to existing theories needs to be clarified. This was of course especially true for quantum mechanics, with the measurement problem, superpositions, and non-locality posing particular challenges.

It is correct, then, that science itself gives rise to interpretive questions of some sort, and that a good scientist needs to be able to respond to such situations of crisis. But this alone does not yet show that an appropriate response requires explicitly philosophical reflection, or rising 'above' the normal practice of science. Debates over interpretation were vivid, precisely because there was a plurality of viewpoints among scientists. It doesn't seem as though we need to describe these debates as involving active 'bracketing' of what particular individuals took to be true, nor did it involve stepping back to a different, non-scientific activity altogether. Far from being bracketed, the different theoretical proposals were vehemently defended. The appeal to thought experiments or methodological principles in the course of these debates should not distract us from the fact that the ultimate goal was to develop an adequate physical theory, not a metaphysical world view.

We should also note that the (re-)solution of these disputes was not a unified metaphysical picture of the world according to quantum mechanics. Instead we find a package of technical solutions, the consolidation of results, and a redirection of physics away from foundational disputes towards the development of quantum field theory and applications of quantum mechanics. It is also clear that philosophers do not think that the debate over the interpretation of quantum mechanics was resolved at this point, and of course difficulties, like the relationship of classical to quantum mechanics, remained unresolved (Bokulich, 2008). We need to acknowledge, however, that closing off further interpretive questions helped physics to move beyond the state of crisis science and to return to normal science under a new paradigm. Not only do scientists need to be able to respond to crisis by stepping back from their beliefs and practices, as van Fraassen rightly observes; they conversely also need to be able to close debate after a new theory emerges. The question, to which we shall return below, is whether naturalists have to side with the scientists who closed off these further questions, or whether they can engage with interpretive questions that arise even after the new paradigm has been established.

6.3.2 *The development of alternative theories*

A more serious challenge arises for naturalism with respect to theories like Bohmian mechanics and GRW (after the developers: Ghirardi, Rimini, and Weber, 1986). Bohmian mechanics is a form of non-local hidden variables theory of quantum mechanics, whereas GRW offers a spontaneous collapse theory. Both views were primarily developed after the initial consolidation of quantum mechanics, and they were offered as alternatives to orthodox quantum mechanics.[4] Van Fraassen moreover points to these two views in particular as interpretations that aided the understanding of quantum mechanics. What motivations might naturalistic natives have to engage with either of these viewpoints?

On behalf of the naturalist one might say that both Bohmian mechanics and GRW are in the first instance alternative physical theories, not (philosophical) interpretations of theories. Just like naturalistic natives could engage in the development of a brand new theory, we should expect them to be able to develop and assess possible rival theories even where one theory is already established. While doing so requires the ability to step back from a particular set of beliefs, it doesn't require stepping outside the framework of science altogether.

The question whether Bohmian mechanics and GRW are alternative theories or interpretations of quantum mechanics is contested. Van Fraassen offers the following distinction (van Fraassen, 1991, 9): an extension of a theory offers new empirical predictions, whereas an interpretation of an extant theory preserves all the predictions of the theory, and does not add any additional predictions. Since both Bohmian mechanics and GRW are developed (in the first instance) to match standard quantum mechanics in the realm of phenomena for which quantum mechanics was first developed, they would seem more like interpretations than like rival theories.

On the other hand, the Bohmian formalism differs from standard approaches to quantum mechanics. Some versions of Bohmian mechanics, for example, treat the

[4] The origins of Bohmian mechanics date back to de Broglie (1928), but the versions currently discussed as competitors to orthodox quantum mechanics were developed later.

Born rule as a theorem, not as an independent postulate (Dürr, Goldstein, and Zanghì, 1992; Valentini, 1991). Moreover, it is not clear what a Bohmian version of quantum field theory should look like (Dürr et al., 2004; Wallace, 2012). So the theory differs both in the role played by certain propositions, and its relation to other theories. Those are relevant theoretical differences, even if no new empirical predictions were forthcoming.[5] Indeed, the apparent lack of prospects for a Bohmian quantum field theory provides a strong reason for physicists to remain sceptical about Bohmian views and a disincentive to take the theory seriously as a rival to quantum mechanics.

Similarly, GRW is an extension of sorts, since it aims to offer a unified dynamics for microscopic and macroscopic systems. Like Bohmian mechanics, it is offered as a theoretical improvement over standard quantum mechanics. Where Bohmian mechanics seeks to improve upon 'orthodox' quantum mechanics by offering a deterministic theory that permits (some forms of) causal explanation, GRW aims to address the 'measurement problem' by offering a dynamics that does not single out measurement processes as primitive elements of the theory.

Neither theory, then, is an extension in the sense of (primarily) offering new testable empirical hypotheses, but each can nonetheless be viewed as a new theoretical development in physics. To make matters even more complicated, not only were Bohmian mechanics and GRW themselves motivated by dissatisfaction with the orthodox interpretation of quantum mechanics, there now also exist further competing 'interpretations' of both Bohmian mechanics and GRW (e.g. Allori, 2013; Egg and Esfeld, 2015; Suárez, 2015). These interpretations are usually interpretations in the third sense (discussed below): they are attempts to find metaphysically palatable ontologies for Bohmian mechanics and GRW respectively. It seems that in whatever sense either Bohmian mechanics or GRW is to count as an interpretation of quantum mechanics, there is also a sense in which they are themselves theories open to interpretation. The distinction between a rival theory or an interpretation of an existing theory does not seem to be clear-cut in these cases, then. I shall argue below that whether these alternatives should be viewed as interpretations or as rival theories depends on how we think understanding of scientific theories can be achieved.

Whether naturalists can plausibly engage with either Bohmian mechanics or GRW depends heavily on how we think these views should be evaluated. If we take them to be interpretations, rather than rival theories, then their evaluation would seem to be a matter of extra-scientific considerations, and hence not something a naturalist can easily engage with. If they are rival views or theoretical developments of an existing theory, on the other hand, they will be evaluated along the same lines as other scientific theories, and insofar as naturalists are not bound to adhere strictly to the theories they 'grew up with', they can certainly entertain and evaluate competing proposals on the same standards they apply to all scientific theories. Unlike the debates during the

[5] The extent to which the theories make differing predictions is controversial. For the sake of the argument I'm willing to restrict the empirical differences to the original domain of quantum mechanics, where GRW and Bohmian mechanics aim to match the predictions of orthodox quantum mechanics. If we instead include the differences in compatibility with relativistic settings among the empirical differences, then both GRW and Bohmian mechanics do differ empirically from orthodox quantum mechanics and the naturalist's suggestion that these are just different physical theories would seem to be even more convincing.

development of quantum mechanics, engagement with rival views to extant theories requires a certain amount of 'stepping back' from the currently accepted theory. But what needs stepping back from here is not science as a whole, but only a subset of beliefs in a particular field. Naturalists may be dogmatic about their stance towards science, but that does not mean they also have to be dogmatic about any particular scientific theory.

6.3.3 *The ontology of theories*

The third project of interpreting quantum mechanics looks more distinctively philosophical: it is to provide a kind of fleshed-out metaphysical picture of what the world would have to be like, if quantum mechanics were indeed true. This is of course the sense of interpretation van Fraassen has in mind, first and foremost. It is this project that both constructive empiricists and realists can engage in when it comes to scientific theories, because both take the semantics of science to be broadly realist—they differ merely in their epistemic attitude towards the theories in question (van Fraassen, 1991, 4). A key feature of interpretations in van Fraassen's sense is that they go beyond the theory on offer, but without being alternative *theories*. All parties to the dispute already accept the theory in question, yet disagree about its interpretation.

Since this project explicitly addresses theory–world relations, it poses the most severe difficulties for the naturalistic native. The naturalistic native is happy to ask whether the evidence speaks in favour of any one of several hypotheses, but questions about how theories *in general* relate to the world, or what possible *overall* ontologies a theory might suggest, are precisely the sorts of questions she wants to resist. Just as she doesn't understand the (on her view) sceptical perspective offered by empiricism, she equally doesn't understand the realists' 'footstamping'.[6] To her, the latter does not add anything over and above the evidence available from within science, and it is on the basis of this evidence, not on account of any global attitude towards science, that she is able and willing to engage ontological questions at all. Maddy suggests that the second philosopher can ask "what does this [successful bit of science] tell us about how the world is?" (Maddy, 2007, 397). But for the second philosopher, the answer to this question will typically be straightforward, with allowances for anything resembling 'interpretation' made only for idealization and mathematization. For Maddy's second philosopher, there is no *general* problem of how to understand scientific theories, nor is there a general, extra-scientific criterion for what there is, or even for what there is according to a particular theory. Instead there will be a long list of particular answers to particular existence questions, based not on 'what science tells us there is', but based on the strength of the evidence offered up for the existence of the putative entities in question (Maddy, 2007, 397).

The second philosopher believes that particles like electrons exist on the basis of specific evidence, not because she is in general committed to believing in the entities 'our best science' says there are. Upon inspection she might, for example, decide to withhold belief in the Higgs boson, if she finds the LHC results insufficient.

[6] For a detailed account of the relationship between Maddy's naturalism and scientific realism, see Wolff (2015).

At the same time she is unlikely to demand interpretation or ontology beyond what is currently offered by the relevant scientific theories. If quantum mechanics suggests that electrons are particles, but that particles do not follow determinate trajectories, and she is persuaded by the evidence in favour of quantum mechanics, then this is the view of particles the second philosopher will adopt, without insisting that we need something else in addition to having a proper interpretation or ontology for the theory.

With this piecemeal approach in mind, it is hard to see how she can find questions as to whether the ontology of quantum theory involves particles (and particles only) in its (primitive) ontology, worth engaging. Similarly, she will be puzzled by questions as to what the ontological status of the wave-function is, or whether quantum mechanics requires a form of holism incompatible with Humean metaphysics. These do seem like extra-scientific considerations and questions. They also seem like considerations that may ultimately rely on different views of science or different background metaphysical commitments, neither of which the naturalistic native is supposed to have. So it seems van Fraassen is right to suggest that this is an interpretive enterprise the naturalistic native will likely reject.

There is a second type of naturalistic response to van Fraassen's interpretive project. Even a naturalist sympathetic to the idea that we should interpret a theory in the sense of providing an ontology for it might disagree with a key element of van Fraassen's characterization of such interpretations. From an empiricist perspective, interpretations are added to the original theory, but make no empirical difference—otherwise they would be alternative theories. Nonetheless, empiricists clearly expect there to be substantive disagreements over the interpretation of theories, at least for theories like quantum mechanics. A naturalist, by contrast, might reject that there really is such a plurality of equally plausible interpretations.

This seems to be David Wallace's understanding of the Everettian 'interpretation': "[T]he unmodified quantum theory can be taken as representing the structure of the world just as surely as any other theory of physics. In other words, quantum mechanics can be taken literally. The only catch is that, when we do take it literally, the world turns out to be rather larger than we had anticipated: indeed, it turns out that our classical 'world' is only a small part of a much larger reality" (Wallace, 2012, 13). Such 'literalism' with respect to quantum theory is of course anathema to van Fraassen's interpretive project. For Wallace, quantum theory is no different from other scientific theories. We expect scientific theories to provide explanations in addition to predictions, so we should not adopt instrumentalism as an outlook on science in the light of quantum mechanics. Moreover, quantum theory, as Wallace sees it, has no more need for a special interpretation than any other scientific theory: "The 'Everett interpretation of quantum mechanics' is just quantum mechanics itself, 'interpreted' the same way we have always interpreted scientific theories in the past: as modelling the world" (Wallace, 2012, 38). At least in Wallace's hands, then, the Everett interpretation is not an interpretation in van Fraassen's sense, even though it arguably provides an ontology for quantum mechanics, because for the Everettian, the competition isn't between an Everettian metaphysics for quantum theory and some other metaphysical reading of quantum theory. Instead, the competition is between a literal reading of the quantum formalism provided by the Everett reading, or else either a revised stance on what we expect from scientific theories (some form of instrumentalism), or an alternative physical theory along the lines of Bohmian theories or GRW. The alternatives between which we have to decide are not alternative 'interpretations', but alternative theories.

Some naturalists, hence, refuse to engage in the project of providing 'an ontology' for scientific theories altogether, as seems to be the case for Maddy's second philosopher. Others provide ontologies for particular theories, but suggest that doing so does not involve an empirically underdetermined choice among equally good alternatives, but instead involves a literal reading of the theory, as Wallace proposes is the case for the Everettian reading of quantum mechanics. Both reactions are rejections of van Fraassen's interpretive project, since they seem to presuppose that a privileged, 'literal' reading of the theory is a available, which is of course precisely what a constructive empiricist would want to deny.

Rejecting the interpretive project as van Fraassen understands it does not by itself pose a problem for naturalism. It is a problem for naturalism only insofar as the benefit derived from interpreting theories is not otherwise available to naturalists and insofar as these benefits are worth having. Since the main purpose or aim of interpretation is greater understanding of scientific theories, we need to ask what kind of understanding one might hope to obtain from the kind of interpretive projects described here. If there is understanding to be gained, and gained only by way of evaluating a range of different candidate ontologies for a given theory, and this understanding is arguably an aim of science, then the naturalistic native will indeed fall short of being a model participant in the scientific enterprise. For then there would be a cognitive achievement not available to her, but available under a competing conception of what it means to be a model participant. On the other hand, if there is no such understanding forthcoming, or if such understanding is not within the remit of science, then it seems the naturalistic native is not losing out by failing to engage in this form of interpretation and can hence not be said to fail by her own lights (or by the lights of the naturalistic philosopher, who offers the native as a stand-in).

6.4 Interpretation and Understanding

6.4.1 Understanding theories

The main purpose of interpreting theories was to increase our understanding of them. We should ask, then, whether naturalists can achieve understanding without engaging in explicitly interpretive projects of the third type as van Fraassen conceives of them, or whether they might have reasons to reject the understanding allegedly provided by such projects as spurious. What do we mean by 'understanding'?

While there is no unique way of classifying understanding, a distinction is commonly drawn between symbolic, objectual, and explanatory understanding (Baumberger, Beisbart, and Bru, 2017, 4). The first refers to understanding in the sense of understanding a language, or more broadly a type of symbolic representation; the second means understanding a particular domain or subject matter, and the third is to understand why something is the case. The relationship between these different forms of understanding is somewhat controversial, but at least prima facie they are quite different. The question is, therefore, which forms of understanding naturalists might aim for, and how they might do so.

Since quantum mechanics is in the first instance treated as a mathematical formalism, understanding quantum mechanics seems to require symbolic understanding (Baumberger, Beisbart, and Bru, 2017). It seems fair to say that the symbolic understanding of quantum mechanics has increased since its first inception in Heisenberg's and Schrödinger's seminal work. All three forms of understanding were initially absent, which perhaps explains the sense of crisis and the intense engagement in interpretation in the early days of quantum mechanics. Once Heisenberg's and Schrödinger's formulations had been shown to be equivalent, a passable level of symbolic understanding had been achieved and the standardization of presentations of quantum mechanics using Hilbert spaces helped to spread a unified understanding of quantum mechanics as a formalism.

After such symbolic understanding had been achieved, it became possible to reject other interpretive projects as superfluous; it was now possible to calculate, and hence possible to 'shut up' about other forms of understanding one might hope to achieve. The attitude many physicists adopted in this period was to accept symbolic understanding as the only understanding necessary for doing physics, and to reject the demands for other forms of understanding as not within the remit of physics. If physics only provides symbolic understanding, then any questions aimed at explanatory or objectual understanding are left unanswered. With this sharp division between symbolic understanding as the only task of physics and all other forms of understanding relegated to 'mere' philosophizing, van Fraassen's suggestion that interpretation involves analysis at a level above science seems quite plausible.

Accordingly van Fraassen sometimes seems to think of the naturalistic native as a scientist who only aims for symbolic understanding, and who, in virtue of being a 'native speaker' of the language of the theory, cannot even comprehend what it would mean to treat the theory as open to interpretation (van Fraassen, 2015, 82). If this is the right way to think of the naturalistic native, naturalists would lose out on two forms of understanding ostensibly provided by interpreting scientific theories: explanatory and objectual understanding. Naturalists might want to reject the claim that such interpretations provide any genuine or useful understanding at all, but to do so requires an independent argument and it is not clear that such an argument can be given. So we might agree with van Fraassen that naturalists would be missing something. Van Fraassen likens this type of naturalistic native to a fundamentalist about scripture (van Fraassen, 2015, 82). For a fundamentalist about scripture, understanding doesn't go beyond faithful recitation of words on the page, and any doubts or questions about the 'text', in this case, about the quantum formalism, are suppressed and extinguished. In particular, a fundamentalist can neither ask how we relate to the 'text' or how the 'text' relates to the world. Even asking for a different kind of understanding is to be made illegal. A fundamentalist literalist of this type will not strike most philosophers as the ideal participant in scientific inquiry, and would indeed seem lost in situations of scientific crisis.

A naturalist, who is already committed to the view that science and philosophy are continuous, might take a different stance. She might suggest instead that explanatory understanding was simply unavailable at the time and that scientific methods were insufficient to choose a more specific theory—one that settled certain explanatory questions—based on the evidence available. So it was reasonable in that context to continue work based on a purely symbolic understanding, bracketing all questions of explanation. She might nonetheless suggest that explanation is in principle an epistemic aim of science, and that providing explanatory understanding is a distinct virtue of a scientific theory. That is to say, she might accept that historically 'shut-up-and-calculate' was the right strategy, while acknowledging that explanatory understanding is still missing. Unlike van Fraassen's empiricist, and unlike the shut-up and calculate physicist, however, she does take such understanding to be an aim of future science. If so, naturalists may not have to forgo explanatory understanding altogether, but will have to achieve it by means other than interpretation in van Fraassen's sense.

The contextualist approach to explanatory understanding suggested by de Regt and Dieks (2005) is particularly helpful for developing this naturalistic response. They suggest that explanatory understanding is *contextual* in the sense that what particular features of a theory are considered explanatory is highly sensitive to historical context. They show, for example, that whether a mechanistic explanation or an explanation involving action-at-a-distance forces counts as a paradigmatic case of an explanatory theory is highly dependent on the historical context. This contextual account is developed in more detail by de Regt (2017). Crucially de Regt insists that understanding, including explanatory understanding, is a central aim of science itself, despite its pragmatic character. De Regt offers the following criterion for understanding a phenomenon: "A phenomenon *P* is understood scientifically if and only if there is an explanation of *P* that is based on an intelligible theory *T* and conforms to the basic epistemic values of empirical adequacy and internal consistency" (de Regt, 2017, 94). Understanding of a domain or phenomenon is hence gained by developing an intelligible (or more intelligible) theory of the domain and by acknowledging that explanation is an epistemic aim of science. Intelligibility of a theory is defined as "the value that scientists attribute to the cluster of qualities of a theory (in one or more of its representations) that facilitate the use of the theory" (de Regt, 2017, 40). What makes a theory intelligible can hence vary from context to context.

Naturalists will find the contextual approach to explanatory understanding congenial for two reasons. First, it does justice to the wide range of demands for understanding that are made of scientific theories, and it does particularly well in cases like quantum mechanics, where not just one, but multiple standards for understanding were invoked to evaluate the different proposals. Quantum mechanics posed a challenge to understanding in numerous ways: the lack of anything like a causal-mechanical story, non-locality, and surprising principles like the uncertainty relation and superposition. Unsurprisingly, then, orthodox quantum mechanics has not been treated as an intelligible theory in the sense given by de Regt.

Second, it makes understanding an epistemic aim of science, despite its acknowledgement of pragmatic aspects of understanding. This is important for naturalists if they want to acknowledge that understanding is important without committing to a further, extra-scientific project of interpretation as a means for providing understanding. Understanding can be achieved as part of developing a scientific theory, not (only) through offering interpretations of the theory. Given that understanding phenomena depends on having an intelligible theory of them, de Regt seems to prioritize explanatory understanding over objectual understanding. We gain the latter by having a theory that provides us with the former. If we think of explanatory understanding as an epistemic goal of science, we should expect that we understand a particular phenomenon or aspect of the natural world by developing a scientific theory for it. Mere symbolic understanding of the quantum formalism will not provide explanatory or objectual understanding, but if explanatory understanding is an aim of science, and objectual understanding is achieved through explanatory understanding of a relevant scientific theory, then both objectual and explanatory understanding are within reach for a naturalistically oriented scientific inquirer.

Bohmian mechanics, and to a lesser degree GRW, can be seen as later attempts to improve upon orthodox quantum mechanics, precisely by offering *explanatory understanding* in addition to merely symbolic understanding of an empirically successful theory. Both theories are supposed to be more intelligible than orthodox quantum mechanics, albeit in different ways. In the context of 'shut-up-and-calculate', explanatory understanding is regarded as a purely philosophical goal, and hence these views continue to be regarded, both by physicists and philosophers, as being 'philosophically' motivated. But the allocation of labour here is secondary; the more important point is that these theories aim to do better than quantum mechanics by offering explanatory understanding in addition to empirically adequate predictions. For a naturalist not trapped in the shut-up-and-calculate paradigm, this means that these theories can be evaluated according to their overall virtues. Since they are meant to be effectively empirically equivalent to quantum mechanics in its original domain, a naturalist will turn to non-empirical virtues to choose between them. Intelligibility, however it may be cashed out in a given context, will be among such virtues for the naturalist. But there are others to be considered as well: simplicity and fruitfulness, for instance. It is with respect to fruitfulness that Bohmian mechanics and GRW fall short in the eyes of many physicists, because they are restricted to non-relativists settings. Crucially, these virtues will be treated as *epistemic* reasons for accepting or rejecting the theory, not merely as pragmatic reasons.

Now it begins to look as though naturalists might have a route to understanding that does not involve interpretation of scientific theories 'from the outside', as it were. Assessing what is gained by accepting a more intelligible theory is traded off against the fruitfulness or simplicity of the theory. These decisions involve pragmatic elements, but they are not *merely* pragmatic and are taken by the naturalist as being a normal part of what scientists do when developing and assessing new theories. This naturalistic response seems a lot more attractive than the dismissive attitude towards explanatory understanding seemingly displayed by the naturalistic native understood as a follower of shut-up-and-calculate. For the naturalist, there will be no sharp distinction between a theory and its interpretation, and no division of labour between developing a fruitful theory and an intelligible one.[7]

This does not mean that naturalists will engage in exactly the same questions as the philosophers who offer interpretations of physical theories. Merely arguing that

[7] Maddy seems to agree; she approvingly cites Marc Lange's rejection of a sharp division between theory and interpretation (Lange, 2002, 250). See Maddy (2007), 409.

a particular theory is or is not compatible with a preferred metaphysical outlook does not by itself enhance the intelligibility of the theory. Instead the naturalist will demand that scientific theories offer explanations, and combine this demand with a 'literalist' reading of the theory's formalism. This may mean adopting new elements into the ontology, but it doesn't mean stepping outside the practice of science. Perhaps we should then think of the naturalistic native not as rejecting all forms of understanding besides symbolic understanding as lying outside the scope of science, but instead as somebody who takes explanatory understanding as an epistemic aim of science and who hence treats intelligibility as an epistemic virtue of a theory. This contrasts of course with van Fraassen's view, who suggests that explanation, and hence presumably also explanatory understanding, are too pragmatic to be properly among the epistemic goals of science (1980).[8] And it is here, perhaps, that we find the real difference between naturalism and empiricism when it comes to the interpretation of theories.

6.4.2 *Interpreting science*

Naturalists, then, can aim to understand scientific theories, even if not by means of extra-scientific interpretations. There is another interpretive project van Fraassen asked naturalists to engage in, namely the project of taking a stance towards science as an activity: "an interpretive stance is open to us and indeed, is *needed* to understand our own situation properly" (van Fraassen, 2015, 85). Whereas the former project was aimed at understanding a particular scientific theory, the latter is aimed at understanding *our relationship* to particular theories and, more importantly, to science as a practice or activity. 'Our situation' here would seem to mean the human condition as a whole, not just the particular qualitative picture appropriate for quantum mechanics. Nothing in what I've said so far shows how a naturalist can engage in this second project of interpretation. One reason for this might be that the two projects, *pace* van Fraassen, can come apart. A theory can be intelligible to a scientist, and hence confer understanding of a phenomenon in the sense of de Regt and Dieks, even while the same scientist is not particularly reflective of her engagement in scientific practice. As *philosophers*, by contrast, we indeed often aim for understanding not only of a particular phenomenon, but of entire practices, like science. What seems puzzling about Maddy's second philosopher is how uninterested she seems to be in questions aimed at this second interpretive project. But that doesn't mean her portrait of the ideal inquirer is inadequate as a portrayal of the model empirical scientist. If second philosophy is dissatisfying as a philosophical stance, it is not dissatisfying because of its view of science and its aims, but because of its view of philosophy and its aims. To say so will not do as an *argument* against naturalism as a stance, since it would beg the question against the key naturalistic commitment—that philosophy is continuous with science—but it helps to explain the dissatisfaction non-naturalists have with that stance.

[8] Compare also: "The reason [questions of interpretation] are often difficult to answer is, in my opinion, that scientific discussion is so thoroughly focused on the question of empirical adequacy alone" (van Fraassen, 1991, 242).

6.5 Conclusion

I've distinguished three different interpretive projects with respect to quantum mechanics and considered three types of understanding we might hope to gain from scientific theories. The first interpretive project occurs in the context of developing a new theory or paradigm, and it involves arguing over different standards of intelligibility for scientific theories, where intelligibility is considered one among several virtues a theory ought to possess. In this context, even symbolic understanding of the new theory is tenuous. The second interpretive project arises after the consolidation of the new theory and after symbolic understanding has been achieved. It aims at a greater explanatory understanding and is pursued through the development of alternatives like Bohmian mechanics and GRW. While empiricists will be inclined to treat these as alternative *interpretations*, naturalists will see them as alternative theories and evaluate them on their overall theoretical virtues. The difference between naturalists and empiricists here comes down to the question whether explanation is an epistemic aim of science, or instead a philosophical add-on. The third type of interpretive project addresses the question of what the world is like if a given theory is true. Answering this project involves providing an ontology for the theory in question. The type of understanding provided by such an interpretation would be something like understanding a domain or subject matter, although it is questionable whether such understanding can be achieved solely by providing an ontology. Instead it seems that having a theory that satisfies criteria of intelligibility and that explains features of the domain in question would be what provides understanding of the domain. While empiricists see a theory as open to a range of different candidate ontologies, naturalists will assume a certain literalism with respect to these theories: there is a preferred ontology for a given theory, and the question whether we should accept the ontology just comes down to whether we should accept the theory. No special philosophical criteria for accepting or rejecting ontological commitments need be employed to gain the understanding the theory provides of its domain.

There are, then, interpretive projects that a naturalist might engage in, although she will typically not conceive of them as distinctively philosophical endeavours. Instead she will pursue them as part of building and evaluating scientific theories, which for her always aim at more than empirical adequacy. In particular, any understanding an empiricist might hope to gain from distinctively philosophical interpretations, naturalists will hope to achieve as part of developing scientific theories. With respect to quantum mechanics, naturalists will not think of it as being more in need of interpretation than any other scientific theory.

References

Allori, V. (2013). Primitive ontology and the structure of fundamental physical theories. In A. Ney and D. Z. Albert (Eds), *The Wave Function: Essays in the Metaphysics of Quantum Mechanics*. Oxford: Oxford University Press.

Baumberger, C., C. Beisbart, and G. Bru (2017). What is understanding?–An overview of recent debates in epistemology and philosophy of science. In S. R. Grimm, C. Baumberger, and S. Ammon (Eds), *Explaining Understanding: New Perspectives From Epistemology and Philosophy of Science*. London: Routledge.

Beller, M. (1999). *Quantum Dialgoue: The Making of a Revolution*. Chicago: The University of Chicago Press.

Bokulich, A. (2008). *Reexamining the Quantum-Classical Relation: Beyond Reductionism and Pluralism*. Cambridge: Cambridge University Press.

de Broglie, L. (1928). La nouvelle dynamique des quanta. In *Institut international de physique Solvay (1928). Electrons et photons: rapports et discussions du cinquième Conseil de physique tenu à Bruxelles du 24 au 29 octobre 1927*, pp. 105–32. Paris: Gauthier-Villars.

de Regt, H. and D. Dieks (2005). A contextual approach to scientific understanding. *Synthese 144*(1), 137–70.

de Regt, H. W. (2017). *Understanding Scientific Understanding*. Oxford: Oxford University Press.

Dürr, D., S. Goldstein, R. Tumulka, and N. Zanghì (2004). Bohmian mechanics and quantum field theory. *Physical Review Letters 93*(9), 090402.

Dürr, D., S. Goldstein, and N. Zanghì (1992). Quantum equilibrium and the origin of absolute uncertainty. *Journal of Statistical Physics 67*(5–6), 843–907.

Egg, M. and M. Esfeld (2015). Primitive ontology and quantum state in the GRW matter density theory. *Synthese 192*(10), 3229–45.

Ghirardi, G., A. Rimini, and T. Weber (1986). Unified dynamics for microscopic and macroscopic systems. *Physical Review D34*(470).

Ladyman, J. and D. Ross (2007). *Every Thing Must Go*. Oxford: Oxford University Press.

Lange, M. (2002). *An Introduction to the Philosophy of Physics*. Oxford: Blackwell.

Maddy, P. (2007). *Second Philosophy*. New York: Oxford University Press.

Rowbottom, D. and O. Bueno (2011). How to change it: modes of engagement, rationality, and stance voluntarism. *Synthese 178* (1), 7–17.

Suárez, M. (2015). Bohmian dispositions. *Synthese 192*(10), 3203–28.

Valentini, A. (1991). Signal-locality, uncertainty, and the subquantum h-theorem. *Physics Letters A 156*(1–2), 5–11.

Van Fraassen, B. (1980). *The Scientific Image*. Oxford: Clarendon Press.

Van Fraassen, B. (1991). *Quantum Mechanics: An Empiricist View*. Oxford: Clarendon Press.

Van Fraassen, B. (2002). *The Empirical Stance*. New Haven: Yale University Press.

Van Fraassen, B. (2015). Naturalism in epistemology. In R. N. Williams and D. N. Robinson (Eds), *Scientism: The New Orthodoxy*, 63–96. London: Bloomsbury.

Wallace, D. (2012). *The Emergent Multiverse: Quantum Theory according to the Everett Interpretation*. Oxford: Oxford University Press.

Wallace, D. (2019). What is orthodox quantum mechanics? In A. Cordero (Ed.), *Philosophers Look at Quantum Mechanics*, 285–312. Dordrecht: Springer.

Wolff, J. (2015). Naturalistic quietism or scientific realism? *Synthese DOI 10.1007/s11229-015-0873-3*.

PART III

Pragmatism about Quantum Theory

7

Pragmatist Quantum Realism

Richard Healey

7.1 Introduction

Quantum theory has long presented a challenge to a would-be scientific realist who maintains that the historical sequence of successful theories in a mature science like physics offer increasingly accurate representations of a physical world that is largely independent of our observational abilities and practices. Over the past century, quantum theory has amassed an unparalleled record of successful applications over an increasingly wide field of science. But heated controversy continues over whether, and if so how, it represents the world as distinct from our observations of it.

Many philosophers seeking to secure the relation between scientific realism and quantum theory have recently adopted overly narrow conceptions of both relata. The general goal of this chapter is to enrich the discussion by noting alternative ways of understanding the project of reconciling quantum theory with scientific realism. My more specific aim is to argue that a pragmatist view of quantum theory facilitates just such a reconciliation.

I begin by building a framework in which to set alternative ways of understanding quantum realism. Besides helping to orient readers who haven't encountered them all, I use this to motivate the pragmatist realism described and defended in the bulk of this chapter. Section 7.3 locates a novel form of physical realism recently advocated by Chris Fuchs within this framework, at the same time explaining why his QBist understanding of quantum theory is appropriately regarded as anti-realist.

Then in Section 7.4, I briefly sketch key features of a pragmatist view of quantum theory I have been developing recently. Scientific realism is often taken to be a view or attitude toward how far, and how well, a scientific theory represents the world. But ultimately it is we, as users of theories, who represent the world. In the case of quantum theory, we do so by applying mathematical models to the world. This application involves claims about a target system that are appropriately evaluated for truth and representational accuracy. But, in this pragmatist view, quantum theory itself makes no claim of this form: and scientific realism, broadly construed, does not require a theory to claim that the target system is (more or less) faithfully represented by the model being applied. Section 7.4 argues that this is not how models of quantum theory are applied.

Richard Healey, *Pragmatist Quantum Realism* In: *Scientific Realism and the Quantum*. Edited by: Steven French and Juha Saatsi, Oxford University Press (2020). © Richard Healey.
DOI: 10.1093/oso/9780198814979.003.0007

One symptom of the poverty of recent discussion of scientific realism (at least in the quantum context) is the implicit assumption that any view of a scientific theory is either (narrowly) realist or instrumentalist. In Section 7.5, I review reasons why the pragmatist view just sketched is not instrumentalist. In the following sections of the chapter I argue for a conception of scientific realism that deems this pragmatist view realist. Realism about the quantum domain is often taken to require provision of an ontological model. In Section 7.6, I argue that quantum theory does offer such a model in the pragmatist view, despite an unsatisfying incompleteness that may motivate attempts to construct a successor theory.

Section 7.7 argues that quantum theory offers genuine explanations in the pragmatist view. In Section 7.8, I consider but reject the view that scientific realism involves a thick notion of correspondence truth that is unavailable here.

Section 7.9 illustrates the limitations of correspondence truth using an interesting recent argument that the universal applicability of quantum theory is incompatible with the assumption that each quantum measurement has a unique, objective outcome. I defend this assumption by appeal to a pragmatist inferential (rather than referential) account of the content of basic representational claims that issue from an application of quantum theory. This account plays a key role in the pragmatist view of Section 7.4 by showing how to eliminate talk of measurement from a precise statement of quantum theory's Born rule without running afoul of 'no-go' theorems that refute a naive realist interpretation of quantum theory.

Some follow Putnam in viewing scientific realism as an empirical hypothesis supported by the history of science. Section 7.10 offers quantum theory as evidence against this hypothesis, suggesting instead that improved symbolic representation of the world be viewed as a long-term scientific aspiration. In conclusion I review the form of quantum realism associated with the pragmatist view of Section 7.4, pointing out where it diverges from what others have taken quantum realism to involve.

7.2 What is Quantum Realism?

7.2.1 Realism

In approaching the varieties of quantum realism, I'll start with realism in general:

> (Realism) The world is the way it is (almost) no matter what anyone may think about it: the only exceptions are processes involved in thinking our thoughts.

A realist about the world of mathematics believes that numbers and their properties exist independently of the existence or thoughts of mathematicians or anyone else. Some mathematical realists take them to exist and have their properties *necessarily*, so (for example)

$$5+7=12$$

would have been true even if there had been no thoughts and even no physical world at all. But if the world contains thoughts, then at least some of its features will depend on whatever is involved in thinking them—hence the qualification.

Quantum realism is concerned with the physical world, not with the world of mathematics. So this gives us

> (Physical Realism) The physical world is the way it is (almost) no matter what anyone may think; the only exceptions are physical processes involved in thinking our thoughts.

Our thoughts affect the physical world at least through the actions to which they lead, and this is one way in which physical processes are involved in thinking them (as when my thought of a drink prompts the raising of my glass). For a physicalist this is not the only way, since he believes there is a sense in which our thoughts *are* physical processes.

7.2.2 *Scientific realism*

Physical Realism is a metaphysical thesis. But we ordinarily assume we learn a lot about the world through science. This adds an epistemological, or at least axiological, component to a scientific realism that incorporates Physical Realism. According to Bas van Fraassen, scientific realism is the view that

> Science aims to give us, in its theories, a literally true story of what the world is like; and acceptance of a scientific theory involves the belief that it is true. (1980, 8)

Such a scientific realist (unlike van Fraassen himself!) takes science to be guided by an epistemic aim that would be achieved only if scientists arrive at literally true theories. Even if the world is ultimately physical, a science like psychology or sociology is not primarily concerned with its purely physical aspects: but physics is. Applied to quantum theory, this kind of scientific realist maintains that

> Physics aims to give us, in quantum theory, a literally true story of what the physical world is like; and acceptance of quantum theory involves the belief that it is true.

In accordance with Physical Realism, such a literally true story would say what the physical world is like: it would have nothing to say directly about anyone's thoughts or experiences of it. Quantum theory is undoubtedly very successful: there are no empirical reasons to withhold acceptance. This leads to one way of characterizing quantum realism:

> (Quantum Scientific Realism$_0$) Quantum theory gives us a literally true story of what the physical world is like.

Scientific realists have a positive epistemic attitude toward the content of our best theories, but the history of physical science cautions against overoptimism. So contemporary scientific realists often restrict their epistemic commitment to the *approximate* truth of our best theories, despite the obscurity of this notion of approximate truth. Such caution suggests this alternative expression of realism about quantum theory:

> (Quantum Scientific Realism) Quantum theory purports to give us a literally true story of what the physical world is like, and we should believe that story is approximately true.

But what is that story? A would-be Quantum Scientific Realist must answer the question 'How could the world possibly be the way quantum theory says it is?': she must provide an Interpretation of quantum theory.[1] Notoriously, this has turned out to be very difficult. Easy answers have been shown to be inadequate, and today the field is littered with a proliferating variety of competing Interpretations of quantum theory in various states of health.

7.2.3 *Naive realism*

Many years ago I formulated arguments against what I called Naive Realism, which tells the following story (Healey, 1979). The quantum domain is supposed to be determinate in that magnitudes always possess definite values on any quantum system. It is held to be objective in that measurement is taken to be merely our way of getting to know what some of these values are. The theory is probabilistic, on this view, because in even the most favorable circumstances there are magnitudes whose values on a given system at a given time are not derivable from the appropriate quantum mechanical description of the system at that time, while this description does yield probability distributions over sets of distinct values. This simple and inviting form of quantum mechanical realism may be reasonably associated with the names of Popper (1967) and Einstein (1949). It may be defined by principles of Precise Values and Faithful Measurement:

> (PV) For any quantum system *s*, and any dynamical variable *A* pertaining to *s*, if *t* is a time lying within the lifetime of *s*, then *A* has a unique real value on *s* at *t*.
>
> (FM) If successful, a measurement of a dynamical variable *A* pertaining to a quantum system *s* reveals the pre-existing value of *A* on *s*.

The arguments were not wholly original: they were based on work by Kochen and Specker (1967) and by Bell (2004b). The arguments rested on two implicit assumptions.

> (Value Independence) The precise value of a dynamical variable is independent of what variable or variables are to be measured.
>
> (Fair Sampling) The probability of success is independent of the value of the measured variable.

The Einstein–Podolsky–Rosen (1935) argument may be seen as an argument *for* Naive Realism, based on their famous Reality Criterion:

> If, without in any way disturbing a system, we can predict with certainty (i.e. with probability equal to unity) the value of a physical quantity, then there exists an element of physical reality corresponding to this physical quantity.

In recent discussions of so-called loophole-free tests of local realism the term 'realism' is often equated to Naive Realism, with or without some of these other assumptions (Giustina et al. 2015; Hensen et al. 2015; Shalm et al. 2015).

[1] I capitalize the initial 'I' to distinguish attempts to answer this question within a broader category of attempts to say how quantum theory should be understood. It is a substantial representationalist assumption that the only way to understand quantum theory is to understand how the world could be the way quantum theory says it is.

The refutation of Naive Realism is not a death-blow to Quantum Scientific Realism, since Naive Realism is an overly narrow understanding of realism about quantum theory. Would-be scientific realists have implicitly recognized this by investigating the prospects of alternative forms of Quantum Scientific Realism.

7.2.4 *Wave function realism*

One alternative has become known as wave function realism. The name is unfortunate. A wave function ψ is a mathematical, not a physical, entity. (Einstein preferred the more abstract term 'ψ-function' while Schrödinger used 'representative,' in each case to highlight the object's status as a piece of mathematics.) What a wave function realist is realist about is not this mathematical entity, but some *physical* entity she takes it to represent. Albert (1996; 2013) takes ψ to represent a physical field in a multi-dimensional configuration space; an adherent to the eigenstate–eigenvalue link might identify the physical referents of ψ with the eigenvalue possessed by each variable represented by a self-adjoint operator for which ψ is an eigenvector; a collapse theorist might take ψ to represent mass density (Ghirardi, 2006), or alternatively flash-propensity (Tumulka, 2006); and so on.

The Wave Function (Ney and Albert, 2013) offers a sampling of recent philosophical work in this tradition that reveals a proliferation of ways of developing, advocating, and criticizing it. As its subtitle reveals, the primary focus of the volume is on metaphysical issues, especially the nature and dimensionality of the space in which the wave function is defined. These are not issues with which physicists are much concerned. As Wallace notes in his contribution to that volume, discussions of the metaphysics of wave function realism are often conducted on the basis of assumptions that are hard to square with the way physicists understand currently fundamental forms of quantum theory. Of course there are those who believe that current formulations of these theories are inadequate, so that the term 'quantum theory' as it appears in the thesis of Quantum Scientific Realism must be treated as open to negotiation in the interests of securing the truth of the thesis.[2] But to arrive at a charitable understanding of the quantum theory we have it seems wise not to adopt a procrustean attitude to that theory in order to fit it to the Interpreter's bed.

7.2.5 *Spacetime state realism*

Contemporary formulations of quantum theory do not take the wave function to be the only way to represent a system's quantum state. Even pure states are typically represented by Hilbert space vectors or rays, while mixed states are represented by density operators. In C^*-algebraic formulations states are represented not in Hilbert space but by positive, normalized, linear functionals on algebras of observables, and the pure/mixed distinction (differently defined) is supplemented by a further distinction, between normal states and non-normal states that cannot be represented by density operators on a Hilbert space (Ruetsche, 2011).

After raising objections to wave function realism, Wallace and Timpson (2010) instead advocate what they call *spacetime state realism* as a form of realism about quantum states that better accords with physical practice. For a quantum field theory on a fixed background spacetime, a quantum state is defined on each open region of

[2] The work of John S. Bell has been particularly influential in this regard: see Goldstein and Zanghì (2013).

spacetime, including arbitrary unions of disjoint regions. Ignoring non-normal states, the state of each region may be represented by a density operator on a corresponding Hilbert space. For a quantum spacetime state realist, to specify the state of every open region of spacetime in this way is to give a complete fundamental physical description of world history.

It is not easy to recognize our familiar world in this description, and a proponent of spacetime state realism has his work cut out to convince critics that it can be reconstructed from such spare and unfamiliar materials. Wallace and Timpson propose to do this by appeal to decoherence within an Everettian multiverse. But they acknowledge other possible routes, following alternative Interpretational strategies based on hidden variables or dynamical-collapse theories.

7.2.6 Should a realist about quantum theory be a quantum scientific realist?

This section has briefly explored a variety of proposed views of quantum theory, each designed to answer the question 'How could the world possibly be the way quantum theory says it is?' They by no means exhaust the Interpretations currently on offer, and we can confidently expect new additions. Notoriously, these Interpretations take quantum theory to offer mutually incompatible stories of what the physical world is like. For a constructive empiricist like van Fraassen this presents a problem for physics only to the extent that different Interpretations fail to be empirically equivalent. But the would-be scientific realist cannot be so sanguine. If quantum theory cannot give us a single, literally true story of what the physical world is like then Quantum Scientific Realism cannot be true. To defend realism about quantum theory it seems that one must develop an Interpretation and convince everyone of its superiority over all others, largely if not completely without appeal to experiment. While the arguments have improved over the years, so have the objections against every proposed Interpretation. Perhaps we should think again about what it means to be a realist about quantum theory?

7.3 Participatory Realism

By identifying realism about quantum theory with what I called Quantum Scientific Realism we were led into the morass of rival Interpretations of quantum theory. If we retrace the steps that led us there we arrive once more at what I called

> (Physical Realism) The physical world is the way it is (almost) no matter what anyone may think; the only exceptions are physical processes involved in realizing our thoughts.

Frustrated by attempts to label his QBist view of quantum theory instrumentalist or anti-realist, Chris Fuchs has recently classified his view as a form of *participatory realism*, along with a number of other views of quantum theory.

> These views have lately been termed "participatory realism" to emphasize that rather than relinquishing the idea of reality (as they are often accused of), they are saying that reality is more than any third-person perspective can capture. Thus, far from instances of instrumentalism or

antirealism, these views of quantum theory should be regarded as attempts to make a deep statement about the nature of reality. (Fuchs, 2016)

By inserting a first-person perspective into the heart of physics, Fuchs portrays QBism as honoring Wheeler's vision of a participatory universe in which 'particles' and agents jointly create a universe whose external reality is manifested by the unpredictable experiences that result from each agent's interactions with it.

I see participatory realism as a novel form of Physical Realism. While the physicalist locates my thoughts as ultimately just elements of physical reality, the participatory realist sets them apart in a way that resists inclusion within any merely third-person perspective. So this is not just a novel form of dualism. It is reminiscent of Putnam's (1981, xi) (subsequently repudiated) internal realist metaphor

The mind and the world jointly make up the mind and the world.

Taken literally, Fuchs's participatory realism is no metaphor, but the radical metaphysical thesis that reality admits of no third-person view—neither materialist, idealist, nor dualist. As he puts it,

these views of quantum theory should be regarded as attempts to make a deep statement about the nature of reality (2016, 1).

But need a quantum realist accept this, or any similar, metaphysical thesis? Science-based arguments for metaphysical conclusions are notoriously controversial, and in this case so are the arguments for QBism or other participatory-realist views of quantum theory.[3] By retreating so far from the morass of Interpretations our would-be quantum realist has backed himself into a dark metaphysical corner!

Moreover, a QBist who endorses this novel form of Physical Realism has totally repudiated the Quantum Scientific Realist demand that quantum theory tell us what the physical world is like. For a QBist, rather than *describing* the physical world, quantum theory is merely a tool each individual may take up and use to better anticipate *that individual's* experiences when acting in the physical world. Even after abandoning Quantum Scientific Realism, a scientific realist should expect the practice of quantum physics to yield more substantive information about an independently existing physical world. Surely there's a better way to be a quantum realist?

7.4 Representation and a Pragmatist Alternative

Van Fraassen's formulation of scientific realism arose in the course of his attempt to avoid metaphysics in an account of scientific practice as directed toward epistemological goals. He went on to propose constructive empiricism as a less metaphysically loaded alternative. Recall the way he formulated scientific realism:

Science aims to give us, in its theories, a literally true story of what the world is like; and acceptance of a scientific theory involves the belief that it is true. (1980, 8)

[3] Healey (2017a) provides an introduction to the controversy, including references to relevant literature.

He inserted the word 'literally' to disqualify positivist and other non-literal accounts of theoretical truth. Van Fraassen endorsed the so-called semantic conception of a scientific theory as defined by a class of models, and identified the truth of a theory with the truth of the statement that the world is faithfully represented by some model in this class.

Quantum theory in its various forms is naturally understood to involve a family of mathematical models, some non-relativistic, others relativistic; some used to model systems of particles, others to model systems of fields, and so on. Following van Fraassen's lead, one could then say that, for a scientific realist, to accept quantum theory is to believe that it is literally true—that some model in this family faithfully represents the world. But to say that is to commit once more to the perilous Interpretative quest pursued by the Quantum Scientific Realist.

Here is where pragmatism can offer the would-be quantum realist a better way, by querying the representationalist assumption that a scientific theory can give us a literally true story of what the world is like only by faithfully representing the world. At first sight the realist may see no alternative to this assumption. But an alternative will emerge once we focus on how mathematical models of quantum theory are applied to the world. The key is to see how, in application, a model can guide us toward a literally true story without itself telling it.

I begin with some seemingly pedantic remarks. It is scientists, not science, who create theories even while pursuing scientific aims. And representation is something scientists and other people do with a model, not something the model does by itself. In pursuit of the aims of science, scientists do many things with the theories they create. They apply theoretical models to the world in various ways, including using a model to represent the world or, more typically, a system of interest in the world. Even in a representational application, the model is almost never claimed to offer a faithful representation of the target system, but only to represent it well enough for the purposes at hand. So claims based on application of a theory are almost never given to us by, or in, that theory.

It is distinctive of quantum theory that, when applied to a target system, its models are *not* applied to represent that system. As I see it, the role of the wave function (or other mathematical representative of a quantum state, a phrase I shall not repeat) is not to describe or represent some physical magnitude, entity, or law, but to provide good advice to any user of quantum theory about the significance and credibility of magnitude claims about physical systems. This is how a quantum model guides us toward the literal truth of some such claims when the model is applied. Certainly a target system is assigned a quantum state, but the primary role of that state is not to represent the system or its properties: it is to permit application of the Born rule to determine what probabilities to assign to various eventualities involving this and similar systems.

I call this the state's primary role because that is the main use of quantum states in physical practice. Physicists usually say that the resulting Born probabilities concern possible measurement outcomes: the demise of naive realism shows they cannot be *uniformly* understood as probabilities of possessed properties.

But we can and should avoid any talk of measurement by using quantum models of decoherence when determining which properties may be meaningfully ascribed to a system in an application of the Born rule. Ascription of a property to system *s* is through a magnitude claim of the form **s has Q** $\in \Delta$, where *Q* is a magnitude such as energy, or a component of position, momentum, or spin, and Δ is a Borel set of real numbers. Models of decoherence may be applied to show that Naive Realism's (PV) is false because there is no physical situation in which, for every such property **Q** $\in \Delta$, s may be meaningfully said either to possess, or to lack, that property.

In this way a quantum model can play an important preliminary role by advising a user on how significant is each magnitude claim about a physical system in a particular situation. To assess its significance, the user may apply (unitary) quantum theory to that system in interaction with its environment. In a quantum model of a system and its environment, an initial wave function assigned just to the system would typically evolve extraordinarily rapidly and robustly into a (reduced state) density operator that is extremely close to diagonal in some 'pointer basis' determined by the nature of the interaction Hamiltonian.[4]

Only claims concerning each magnitude represented by an operator near-diagonal in the pointer basis are thereby selected as having enough content to be assigned a Born probability, resulting in significant sets of mutually exclusive and jointly exhaustive magnitude claims suitable for application of the Born rule. Anyone who accepts quantum theory should set credence (only) in each *significant* magnitude claim equal to its Born probability, confident that exactly one claim in the set for each such magnitude is true while the others are false. No magnitude claim that lacks significance in that environmental context is worthy of credence—such meaningless claims cannot even be entertained.

It is on significant magnitude claims that a quantum realist can base a literally true story of what part of the world involving the target system is like, though which story is true will depend on which eventualities actually come to pass. Quantum theory does not tell that story. But we can use quantum theory to reassure ourselves that there is some true story to be told, in terms of true significant magnitude claims and other statements whose truth these claims determine, such as Bell's (2004b, 52) settings of switches and knobs on experimental equipment, the currents in coils, and the readings of instruments. And we can use quantum theory as a source of good advice on what to expect the true story to be.

The reader will have noticed a certain 'fuzziness' in my statement of the pragmatic rules governing the use of the wave function. This is as unproblematic as it is inevitable. Bell began his "Against 'measurement'" (Bell 2004a) with the complaint that surely by now we should have an exact formulation of some serious part of quantum mechanics. This may be achieved simply by dropping von Neumann's notorious 'projection postulate' (collapse of the wave function on measurement) and removing any reference to measurement, observation, apparatus, classical system, etc. from a statement of the Born rule in the way I have indicated.

Pragmatic rules governing the use of the wave function should not appear in the resulting exact formulation: They concern the *application* of the theory so formulated. No matter how exactly or precisely a scientific theory is formulated, its application

[4] For further details, see Healey (2012a).

always requires skill and judgment that cannot be made fully explicit. Any pragmatic rule guiding that application remains subject to interpretation by the skilled practitioner. A good physicist is able to judge when it is permissible to apply the Born rule even without deploying a model of environmental decoherence—fortunately, because environments are typically complex open systems for which there are few tractable quantum models of decoherence, and even in these few, completely robust and irreversible diagonalization of system density operator is never more than a *very* good approximation.

7.5 Is this Instrumentalism?

There are realists who will dismiss the pragmatist view briefly introduced in the previous section (and developed further in Healey, 2012b; 2017b) as an unacceptably instrumentalist view of quantum theory. Before going on to argue that this pragmatist view offers a novel form of quantum *realism*, I'll first respond to this charge of instrumentalism.

According to current usage, instrumentalism in the philosophy of science is the view that a theory is merely a tool for systematizing and predicting our observations: For the instrumentalist, nothing a theory supposedly says about unobservable structures lying behind but responsible for our observations should be considered significant. Moreover, instrumentalists characteristically explain this alleged lack of significance in semantic or epistemic terms: claims about unobservables are meaningless, reducible to statements about observables, eliminable from a theory without loss of content, false, or (at best) epistemically optional even for one who accepts the theory.

But the pragmatist view sketched in the previous section makes no use of any distinction between observable and unobservable structures, so to call it instrumentalist conflicts with current usage. In this pragmatist view, quantum theory does not posit novel, unobservable structures corresponding to quantum states, operators, and Born probabilities: these are not beables represented in quantum models. Nevertheless, claims about them in quantum theory are often perfectly significant, and many are true. This pragmatist view does not seek to undercut the semantic or epistemic status of such claims, but to enrich our understanding of their non-representational function within the theory and to show how they acquire the content they have.

As we saw in Section 7.4, wave function realists take the wave function to represent a novel physical structure—the quantum state—whose existence is evidenced by the theory's success. In this view, a wave function represents a physical structure that either exists independently of the more familiar physical systems to which magnitude claims pertain or else grounds their existence and properties. From this realist perspective, it may seem natural to label as instrumentalist any approach opposed to that account of the quantum state.

But a pragmatist should concede the reality of the quantum state—its existence follows trivially from the truth of quantum claims ascribing quantum states to systems. What he should deny is that a quantum state assignment is true independently of or prior to the true magnitude claims that (in his view) provide the backing for it. Note that any such state assignment is relational, since these are the claims about values of magnitudes accessible from the physical situation of an actual or hypothetical agent that would make this the correct quantum state to assign, relative to that situation. But although they are relational, quantum state assignments are perfectly objective, since this agent-situation is *physical* and not merely epistemic. (Relational) quantum states could exist even in a world without agents, as long as that world contained physical situations that some agent *might* have occupied. Under a reasonably expansive conception of agency that will include most or all spatiotemporal worlds.

The truth of a quantum state assignment trivially implies that a wave function represents *something* we call a quantum state. It does not imply that this quantum state is a beable of quantum theory—a purported element of physical reality that it is the job of the wave function in a quantum model to represent. The previous section described the non-representational roles a wave function plays when a quantum model is applied. In its primary role the wave function offers advice on how strongly to believe magnitude claims. But when a model of quantum theory is applied it is the function of magnitude claims to represent elements of physical reality. These are the claims that underlie statements about the outcomes of quantum measurements, and they play a crucial role in representing what quantum theory is used to predict and explain.

A radical pragmatist might deny that representing reality is ever the primary function of any claim. Such a radical pragmatist could reject the representationalist presupposition of this realist/instrumentalist dilemma—the assumption that mere representation could be a (key) function of an element of theoretical structure that figures centrally in an account of its content. The pragmatist view of quantum theory briefly sketched in Section 7.4 does not require this denial or rejection. But as we'll see in Section 7.9, acceptance of this view of quantum theory does require re-examination of what it is for a magnitude claim to represent physical reality.

Not only does quantum theory inform us about the unobserved: it helps agents improve their beliefs about microscopic phenomena that are often considered unobservable. Many magnitude claims are about properties of systems that are unobservable by unaided human senses. This is to be expected, since quantum theory was initially developed as a theory of the microworld, where classical physics was first seen to break down. For the pragmatist, unlike the constructive empiricist or traditional instrumentalist, the observable/unobservable distinction is of no special semantic, epistemic, or methodological significance. The use of quantum theory to adjust credences in magnitude claims about microscopic phenomena is not only compatible with the present pragmatist view, but plays an important role in helping us explain regularities they exhibit (see Section 7.7).

Some may concede that the present pragmatist view of quantum theory is not instrumentalist in the classical sense because it abjures any observable/unobservable distinction. But they may wish to count it as instrumentalist in a more general sense, because it relies on a parallel distinction between quantum and classical, or between quantum states and the magnitudes about whose values these are taken to offer advice.

Certainly, in this pragmatist view, quantum theory itself distinguishes between these magnitudes and quantum states. But the distinction is functional, not epistemic or semantic: and it does not rely on any problematic notion of classicality.

The progress of physics has revealed magnitudes unknown to classical physics such as strangeness and (the value of) the Higgs field. Quantum states may offer advice on these magnitudes as well as classical magnitudes such as position and momentum. In an application of a model of quantum field theory, a quantum state can also offer advice on the circumstances in which it is legitimate to make claims about entities, such as Higgs particles or the Higgs field as well as photons and the (classical) electromagnetic field. But quantum states are themselves neither magnitudes nor entities, even when they are real (as follows from the truth of the corresponding quantum state assignments).

Here is how quantum theory relates to the physical world, in the present pragmatist view. In accordance with Physical Realism, the physical world exists and has its properties (almost) independently of the existence, thoughts or activities of human or other agents situated within it. Because of their physical situation, agents like us lack a great deal of information about the world, and especially about how it will develop in our future. By creating physical theories we have been able to improve our imperfect epistemic situation. A theory in classical physics was characteristically applied in claims to the effect that mathematical structures in its models represented physical structures in the world (well enough for the purposes at hand). In this way, the theory was itself taken to represent the world, as containing entities with particular values of physical magnitudes. But an application of a model of quantum theory works differently. The goal is still improved beliefs about the values of physical magnitudes on physical entities, but the theory itself does not represent magnitudes as having values. Instead, it offers advice on which entities can meaningfully be assigned values for particular magnitudes, and what credence should be assigned to possession of different possible values for those magnitudes.

The wave function in a quantum model plays its primary role in issuing this advice by serving as input to the Born rule when the model is legitimately applied. The resulting advice is good just when the correct wave function is assigned, in which case the wave function represents the real quantum state of the system to which it is assigned. But correctly representing the quantum state of the system is incidental to the purpose of providing good advice to physically, and so epistemically, limited agents on the values of magnitudes on physical entities. To put it metaphorically, quantum states, though real, would be of no particular interest to an omniscient God with direct epistemic access to the values of all magnitudes on all physical entities throughout world history.

7.6 No Ontological Model?

Even if the arguments of the last section convinced you that the pragmatist view briefly sketched in Section 7.4 is not instrumentalist, you may be unwilling to call it realist. In recent foundational discussions, realism has been cast in the framework of ontological models. Here is an example from one recent paper.

(Single-world) **Realism**: The system has some physical properties, a specification of which is called its *ontic state*, denoted λ. Ontic states take values in a (measurable) set called the *ontic state space*. (Leifer and Pusey, 2017, 6)

In non-Everettian versions of wave function realism, a system's ontic state is specified by its wave function ψ. In the more general ontological models framework λ may or may not include ψ, but it may include 'hidden' variables (as Bohmian mechanics includes position variables). The motivating thought here is that a realist must be able to tell a story about what is going on in the world in situations where quantum theory makes correct predictions, even though no such story emerges from quantum theory itself (as usually understood).

For one who takes the pragmatist view of Section 7.4, there is a meaningful story to be told about the values of various magnitudes in circumstances when the content of claims about them is well enough defined. But these magnitudes are not always well defined, and so there is no ontic state space in which even one of them always has a precise value. So this view permits only a 'gappy' story of what is going on in the world that does not conform to (Single world) **Realism**. For example, when Bell inequalities are violated in photon experiments there is nothing significant one can say about the polarization properties of each entangled photon prior to detection.

But how reasonable is it to impose (Single-world) **Realism** as a condition on anything that could count as quantum realism? A 'gappy' story is still a story about what the physical world is like, independent of the existence or activities of agents or observers. Accepting this view of quantum theory does not commit one to the belief that the world is mind-dependent in any way that would conflict with Physical Realism.

The story provided by Newton's theory was not rejected as anti-realist despite the fact that it included no mechanism filling the gaps between the sun and the earth on which it exerted a gravitational attraction. Admittedly, its successor (Einstein's theory of general relativity) gave a more complete story with its dynamic spacetime permitting continuous propagation of (now confirmed) gravitational waves. A realist may hold out the hope of a similar completion of the 'gappy' quantum story, perhaps involving a kind of retrocausation that would undermine the independence condition required to rule out Naive Realism (Price and Wharton, 2016).

7.7 Is this Pragmatist View Explanatory?

For a realist, the demand for an ontological model is closely associated with the need to provide genuine understanding. Bell, for example, maintained that only if reformulated precisely in terms of a clear ontology of 'beables' could quantum theory supply the kind of explanations we need to understand the big world outside the laboratory. How genuine is the understanding provided by a 'gappy' story that is all quantum theory permits on the pragmatist view?

As I have argued elsewhere (Healey, 2012b; 2015; 2017b), in this pragmatist view quantum theory helps us to explain a host of otherwise puzzling phenomena by showing that they were to be expected and what they depend on. The primary target of explanation is not individual events but what I call *probabilistic phenomena*. A probabilistic phenomenon is a probabilistic data model of a statistical regularity. To explain a probabilistic phenomenon using quantum theory one locates it within a general class of similar phenomena and shows how the probability distributions in each case are a consequence of a similar legitimate application of the Born rule to a quantum state that is correctly assigned to the systems concerned. This unifies all phenomena in this class by providing something close to what Railton (1978) called a deductive-nomological probabilistic (DNP) explanation of each phenomenon.

Here are my reasons for the qualification. Unlike Railton, I consider the primary *explanandum* in each case to be not an individual chance event but the probabilistic phenomenon such events manifest. For Railton, a DNP explanation of an individual event must allude to a (probabilistic) causal mechanism that gives rise to that event. But while use of quantum theory to explain a probabilistic phenomenon must display an appropriate dependence of events manifesting it, the form of that dependence need not be causal; and even when it is, the term 'mechanism' does not seem apt. Here are two examples to illustrate each point.

That individual hydrogen atoms are stable against spontaneous collapse is a (probability 1) example of a probabilistic phenomenon. We can use quantum theory to explain this phenomenon by showing that the expectation value of a hydrogen atom's energy has a finite lower bound, so the probability is zero for it to have the arbitrarily large negative energy associated with collapse. No causal mechanism is appealed to in the explanation. That hydrogen atoms manifest this phenomenon (by not collapsing) does not depend on a causal mechanism that produces it but on their constitution: each is constituted by an electron and a proton interacting through an electromagnetic potential.

A violation of a Bell inequality is a probabilistic phenomenon. We can use quantum theory to explain this phenomenon by deriving the relevant probability distributions from the Born rule, legitimately applied to the appropriate polarization-entangled state of photon pairs whose detection manifests the phenomenon. The events of polarization detection in an individual pair depend causally on whatever events (described by magnitude claims about preparation devices) back the assignment of that entangled state, since interventions on those devices would alter their probabilities. But the distant events of polarization-detection in each instance are causally independent. Moreover, quantum theory has nothing to say about any continuous causal process mediating the causal dependence of an individual detection event on its backing conditions, and Bell's theorem shows that these events can be produced by no causal mechanism of a kind we have previously encountered.

Some realists may not be satisfied by such explanations on the grounds that they yield only partial understanding. But this is not sufficient reason to reject the view of quantum theory that lay behind them as not realist. In this pragmatist view, quantum theory helps us explain these phenomena by appeal to what is going on in the physical world with no reference to observers, agents, or minds. A realist who hankers after more may be compared to Newton, who remained dissatisfied with explanations using his own theory of universal gravitation and continued to search fruitlessly for 'the cause of gravity.' Over 200 years passed before Einstein's general relativity permitted the reconceptualization that since gravity is not a force it needs no 'cause.' Dissatisfaction with the good explanations quantum theory helps us to give is not a reason to reject the pragmatist view of quantum theory, though it may motivate the search for a theory that can help us give more satisfying explanations.

7.8 Truth

In this pragmatist view, quantum theory helps us describe the physical world by means of magnitude claims, on whose significance and credibility it offers us advice. These are what we use to make statements about physical reality when applying a quantum model, and their truth or falsity is what we ultimately care about. Realism is often associated with a correspondence theory of truth, so it is important to address the objection that the way the pragmatist view treats magnitude claims is incompatible with correspondence truth and therefore also quantum realism.

Recall that, in this view, the significance of a magnitude claim about a system depends on the system's environment. Consider the claim *C*: **s has** $\mathbf{Q} \in \Delta$. This attributes property $\mathbf{Q} \in \Delta$ to *s*. As Tarski insisted, *C* is true if and only if *s* has $\mathbf{Q} \in \Delta$, and that will be so just in case '*s*' refers to *s* and *s* satisfies '$\mathbf{Q} \in \Delta$.' The pragmatist readily assents to this Tarskian demand, and if that is all that correspondence truth comes to then correspondence truth is part of this pragmatist view. But some scientific realists demand more of correspondence truth. Field (1972) further demanded a physicalist reduction of reference relations, and even self-avowed pragmatist Philip Kitcher once argued that

> Reference relations are causal relations between mind-independent entities and linguistic tokens. (2002, 347)

However, there seem to be cases in which linguistic tokens succeed in referring to mind-independent entities incapable of bearing causal relations, as when I use the word 'one' or the numeral '1' to refer to the number 1. Of course nominalists deny the existence of numbers and other non-physical entities. But one who adopts this pragmatist view of quantum theory is already committed to the reality of quantum states and Born probabilities to which we succeed in referring even though these don't enter into causal relations.

I agree with Stephen Leeds (2007) that a quantum realist can and should reject a 'thick' notion of correspondence truth that requires reference to be understood as a causal relation in favor of a more minimal account of truth and reference. This is not the place to survey the several deflationary options currently on offer. I merely insist that whatever version of minimalism is adopted should be accompanied by an account of the wider function of attributions of truth and reference capable of explaining the importance of these concepts in our discourse. This seems especially important in an era of 'alternative facts'!

In quantum theory, claims about quantum states and Born probabilities guide belief about the magnitude claims that are basic to the theory's use in predicting, explaining, and controlling phenomena. From this perspective it is natural to think that terms appearing in magnitude claims refer to their subject matter in a way that is somehow more immediate or concrete than terms like 'probability' or 'wave function.' But I think it would be a mistake for a realist to be misled by this thought into adopting a causal account of how reference works in the case of magnitude claims. Indeed, in the next section I will show how a recent argument undermines any such 'thick' notion of correspondence even for claims about the outcomes of quantum measurements whose truth some true magnitude claims determine.

7.9 The Limits of Quantum Objectivity

Realists may disagree about whether wave functions represent something physically real, whether electrons have precise positions and momenta, and whether the world is non-local. But on one point all (except Everettians) agree: quantum measurements each have a unique, physically real outcome and their statistics are correctly predicted by quantum theory. A recent argument by Frauchiger and Renner (2016) seeks to show that this assumption is inconsistent with the universal applicability of quantum theory itself. The original argument is too long to repeat here, so instead I shall sketch a simpler version I first heard in a talk by Matthew Pusey.

Consider the following (completely unrealizable!) thought-experiment. Suppose that Alice and Bob decide to conduct measurements of various polarization components on a large number of pairs of photons, where each pair is correctly assigned the same polarization-entangled state. Being lazy, they do not at first perform any measurements themselves, but delegate that task to their friends, Carol and Dan respectively, each of whom performs the required measurements in his or her otherwise completely physically isolated laboratory. For each pair of photons, Carol measures polarization of one photon with respect to axis c while Dan measures polarization of the other photon with respect to axis d. By assumption, each of their measurements has a unique, physically real outcome (as registered in their notebooks or stored in their computers): and quantum theory correctly predicts the correlations between these outcomes from the joint probability distribution $P(c, d)$ calculated by application of the Born rule to the entangled state assigned to the pairs (where c, d are random variables whose values indicate the outcome of the respective polarization measurement).

After each photon pair is measured by Carol and Dan, instead of asking them what outcomes they observed, Alice and Bob apply very carefully tailored interactions to the entire contents C, D of their respective laboratories (including Carol and Dan themselves). They do this repeatedly, for each photon pair independently. Quantum theory then predicts the effect of these interactions is to restore each photon pair to its original entangled state and to restore each lab+occupant C, D to its state prior to the polarization measurement, thus permitting Carol and Bob to continue their measurements. Finally, Alice measures polarization of one photon in each pair with respect to axis a while Bob measures polarization of the other photon with respect to axis b.

By assumption, each of Alice's and Bob's measurements also has a unique, physically real outcome (as registered in their notebooks or stored in their computers): and quantum theory correctly predicts the correlations between these outcomes from the joint probability distribution $P(a, b)$ calculated by application of the Born rule to the same entangled state assigned to the pairs. Given our working assumption, quantum theory also correctly predicts the correlations between Carol's outcomes and Bob's from the probability distribution $P(b, c)$, and between Alice's outcomes and Dan's from the probability distribution $P(a, d)$, each of which may again be calculated by applying the Born rule to the same entangled state assigned to the pairs.

If the entangled photon state and the axes a, b, c, d are chosen appropriately, the probabilistic correlations predicted in this way by quantum theory will violate a

Bell inequality (the so-called CHSH inequality). But since they constitute a joint distribution over all four measured variables the statistics of these assumed real outcomes will always conform to the inequality. We have a contradiction. So the assumption is false: quantum measurements *do not* always have unique, physically real outcomes whose statistics are correctly predicted by quantum theory. But predictions of quantum theory have always been confirmed by the statistics of measurement outcomes. So we cannot assume that these measurement outcomes are uniquely physically real!

For one who adopts the pragmatist view of Section 7.4, a quantum measurement has a unique, physical outcome. A statement about that outcome has a determinate, mind-independent truth-value: its truth hinges on that of magnitude claims about physical systems involved in the measuring apparatus. This pragmatist view not only escapes refutation in the imagined scenario, but also receives support from its deeper analysis. The key point is that, in this view, acceptance of quantum theory modifies the *content* of a statement about the outcome of a quantum measurement by restricting what inferences may legitimately be made from its truth. The restriction effectively relativizes that content to the environmental context of the system to which quantum theory may be applied. That content then becomes a function of the physical environment within which the system is located. The way content depends on physical environment may be modeled by a quantum model of decoherence.

In all practicably realizable circumstances the environmental context is appropriately modeled by massive decoherence of the relevant quantum state, so that all physically situated agents (not only human agents like Alice, Bob, and friends) are able to neglect the fact that content may depend on environmental context and successfully attribute a context-independent content to a statement about any measurement outcome. But in the (completely practicably unrealizable) circumstances described in the thought-experiment this is not so, since Alice, Bob, and friends do not share a single environmental context. In these circumstances it is investigators' labs that provide the relevant environmental context. Here context-relativity can be indicated by an appropriate subscript. For example, the content of a magnitude claim M reporting the outcome of one of Carol's measurements on a system in her lab C may be represented as $[M]_C$. In their situation they (and we) may continue to agree that there are true statements about their unique physical measurement outcomes with objective, mind-independent content. But that content is not context-independent since it does not license reliable inferences between different environmental contexts.

Decoherence confined to each of their individual laboratories models the environmental context underlying the content of each claim about the outcome of a measurement in that lab. For Carol and Dan physically to have exchanged information they would have had to physically combine their environmental contexts to form a unified context $C \cup D$ into which their statements about their outcomes could have been reliably exported. Alternatively, either Carol or Dan might have physically exchanged information with Alice or Bob without first exchanging information with each other, permitting each reliably to export statements about his or her outcomes into that incompatible larger context. But in fact no such physical interactions occurred in the imagined scenario, in which the environments C, D remained sealed off from each other and also from the environments A, B of Alice and Bob. The upshot is that while the content of each statement about the unique outcome of every measurement on each of an entangled pair of photons may be treated here as perfectly well defined within an environmental context, there is no such context in which a statement about all these outcomes has well-defined content.

When Alice and Bob combine their results in their physically unified joint environment $A \cup B$, statements of the outcomes a, b modeled by the joint distribution $P(a, b)$ are significant in $A \cup B$, and the Born rule may legitimately be applied to (correctly) predict this joint distribution. But the absence of the required unified contexts renders illegitimate any application of the Born rule to predict the joint probabilities $P(c, d)$, $P(b, c)$, $P(a, d)$ that also appear in a statement of the CHSH inequality. Indeed, by applying quantum theory in this scenario we can see that the CHSH inequality cannot be derived here since there is no environmental context in which its constituent probabilities are all well defined. So statements of the unique, contextually well-defined, physical outcomes of all quantum measurements in the imagined scenario violate no legitimately derivable Bell inequality.

There is a common philosophical view of how content depends on context that may appear to be in tension with the idea that content can depend on context in this way. In this view, a statement has significant content if and only if it expresses a determinate proposition. While what content a statement expresses may depend on the context to which it relates (loosely, to the context in which it is made), context merely determines what proposition a claim expresses. Any variation of content with context can be represented by a function from context into proposition expressed.

In this view, an adequate analysis of a statement's content must then supply an account of the content of each proposition in the range of that function in a referential semantics that provides its truth conditions: if the function is only partial, the claim has no content in a context in which it expresses no determinate proposition. So an adequate analysis of the content of a statement S reporting the outcome of a measurement will either assign it some specific content (varying from context to context) or no content at all (in other contexts). It follows that no analysis is adequate according to which what varies with context is not simply the specific content of the statement but also how much content it has.

This philosophical view provides an idealized model of content that is helpful in elucidating the meaning of indexicals like 'I' and 'now,' for which the context in which a statement is made seems readily specifiable (by saying who made it and when). It becomes problematic in circumstances where contextual elements are harder to pin down. As an example, consider the statement K, when uttered on a road trip across Kansas

K: Kansas is flatter than a pancake.

What determinate proposition does this express? K may be taken literally or as a metaphor for the literal claim

V: Kansas is very flat.

Taking it literally, intrepid researchers compared a geographic profile of the state based on a digital elevation model provided by the United States Geographic Survey to the profile of a particular pancake from the International House of Pancakes using a confocal laser microscope (Fonstad, Pugatch, and Vogt, 2003). They chose as a measure of flatness the deviation from sphericity of an ellipsoid, and estimated this in each case from a best fit to two chosen orthogonal transects of the surface. On this measure, Kansas proved to be much flatter than the pancake. The authors concluded that:

> The calculated flatness of the pancake transect from the digital image is approximately 0.957, which is pretty flat, but far from perfectly flat. Kansas's flatness is approximately 0.9997. That degree of flatness might be described, mathematically, as "damn flat."

The latter claim presumably establishes the truth of *V*. But on an alternative, qualitative local measure of flatness the authors commented that

> When viewed at a scale of 50 mm, a pancake appears more rugged than the Grand Canyon.

One might quibble that Kansas cannot be flat since its elevation varies from 4039 feet to 679 feet above sea level, or that (as noted in one YouTube video taken from a speeding car) you can see the curvature of the earth. You see the problem: When uttered on a road trip across Kansas, the context fails to pick out any well defined proposition expressed by statement *K*, even though that statement clearly conveys something that is both significant and true.

It may be tempting to dismiss such difficulties in specifying the context of utterance for claims like *K* and *V* as arising from their vague or metaphorical language. But on closer examination the same kind of problems afflict even the paradigm case of the indexical 'now.' For what exactly is it to specify the time at which a statement is uttered? Any actual utterance is not an event but a process extending over an interval of time. But even precisely to specify an instant within this process either (falsely) presupposes a universal absolute time or (assuming relativity theory) requires a further specification of a state of motion associated with a local inertial frame and/or a spacelike hypersurface including some point-event in the utterer's vicinity.

One committed to a propositional model of content might acknowledge the resulting indeterminacy in exactly what proposition is expressed by utterance of a sentence including the indexical 'now,' while maintaining that essentially the same content is conveyed by the utterance for every admissible way of resolving this indeterminacy. But that would commit her both to a non-propositional notion of content and to the task of explaining why the contingent circumstances of human communication render this propositional indeterminacy harmless. Such contingencies have been explored by both philosophers and physicists.

Butterfield (1984) notes the importance of several physical features of the environment in which we generally communicate, including the proximity of the parties and the slow rate at which the timely topics of their communication change, compared to the speed of their communication. By considering relativistic physics, Hartle (2005) and Penrose (1989, 392–3) note the further importance of the slow relative speed of communicators. Any attempt *precisely* to specify what proposition is expressed by utterance of a sentence including a temporal indexical like 'now' would require careful application of physical theories to the environment of utterance. But an account of its content does not require this. Even a qualitative analysis generally suffices to explain how such a tensed utterance is able to convey content that is sufficiently well defined for practical human purposes.

Returning to the imagined scenario of the thought-experiment, what proposition is picked out by Alice's statement that the outcome of the 100th of her measurements was a detection of a photon horizontally polarized with respect to axis *a*? In the pragmatist view of Section 7.4, the truth value of this statement in context is determined by that of some magnitude claim of the form *M*: **s has Q** $\in \Delta$. One can give an account of the truth conditions of a claim of the form *M*: **s has Q** $\in \Delta$ but this is trivial. For example: *M* is true if and only if the system to which '*s*' refers has a value for the magnitude to which '*Q*' refers that lies in the set of real numbers to which 'Δ' refers. Once the (tensed) claim is somehow(!) indexed to a time, these truth conditions are independent of context, since the claim contains no explicit indexical elements.

The problem with this referential approach is not that it is wrong but that, once one has despaired of a physicalist or causal account of reference, it is too shallow to be helpful. The approach fails to illuminate the different ways a claim of the form *M* functions when s is in different environments. The claim functions within a web of inferences, and the extent of its content depends on the context provided by the presence of other claims in the web—here, an assumption about *s*'s quantum state is required in assessing the content of a claim of the form *M* about *s*, since assignment of a (reduced) quantum state reflects the extent and nature of environmental decoherence in a quantum model of Alice's environment *A*.

The referential approach to content sometimes provides a useful analytic model of content, but it lacks the resources to account for how content accrues to a statement. To understand quantum theory one needs to adopt a better account of what gives a statement content. By modifying inferential relations involving magnitude claims quantum theory affects their content, rendering this contextual. Philosophers customarily regard a claim as meaningful if and only if it expresses a definite proposition when made in an adequately specified context: otherwise it is taken to be meaningless. An improved pragmatist inferentialist approach to the content of an empirical claim accepts a role for context but replaces this "digital" view of content with an "analog" view (Brandom 1994; 2000). Content accrues by degrees as links are added to the inferential web within which statements are located. By making the content of a magnitude claim about a system a function of the environment, acceptance of quantum theory cautions one against taking that claim to attribute an intrinsic property to an object independently of environmental context, even while insisting on the objective content of the claim.

It is only because the situation of agents like us in the physical world is such that we will always inevitably share a single 'decoherence environment' that we are able to ignore the implicit dependence on our physical situation of the contents, not only of statements about the outcomes of quantum measurements, but of practically all claims about macroscopic, and most claims about microscopic, systems.

7.10 Scientific Realism as an Empirical Thesis

Hilary Putnam (1978; 2012) viewed scientific realism as an empirical hypothesis that we should accept as the best, if not the only, explanation of the success of science. He endorsed a formulation of Richard Boyd as influencing his conception of scientific realism—that terms in a mature science typically refer, and theories accepted in a mature science are typically approximately true. His famous 'no miracle' argument took the success of science and the preservation of terms like 'electron' through scientific theory change as evidence for scientific realism, so conceived. The realist explanation of these features of science (success and reference preservation) is that scientists mirror the world—in the sense of constructing symbolic representations of their environment—and that science succeeds in the way it does because these symbolic representations become increasingly accurate as science progresses.

Faced with potential counterexamples from the history of science, this empirical hypothesis has been clarified or modified in two ways. Some have sought to distinguish terms for a theory's working posits, whose successful reference is supported by its success, from other non-referring terms. Supporters of structural scientific realism have taken refuge from apparent counterexamples in an appeal to theoretical preservation of representations of structure, rather than (the nature of) objects, with the progress of science. A recent paper by Steven French adopts this perspective toward quantum theory.

> Which features of a scientific theory should a scientific realist take to represent the world? Answer: those features that are responsible for the theory's explanatory successes. When the theory is quantum mechanics, the wave function is surely one of those features. (2013, 76)

But, in the present pragmatist view, what is distinctive about the success of quantum theory is precisely that it is *not* due to introduction of new symbols (for beables) to represent novel structures in the physical world. Quantum theory introduces terms like 'wave function' and 'observable' with a different function. They are not intended to mirror the physical world but to guide scientists and other situated agents in better deploying representational resources they already have or are engaged in developing. Even if an important long-term scientific aim is improved symbolic representation of the physical world, science may at times progress faster by introducing key terms *without* that representational function. Language and other symbolic systems provide scientists and the rest of us with wonderful tools for achieving our goals. But these tools don't always function in the same way. Perhaps the central pragmatist moral of quantum theory is that scientists may find new ways of furthering long-term scientific aspirations by creating theories whose key terms do not function as representations of physical reality.

7.11 How to be a Quantum Realist

I have advocated a certain conception of scientific realism and argued that by taking a pragmatist view one can see that quantum theory fits this conception. A scientific realist should at least accept the existence of a physical world that is largely independent of how we think of or observe it. So scientific realism incorporates what I called Physical Realism. Quantum theory is compatible with Physical Realism, in the pragmatist view of Section 7.4. While agreeing that there is a physical world whose existence is independent of how we think of it, a Fuchsian participatory realist takes its development to be sensitive to our thoughts through our intentional actions that precipitate an unpredictable observed response. While I see this as a novel form of Physical Realism, others may count it a rejection of this minimal scientific realist requirement.

As an epistemological optimist, a scientific realist should further maintain that through our best science we have learned a lot about what the physical world is like, including many of its features that we cannot observe through our unaided senses. Applications of quantum theory have certainly led to successful predictions and explanations of physical phenomena, many of which are in this sense unobservable. The pragmatist view of Section 7.4 helps us understand how we apply quantum theory in deepening our knowledge of these phenomena without taking wave functions or other elements of its mathematical models to represent beables. But for a QBist, quantum theory has taught us little or nothing about what the physical world is like: it has merely provided each agent with a tool that is valuable in anticipating that agent's future experiences when acting in the physical world.

The most direct way a scientific theory could teach us about the physical world is by itself describing it, or offering mathematical models by which to represent it. For a Quantum Scientific Realist, that is the only way we could learn from quantum theory about the physical world. Adopting this restricted conception of quantum realism has led to a proliferation of rival Interpretations of quantum theory, each eager to extract a literally true story of what the physical world is like from quantum theory itself. But pragmatism encourages a wider perspective on how we might gain knowledge of the physical world. A pragmatist who takes the view of Section 7.4 rejects Quantum Scientific Realism. She sees quantum theory as a radically different kind of theory that teaches us about the world not by offering models by which to represent it, but by advising us on *how* it may meaningfully be represented, and how likely is each meaningful representation to be true. This is how quantum theory has taught us a lot, not only about what the world is like but also about the scope and limitations of our representations of it.

A scientific realist should reject a 'thick' correspondence view of truth in favor of a deflationary account, and regard referential semantics as offering an analytic model of content rather than a substantive account of its origins. These pragmatist attitudes to truth and reference underlie Section 7.4's pragmatist view of quantum theory. By adopting an inferentialist account of how content accrues to a statement, this pragmatist view shows how to eliminate all Bell's (2004a, 215) proscribed words from a formulation of quantum theory with no measurement problem while yielding a statement of the Born rule compatible with the "no-go" theorems of Bell (2004b), and Kochen and Specker (1967). It also shows how to reconcile the existence of a unique, objective, physical outcome of each successful quantum measurement with the argument of Frauchiger and Renner (2016).

William James presented his pragmatism as a conciliatory view in philosophy. I think of Section 7.4's pragmatist view of quantum theory as offering to reconcile the views of Bohr and Einstein. If Bohr was right that acceptance of quantum theory requires acknowledgment of the limits this puts on our abilities to speak meaningfully about the physical world, perhaps Einstein was right to hold out the hope that these limits may be transcended as quantum theory is succeeded by an even more successful theory that gives us an approximately true, literal story of what the physical world is like.

Acknowledgments

Thanks to David Glick, Doreen Fraser, an anonymous referee, and the editors. I hope your comments and queries have helped me express the views of this chapter more clearly even where they differ from your own!

References

Albert, D. Z. (1996). Elementary quantum metaphysics. In J. Cushing, A. Fine, and S. Goldstein (Eds), *Bohmian Mechanics and Quantum Theory: An Appraisal*, pp. 277–84. Dordrecht: Kluwer.

Albert, D. Z. (2013). Wave function realism. In A. Ney and D. Z. Albert (Eds), *The Wave Function: Essays on the Metaphysics of Quantum Mechanics*, pp. 52–7. New York: Oxford University Press.

Bell, J. S. (2004a). Against 'measurement'. In *Speakable and Unspeakable in Quantum Mechanics*, pp. 213–31. Cambridge: Cambridge University Press.

Bell, J. S. (2004b). *Speakable and Unspeakable in Quantum Mechanics*. Cambridge: Cambridge University Press.

Brandom, R. (1994). *Making it Explicit*. Cambridge, MA: Harvard University Press.

Brandom, R. (2000). *Articulating Reasons: An Introduction to Inferentialism*. Cambridge, MA: Harvard University Press.

Butterfield, J. (1984). Seeing the present. *Mind* 93(370), 161–76.

Einstein, A. (1949). Reply to criticisms. In P. Schilpp (Ed.), *Albert Einstein, Philosopher-Scientist*, pp. 665–88. La Salle, IL: Open Court.

Einstein, A., B. Podolsky, and N. Rosen (1935). Can quantum-mechanical description of physical reality be considered complete? *Physical Review* 47(10), 777–80.

Field, H. (1972). Tarski's theory of truth. *The Journal of Philosophy* 69(13), 347–75.

Fonstad, M., W. Pugatch, and B. Vogt (2003). Kansas is flatter than a pancake. *Annals of Improbable Research* 9(3), 16–18.

Frauchiger, D. and R. Renner (2016). Single-world interpretations of quantum theory cannot be self-consistent. *arXiv 1604.07422 [quant-ph]*.

French, S. (2013). Whither wave function realism? In A. Ney and D. Z. Albert (Eds), *The Wave Function: Essays on the Metaphysics of Quantum Mechanics*, pp. 76–90. New York: Oxford University Press.

Fuchs, C. A. (2016). On participatory realism. *arXiv: 1601.04360v3 [quant-ph]*.

Ghirardi, G.-C. (2006). Collapse theories. *The Stanford Encyclopedia of Philosophy* (Fall 2018 Edition), ed. Edward N. Zalta: https://plato.stanford.edu/entries/qm-collapse/.

Giustina, M. et al. (2015). Significant-loophole-free test of Bell's theorem with entangled photons. *Physical Review Letters* 115(25), 250401.

Goldstein, S. and N. Zanghì (2013). Reality and the role of the wave function in quantum theory. In A. Ney and D. Z. Albert (Eds), *The Wave Function: Essays on the Metaphysics of Quantum Mechanics*, pp. 91–109. New York: Oxford University Press.

Hartle, J. B. (2005). The physics of now. *American Journal of Physics* 73(2), 101–9.

Healey, R. (1979). Quantum realism: naïveté is no excuse. *Synthese* 42(1), 121–44.

Healey, R. (2012a). Quantum decoherence in a pragmatist view: dispelling Feynman's mystery. *Foundations of Physics* 42(12), 1534–55.

Healey, R. (2012b). Quantum theory: a pragmatist approach. *The British Journal for the Philosophy of Science* 63(4), 729–71.

Healey, R. (2015). How quantum theory helps us explain. *The British Journal for the Philosophy of Science* 66(1), 1–43.

Healey, R. (2017a). Quantum-bayesian and pragmatist views of quantum theory. *Stanford Encyclopedia of Philosophy* (Spring 2017 Edition), ed. Edward N. Zalta: https://plato.stanford.edu/entries/quantum-bayesian/.

Healey, R. (2017b). *The Quantum Revolution in Philosophy*. Oxford: Oxford University Press.

Hensen, B. et al. (2015). Loophole-free Bell inequality violation using electron spins separated by 1.3 kilometres. *Nature* 526(7575), 682–6.

Kitcher, P. (2002). On the explanatory role of correspondence truth. *Philosophy and Phenomenological Research* 64(2), 346–64.

Kochen, S. and E. P. Specker (1967). The problem of hidden variables in quantum mechanics. *Journal of Mathematics and Mechanics* 17(1), 59–87.

Leeds, S. (2007). Correspondence truth and scientific realism. *Synthese* 159(1), 1–21.

Leifer, M. S. and M. F. Pusey (2017). Is a time symmetric interpretation of quantum theory possible without retrocausality? *Proc. R. Soc. A* 473(2202), 20160607.

Ney, A. and D. Z. Albert (2013). *The Wave Function: Essays on the Metaphysics of Quantum Mechanics*. New York: Oxford University Press.

Penrose, R. (1989). *The Emperor's New Mind*. Oxford: Oxford University Press.

Popper, K. R. (1967). Quantum mechanics without 'the observer'. In *Quantum Theory and Reality*, pp. 7–44. Berlin: Springer.

Price, H. and K. Wharton (2016). Taming the quantum spooks. *Aeon*: https://aeon.co/essays/can-retrocausality-solve-the-puzzle-of-action-at-a-distance.

Putnam, H. (1978). *Meaning and the Moral Sciences*. London: Routledge & Kegan Paul.

Putnam, H. (1981). *Reason, Truth and History*. Cambridge: Cambridge University Press.

Putnam, H. (2012). *Philosophy in an Age of Science*. Cambridge, MA: Harvard University Press.

Railton, P. (1978). A deductive-nomological model of probabilistic explanation. *Philosophy of Science* 45(2), 206–26.

Ruetsche, L. (2011). *Interpreting Quantum Theories*. Oxford: Oxford University Press.

Shalm, L. K. et al. (2015). Strong loophole-free test of local realism. *Physical Review Letters* 115(25), 250402.

Tumulka, R. (2006). Collapse and relativity. *Quantum mechanics: Are there quantum jumps? and On the present status of quantum mechanics. AIP Conference Proceedings* 844(1), 340–51.

Van Fraassen, B. C. (1980). *The Scientific Image*. Oxford: Clarendon Press.

Wallace, D. and C. G. Timpson (2010). Quantum mechanics on spacetime I: spacetime state realism. *The British Journal for the Philosophy of Science* 61(4), 697–727.

8

Can Pragmatism about Quantum Theory Handle Objectivity about Explanations?

Lina Jansson

8.1 Introduction

The idea that one of the aims of science (and one that the sciences sometimes succeed in) is to offer objectively correct explanations of phenomena by accurately (enough) representing the system of interest has close ties to scientific realism. The connection is twofold. First, realism is often formulated in opposition to instrumentalism where, tellingly, the primary epistemic goal of science is not explanation but prediction. Second, a common form of scientific realism has a commitment to inference to the best explanation and arguments from explanatory indispensability at its core.[1]

Healey's (2017) pragmatist approach to quantum theory takes the role of quantum theory to be one of guiding an agent's beliefs, instead of representing quantum states and their properties:

> The most significant break marked by acceptance of quantum theory is a novel, indirect use of models to further the aims of fundamental science. This, above all else, is what makes quantum theory radical... Quantum theory is revolutionary not because it represents new and unfamiliar physical things and processes in the universe, but because of the way it improves our use and understanding of representations of the universe we could offer without it. (121)

This would naturally seem to be classified as an instrumentalist attitude towards quantum theory. However, Healey also takes quantum theory to be able to objectively explain phenomena, fitting a more typically realist attitude. When explanation is addressed within anti-realism, it is typically taken to be a pragmatic virtue (e.g. van Fraassen, 1977; 1980). In the debate over realist versus anti-realist approaches to quantum theory, we similarly find arguments that the crucial question hinges on the availability of objective explanations (e.g. Timpson, 2008). If a position that holds both

[1] Chakravartty (2017) calls the strand of realism that emphasizes explanatory indispensability *explanationism*. I use the term here to cover realist arguments that proceed from explanatory commitment to ontological commitment generally. Saatsi (2018) discusses the centrality of explanatory reasoning to realism (and its limitations).

Lina Jansson, *Can Pragmatism about Quantum Theory Handle Objectivity about Explanations?* In: *Scientific Realism and the Quantum*. Edited by: Steven French and Juha Saatsi, Oxford University Press (2020). © Lina Jansson. DOI: 10.1093/oso/9780198814979.003.0008

that the role of quantum theory is to guide agents' beliefs and that quantum theory can offer objective explanations can be maintained, it occupies a middle ground between explanation seeking realism and prediction focused instrumentalism. Moreover, it is an attitude that could be very attractive since it seems to sidestep many of the traditional interpretative challenges for realist approaches to quantum theory without subscribing to an ideal of scientific success that sidelines scientific explanation.

In this chapter I will argue that there is such a middle ground to be had. However, it comes with some significant costs. In order to outline the difficulty, let me turn to the aspect of the (typical) realist attitude towards explanation that is the source of the problem. Namely, that an aim of science is to offer an accurate account of the world, and, in particular, that successful scientific explanations offer accurate representations of systems of interest.[2] Here we seem to have an immediate conflict with Healey's pragmatist approach. Quantum theory is not, on this view, in the business of representing the phenomena of interest (accurately or otherwise). It is in the business of providing advice about what to believe. In Section 8.2, I will spell out the realist requirement that explanations offer accurate representations and briefly consider its plausibility. In Section 8.3, I will raise a challenge to the pragmatist approach to quantum theory that stems from giving up the requirement of accurate representation in providing explanations. I will suggest a way to tackle this problem that allows for (a limited use of) arguments from explanatory indispensability and inferences to the best explanation in Section 8.4. Finally, in Section 8.5, I will illustrate how this suggestion, while offering a middle way between instrumentalism and realism, carries real explanatory costs.

8.2 IBE, Explanatory Indispensability, and a Putative Truism about Explanation

Many realists argue for our commitment to the existence of at least certain entities, facts, or structures based on inferences that assume a putative truism about explanation: falsehoods do not explain. For example, we find this truism behind inferences to the best explanation, in the reliance on objective explanatory dependencies, and in arguments from explanatory indispensability.[3] The idea is that statements of genuine explanans have to be true and that truth is to be understood in a non-deflationary way.[4]

For example, here is Leng's (2005, 173) particularly clear summary of explanatory indispensability arguments (that she does not endorse) for the existence of mathematical objects. Here the truism appears as part of P1 (coupled with a non-deflationary account of truth).

[2] I am purposefully leaving it open *how* accurate a representation has to be in order to count as accurate (enough). I explain why I think that this issue can be left unsettled for the purpose of this chapter at the end of Section 8.4.3.

[3] See, for example, Baker (2005), Berenstain and Ladyman (2012), Colyvan (2001; 2012), Kitcher (1993), and Psillos (1999).

[4] This is common in many accounts of explanation. See, for example, Hempel (1965) and Strevens (2008).

(P1) Genuine explanations must have *true* explanans. In particular, then, the objects posited by those explanans must exist.
(P2) There are some genuine mathematical explanations of some physical phenomena.
Therefore
(C) The mathematical objects posited by these explanations must exist.

The same style of argument could be given from scientific explanations of physical phenomena to the existence of the unobservable entities or structures posited in those explanations.[5]

One of the reasons why (P1) is a putative truism is that it accounts for why a whole range of putative explanations *fail* to be genuine explanations. Consider for example why the examples below fail.

1. Why did the candle in the jar stop burning? Answer: the air was unable to absorb more phlogiston.
2. Why does Mars display retrograde motion? Answer: the epicycles in the orbit of Mars means that it periodically 'reverses' direction.
3. Why did children in Salem display strange behaviour? Answer: they were afflicted by the casting of malicious spells by witches.
4. Why did I go to the fruit stall? Answer: I wanted to buy durian fruits.

In all of these cases we can—if we accept the putative truism—give the same simple answer as to why these explanations fail to be genuine explanations. The explanantia in question are false. Moreover, the truism seems to give the correct prediction of when we would be inclined to retract our endorsement of an explanation. I cannot both claim that the explanation of the behaviour of the children in Salem is that they have been afflicted by spells cast by witches *and* that there are no witches and no such spells. If I previously endorsed the explanation of the behaviour of the children in Salem above, once I come to believe that there are no witches and no malicious spells I must retract my endorsement.

This might seem to make the case overwhelming in favour of the putative truism. However, there are also strong considerations that push in the other direction. This is why statements of the putative truism are nearly always followed by a weakening clause (even if little attention is given to its importance). For example, the description of the laws that an explandum phenomenon depends on have to be true *or approximately so*, and the causal mechanisms that we describe have to be true causal mechanisms *or approximately so*, etc.[6] At least some theories are in common explanatory use even after they have come to be regarded as (strictly speaking) false.[7] For example, take the putative explanation of why the escape velocity on the surface of the earth is roughly 25,000 mph. A standard explanation of this makes use of conservation of energy and Newton's law of gravity. Yet, we do not take Newton's law of gravity to be strictly true.

[5] I think that the realist's case is best put in terms of an argument from a *commitment* to the genuineness of certain scientific explanations to a *commitment* to the existence of the unobservable entities or structures posited in those explanations. However, I think that Leng's summary better captures the state of the debate.

[6] See, for example, Craver (2006), Hempel (2001), and Woodward (2003, section 3.3).

[7] And, depending on our attitude towards our current best scientific theories, we might take them all to be, strictly speaking, false.

For a realist who argues from explanatory commitments to ontological commitments these types of examples raise a challenging question: how can we defend such arguments in light of the seemingly indispensable use of idealizations, distortions, and fictions within scientific explanations?[8]

For those who deny the putative truism about explanation a different challenge looms large: what distinguishes acceptable uses of idealizations, distortions, and fictions in explanations from unacceptable ones (e.g. in 1–4 above)? Or, when is accurate representation of the system of interest not required for explanation?

A realist who accepts some version of the explanationist strategy has a particularly difficult middle road to steer here. For example, inferences to the best explanation lose their force if accurately representing the target system is not among the features that makes an explanation the 'best' of the alternatives. If, in comparing competing explanations, we can judge one that we take to less accurately represent the explanatory target as better than a more accurate competitor, then the plausibility of inference to the best explanation has been undermined. If the truism has to be modified, it has to be modified in some way such that what distinguishes explanatory from non-explanatory uses of idealizations, distortions, and fictions is (i) recognizable in the internal workings of explanations, and (ii) tracks ontological commitment.

For example, arguments from explanatory indispensability would lose their force if explanatory indispensability (suitably understood) did not track ontological commitment, or if the only way to spell out what 'suitably understood' explanatory indispensability is forces us to rely on resources not recognizable internal to the explanation. If we have to appeal to criteria external to the explanation then it is not *explanatory* indispensability that is doing the work in determining ontological commitments. Similarly, in inferences to the best explanation, the comparison between competing explanations has to be such that we can make it without appeal simply to ontological commitments independent of the explanations at hand (since this would undermine the need for the inference to the best explanation). Moreover, the relevant notion of explanatory goodness has to track ontological commitment. If we can distinguish explanatory from non-explanatory uses of idealizations, distortions, and fictions in a way that is internal to the explanations in which they occur and such that ontological commitments are tracked, then we have opened up a strategy for reinstating explanationist arguments for realism.

However, there is one family of accounts of explanation where this precisifying project looks particularly unpromising. On pragmatic approaches to explanation, it is natural to take the difference between explanatory and non-explanatory uses of idealizations, distortions, and fictions to vary by context in a way that resists generalizations based on criteria internal to explanation. This is, for example, Bokulich's (2012; 2016) view. These types of accounts do not naturally support explanationist arguments for realism. On these views, what distinguishes explanatory from non-explanatory uses of idealizations, distortions and fictions is neither recognizable internal to explanation nor such that it tracks ontological commitment. The project of precisifying the

[8] Maddy (1992; 1995) has long noted that idealizations, distortions, and fictions pose a challenge for indispensability arguments. Here that challenge is extended to explanatory versions of indispensability arguments.

relevant explanatory goodness or the relevant explanatory indispensability cannot get off the ground.

If Healey's pragmatist approach to quantum theory is to strike a middle road between instrumentalism and realism (of the explanationist type), then a new solution to the two challenges above is needed. Moreover, it cannot come from drawing on pragmatic approaches to explanation. Of course, it is possible to reject the arguments from explanatory commitment to ontological commitment and instead adopt another reason for accepting (a weakened version of) realism. However, Healey does not adopt a pragmatist account of explanation and this leaves open the possibility that standard explanationist arguments for realism could apply (but in a more restricted form).

In the rest of this chapter, I will argue that there is a way forward available to the pragmatist about quantum theory (albeit at a cost) that keeps the standard realist inferences from explanatory commitment to ontological commitment intact (but restricted). The key to a solution lies in distinguishing different explanatory roles.[9]

8.3 Pragmatism about Quantum Theory without (Pure) Pragmatism about Explanation

A distinctive aspect of pragmatism about quantum theory is that quantum models are not taken to have the task of (accurately or otherwise) representing quantum features of physical systems. This is not to deny that applications of quantum models to target systems take the systems modelled to be real. Rather, the idea is that the role of a quantum model is to prescribe how an agent should update their beliefs regarding non-quantum features.

> I'll call the system(s) to which quantum theory is applied in any instance the *target* of that application. The target is physical, and any actual (rather than merely hypothetical) application takes it to be real—an element of physical reality. Recall how Bell introduced beables as what a theory takes to be physically real... But it does not follow that the quantum model applied *represents* the target of its application or that the target is a beable of *quantum theory* ...Things whose physical existence is presupposed by an application of quantum theory deserve a name of their own: I'll call them *assumables*.
>
> The target of any actual application of quantum theory is clearly assumable in this sense, whether or not the model applied is taken to represent it. Whatever entities and magnitudes back assignment of a quantum state in this model are also assumables, such as experimental equipment including Bell's settings of switches, knobs, and currents. ...The readings of instruments recording such outcomes must also be counted among quantum theory's assumables. Clearly the application of a model of quantum theory assumes a lot about how the physical world can be represented. But examination will show that no element of a quantum model has the function of representing any beable that is novel to quantum theory. (Healey, 2017, 127–8)

[9] The need to distinguish different explanatory roles is also the general suggestion that Baron (2016) and Saatsi (2016) make in recent work. However, the focus of Saatsi's work is to survey how different accounts of explanation interact with arguments from explanatory indispensability. I will provide a solution to one of the problems that Saatsi notices for broadly Woodwardian accounts. My account will differ from Baron's in providing a criterion internal to explanation.

In a first departure from the traditional realist/anti-realist debate, the assumables do not have to concern observable aspects or quantities nor do they have to be capable of being described classically.

When it comes to explanation, a quantum model explains by telling us what to *expect* and what the explanandum *depends* on. The Born rule plays a crucial role in this by relating the mod square of the amplitude of the wavefunction to probabilities of outcomes. This gives us a way to move from features of the quantum model to information about what we should expect (see, e.g. Healey 2017, 138). The Born probabilities are also given a pragmatist interpretation. They do not, for example, represent physical propensities and are not physical magnitudes on this view. Rather, they are guides for how an agent should adjust her credences.

To illustrate this feature of the pragmatist approach, let us consider the situation in an EPR-style set-up. For example, let us say that Alice and Bob are in two spacelike separated regions 1 and 2 (respectively) measuring the linear polarization of photons from a source where each photon pair is assigned an entangled state $|\eta\rangle$. Alice and Bob can both manipulate the settings of the polarization measurement (a and b, respectively) and can be taken to set the measurement parameter just before some period of time T during which the measurement events take place. The measurement is recorded by some outcome event A at 1 and some outcome event B at 2.

In this set-up the probability of some outcome A is not identical to the probability of A conditional on B. However, on the pragmatist approach this simply indicates that *if* Alice had available the information about the setting b and the outcome B (in addition to the information about $|\eta\rangle$ and a), she should now, say, give outcome A the probability 1 rather than 1/2. However, in the set-up described Alice does not have access to this information and, importantly, there is no probability here that is *the* objective chance of A of which we can ask if it is, say, 1 or 1/2.

So far, this looks very much like a through-and-through non-realist account where quantum theory is applied in order to tell us what to expect but not taken to directly be about the way the world is (probabilistically or otherwise). Yet, the pragmatist approach is not supposed to be an instrumentalist position. In a second departure from the traditional divisions in the realism/anti-realism debate, on the pragmatic approach quantum theory is taken to be able to explain, to do so well, and, although dependent on physical situation, to allow for objective explanations in the sense that whether the explanans accounts for the explanandum does not vary with context (as in pragmatic accounts of explanation).

> Anyone can apply quantum theory from the perspective provided by the local situation of Alice or of Bob to show that the correlations of outcomes recorded at 1, 2, as well as the statistics separately recorded at each of 1, 2, were just as expected for an agent in that situation. These applications also make it clear what both these non-localized correlations and the localized outcomes that constitute them physically depend on. *By showing that they were to be expected and what they depend on, quantum theory helps us to explain the non-localized correlations it successfully predicts.* The correlations depend on whatever physical conditions make $|\eta\rangle$ the correct quantum polarization state to assign to the systems involved; the settings a, b; and the physical conditions at 1, 2 in T necessary for events localized in each region to count as recordings of the outcome of the relevant linear polarization measurement there.
>
> (Healey, 2017, 74, my emphasis)

The first thing to note is that on the pragmatist approach to quantum theory the explanatory dependence spelled out above does not directly involve the quantum state (since this too is given a pragmatist understanding). Rather, the dependence holds between the *assumables* of the quantum model.

This approach denies the putative truism of Section 8.2. At first glance, it also offers an easy solution to the challenge posed to deniers of this truism. The answer to what distinguishes acceptable uses of falsehoods, idealizations, or otherwise non-accurate representations of the target system from unacceptable uses is simply pragmatic. However, Healey does not pursue this answer. His pragmatist approach is supposed to offer an alternative between traditional realist and anti-realist views in two respects. First, by not focusing on the observable/unobservable distinction as the crucial one. Second, by keeping a commitment to the importance of objective explanations within quantum theory. The easy answers for deniers of the putative truism threatens the latter aspect.

To see how the pragmatic account fares with this problem we need to say a little more about what is required for successful explanation. So far, we have identified that quantum models can explain, on this view, by showing that the explanandum was expected and by showing what the explanandum depends on. The expectability requirement is clear but also faces a number of well-rehearsed challenges if taken as the sole criterion of explanation.[10] The dependence requirement goes some way towards addressing these shortcomings. However, not all ways of spelling out the dependence requirement will fit with the goal of allowing for objective explanations.

A prominent way of spelling out the dependence requirement is in terms of the ability to answer *what-if-things-had-been-different* questions (*w*-questions for short) as developed by Woodward (2003). This account also yields the desired objectivity of explanations and resists taking the explanatory relation to be context relative.

> On my analysis, interest relativity enters into what we explain but not into the explanatory relationship itself. What we try to explain depends on our interests, but it does not follow that for a fixed explanandum M and fixed explanans E, whether E explains M is itself interest-dependent. Obviously, it is not puzzling and no threat to the "objectivity" of explanation that the explanans E may explain M but a different explanans E' may be required to account for M'.
> (Woodward, 2003, 230)

Several of the central motivations for this account—namely, that explanation "is a matter of exhibiting systematic patterns of counterfactual dependence" (Woodward, 2003, 191) and the idea that the dependencies that matter have their roots in interventions—are attractive on a pragmatist approach to quantum theory. The focus on interventions is tied (at least conceptually) to changes that agents can bring about in a system (even if these are abstractly formulated and not tied to what is physically or even nomologically possible for an agent to do) which fits nicely with the

[10] In particular, this is the basis of familiar objections to the deductive-nomological account of explanation as presented by Hempel and Oppenheim (1965). I argue that some of these objections can be addressed as to not count against nomological accounts of explanation in Jansson (2015) but this requires moving beyond a mere (nomological) expectability account of explanation.

relativity to physical situatedness in the pragmatist approach.[11] However, one of the central commitments of Woodward's account, namely, the focus on *causal* accounts of explanation and causal interventions has to be relaxed.[12] It is easy to see why. The quantum state plays a key role in explanations but is not taken to have the job of describing the target system.

> The account presented here of how we use quantum theory to explain regularities makes no attempt to portray these explanations as causal. But it is consistent with the intuition that such explanations display two kinds of dependence of the regularities on the conditions in which they are manifested.
>
> The regularity depends on the quantum state or states that figure in the explanation: these states are what give one reason to expect the regularity to obtain. This kind of dependence is certainly not causal, as a quantum state does not represent or describe the momentary condition of a physical system to which it is ascribed. But it is still physical, in the sense that the correct quantum state ascription supervenes on the (non-quantum) physical conditions that ground it. (Healey, 2015, 38)

On the pragmatic approach the relevant *w*-questions must allow for dependencies that cannot be understood in causal terms. There are challenges to address here (and I think that they can be addressed) but for now, the relevant point to highlight is that this relaxation of the demand for *causal* dependencies and causal interventions blocks an otherwise tempting avenue towards distinguishing the failed explanations in Section 8.2 from appropriate uses of idealizations, distortions and fictions. The tempting, but no longer available, answer is that the failed explanations in Section 8.2 fail to get the causal dependencies right. On a causal dependence account, this would otherwise be an attractive weakening of the truism (accurate representation of the causal dependencies matter). However, we cannot simply repeat this move once we have relaxed the sense of explanatory dependence. We now need to specify more precisely which dependencies matter. At least the first two of the failed explanations in Section 8.2—the candle ceasing to burn due to the saturation of phlogiston in the air and the geocentric explanation of the retrograde motion of Mars—are capable of successfully answering a range of *w*-questions and supporting counterfactual dependencies in a broader sense. For example, what would have happened had the candle been in an open rather than a closed jar? Answer: it would have kept burning.

While there is scope to argue about the placement of specific cases, in order to keep a middle ground between mere prediction (in the instrumentalist sense) and objective explanations with a commitment to the putative truism (as is typical of realist approaches) something more—beyond an appeal to the ability to correctly

[11] In particular, it fits especially nicely with the motivation for taking applications of quantum theory to explain the correlations predicted in cases of entangled quantum states without invoking causal action at a distance. There are, however, questions about the extent to which this argument relies on taking the explanatorily relevant counterfactuals to represent nomologically possible interventions for the agent. Such a requirement would mean that an account in a Woodwardian spirit would need to be significantly altered to fit the pragmatic approach to quantum theory. I think that such a project could be undertaken, but I will put this concern aside for another time.

[12] This is key also to the treatment of the correlations predicted in cases of entangled quantum states.

answer non-causal *w*-questions—has to be said to rule these cases out. The next section will spell out a suggestion for what can be said.

8.4 Explanatory Roles and Ontological Commitment

In the previous section, I outlined the commonalities and some key differences between Healey's (2017) discussion of explanation and Woodward's (2003) account. Before I can suggest a way forward, I need to spell out why I nonetheless think that an account in a Woodwardian spirit is the right type of account for a pragmatist approach to quantum theory to focus on.

In this section I will introduce the idea of an epistemic account of explanation and argue that this type of account is well suited to Healey's goal of establishing pragmatism about quantum theory as an option between realism and instrumentalism. An epistemic dependence account allows for different types of dependence, and in Section 8.4.2 I will argue that different types of dependence are associated with different explanatory roles. In Section 8.4.3, I apply the account from Section 8.4.2 to address the cases discussed in Section 8.2.

First of all, we need to make some broad distinctions between different ways of thinking about explanation. Sometimes explanations are assumed to be *in the world*; one aspect of the world explains some (other) aspect of the world. I will call this construal *ontic* explanation.[13] However, explanations can also be assumed to be the kinds of things that we—as epistemic agents—seek, give, and receive. Here explanations are often understood primarily as acts of communication where the goal is to increase understanding. If we take an approach that focuses on subjective notions of understanding, then we get an account according to which explanations are—as Jenkins (2008) puts it—*in the mind*.[14] There is also a third, mixed, option. I will call this notion of explanation *epistemic*, since it—like accounts of knowledge—has components reflecting constraints based on what the world is like and components reflecting the mental states of agents.[15] Explanations are (by definition) something that we, as epistemic agents, seek, receive, and give, but whether we succeed in having an explanation once we have grasped a putative explanation is not determined by factors relating to our mental states—collectively or individually—rather, it is determined by the way the mind-independent world is.[16] A successful explanation is one where our beliefs adequately (whatever this turns out to mean) reflects some (to be identified) aspect of the world. In the recent literature, Woodward's (2003) interventionist account of causal explanation (and developments in, for example, Hitchcock and Woodward, 2003 and Ylikoski and Kuorikoski, 2010) stands out as an example of such an account.

[13] Salmon (1978) offers a causal account of this kind. Bokulich (2016) argues for greater clarity in whether an account is ontic or merely has a worldly constraint on successful explanation.

[14] I take Faye's (2014) account to be of this kind.

[15] The closest related notion is Jenkins's (2008, 67) genuine-understanding explanations. Within a specific framework of focusing on *w*-questions and counterfactual dependence this idea is developed by Hitchcock and Woodward (2003) and Ylikoski and Kuorikoski (2010).

[16] I take this category to not include pragmatic accounts of explanation such as that of van Fraassen (1977; 1980), since on these accounts whether a putative explanans E is an explanation for some explanandum M is context dependent and so, plausibly, dependent on factors related to mental states.

Epistemic accounts are particularly friendly to the goals of Healey's pragmatic approach to quantum theories. As I have already emphasized, Healey's account of explanation needs to reject the putative truism of Section 8.2. Epistemic accounts naturally make room for weakening the putative truism while retaining the objectivity of explanation. If the goal is to have our beliefs adequately reflect some aspect of the world, then we can expect to find tradeoffs between favouring features of explanations that allow us to successfully form beliefs and features that increase the accuracy of the representations involved.

For the rest of this chapter I will focus on epistemic accounts in a (very broadly) Woodwardian spirit. However, I think that in principle other accounts in the same broad family could do the job. I will take the crucial notion of explanatory connection to be one of *dependence*, following Woodward in his focus on *what-if-things-had-been-different* questions, and I will formulate the argument in these terms. The challenge is how to combine the three desiderata articulated so far. We would like to:

1. retain an objective account of explanation (and in particular, allow for objective grounds to rule out the failed explanations of Section 8.2).
2. weaken the putative truism in Section 8.2.
3. allow non-causal dependencies to be explanatory.

So far, I have argued that the third desiderata makes the combination of the first two more challenging. In the next section, I will suggest how, nonetheless, I think that all three can be satisfied. A tempting route towards weakening the putative truism on any epistemic dependence account of explanation is to distinguish the worldly relations of dependence from what allows us to grasp those dependencies. As a first, and very rough approximation, what really matters is what the explanandum *actually* counterfactually depends on.[17] The next section will spell this out.

8.4.1 Epistemic accounts of explanation and types of dependence

The explanations that Healey (2017) discusses are explanations provided by modelling a system of interest. If we categorize the possible types of dependence present in an explanatory model according to their relata, then we get four possible types of dependence (Figure 8.1):

Type 1: Dependence relating features of the model to features of the world.
Type 2: Dependence relating features of the model to other features of the model.
Type 3: Dependence relating features of the world to features of the model.
Type 4: Dependence relating features of the world to other features of the world.

So far, I have not said anything about *which* broad type (1–4) of dependence is the *sine qua non* for explanation. On an ontic dependence account, it is natural to focus on the fourth type. The role of the other dependencies between the world and the model, within the model, and between the model and the world are all easily understood as

[17] This is, roughly, the suggestion by Baron (2016) and Saatsi (2016). However, I will not develop the idea closely along the lines suggested there. In particular, I think that Baron's suggestion is not friendly to arguments from explanatory indispensability.

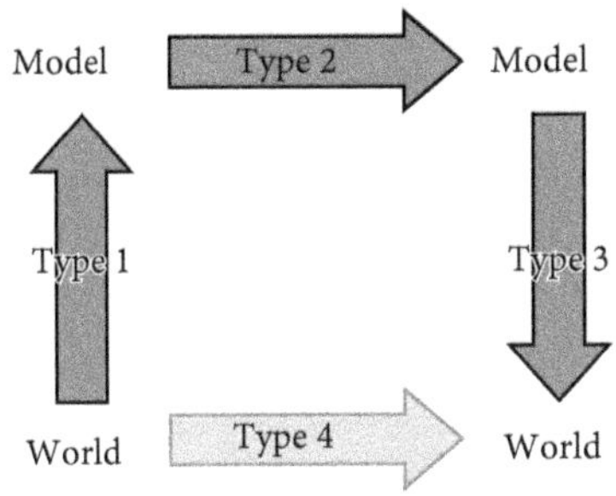

Figure 8.1 Types of dependence

part of the practice of providing information about the worldly dependence.[18] Types 1–3 look like explanatory dependence analogues to generally recognized important steps of modelling. When we try to model a target system we first have to describe the aspects of interest in a way appropriate to the modelling scenario; second, we can reason within the model; and, finally, we need to know how to translate the reasoning from within the model back to information about the target system.[19]

For explanation it is not enough to reason predictively correctly about the target system. Within epistemic dependence accounts (such as Woodward's), we also need to be able to answer questions about how the explanandum would have been different had some aspects of the target system been different. In order to answer such questions, the modelling steps have to support some information about dependence at each step of the modelling process. However, this only tells us that the dependencies of type 1–3 are inferentially robust. It does not turn them into worldly dependencies. On this view, the model is naturally understood as 'mediating' the reasoning about the fourth type of dependence.[20] Epistemic dependence accounts can allow that all types of dependence are crucial in model explanations. However, the type of dependence that *really matters*—what the explanandum actually depends on—must be of type 4.

This has not yet given us an answer to the challenges from Section 8.2. However, in order to do so we can combine the different types of dependence with the roles that they can occupy in explanations.

8.4.2 *Explanatory roles*

The basic explanatory roles that the traditional accounts of explanation recognize start off fairly coarse-grained; there is what is doing the explaining and what is being explained. Quickly, however, even the traditional accounts will provide more fine-grained roles within the explanans. There is typically a privileged place for causal relations or nomological relations, so that we can distinguish between the

[18] I will continue the discussion simply in terms of the model/world relation and worldly dependencies. However, this is a simplification since what the model takes as input is very typically a theoretical description of the system of interest that is available from another model or theory. What is crucial for my discussion is that this input is external to the explanatory model in question.

[19] For a discussion of mathematical models in these terms see, for example, Bueno and Colyvan (2011, 353).

[20] This is in line with the suggestion from Morrison (1999, 63).

causal relations or nomological relations required for explanation and the relata of these relations that may figure in the explanans. For example, within the deductive-nomological account we can distinguish between the laws and the particular facts.[21] On dependence accounts of explanation, such as Woodward's (unlike the deductive-nomological account), it becomes particularly clear that these components have different explanatory roles.

On a dependence account, the crucial distinction is not between particular facts and laws. Rather, it is between what is specified as the initial input—all explanations have to start somewhere—and what allows us to make *inferences* about the counterfactually robust links between the explanans' initial input and the explanandum. In more familiar terms, in causal accounts we distinguish between endogenous variables (with values determined by the model), the exogenous variables (with values assumed to be determined from outside of the model and acting as initial input into the model), and structural equations capturing how exactly the endogenous variables depend on the exogenous ones and each other.

Many structural equations capture what we typically take to be laws. However, Woodward (2003, ch. 6) points out that explanatory generalizations do not have to be exceptionless laws. The crucial role of the explanatory generalizations, on Woodward's account, is to allow us to reason reliably about how the explanandum depends on the initial input into the explanans. I will take this point on board.[22] What will matter for the way that I will carve up the explanatory roles within the explanans is simply the distinction between the initial input into the explanans versus what allows us to reason reliably about how the explanandum depends on the initial input. This way of distinguishing the initial input from the inferential resources is relative to a specific explanation. An inferential resource relative to one explanation might be an initial input relative to another explanation. However, relative to an explanation, it is an internal matter whether something is an initial input into the explanans or specified as an inference supporting principle, such as a law. In order to present a dependence explanation, we have to specify what is playing which role.

On dependence accounts of explanation we are looking for information about what in the world the explanandum phenomenon depends on. Successful explanations will, at minimum, show that the explanandum depends on the initial input into the explanans in some way. The 'some way' needs to be specified in more detail

[21] As Hempel and Oppenheim (1965) do.

[22] There is a slight complication here. If the crucial role is enabling us to reason reliably about robust dependence relations between the initial input into the explanans and the explanandum, then there is no particular reason to restrict ourselves even to *generalizations*. Woodward is focused on causal models, and he focuses on the explanatory project as starting once we have selected a certain model (and accepted it as apt). With this in place, the explanatory generalizations allow us to do the required inferential work. I think that this way of conceptualizing scientific explanation hides the inferential importance of dependence of types 1 and 3, and risks conflating dependence of type 2 with dependence of type 4. In particular, it is not simply the generalizations *within* models that allow us to reason inferentially with the model. The dependence relations between the world and the model and back to the world from the model are equally important to the explanatory use of models. I do not think that Woodward would deny the importance of these aspects of explaining with models, but in his account of explanation they are simply assumed to be in place before the account of explanation kicks in. To make sure that we keep these types of dependence on the table, I will not restrict the inferential resources to generalizations.

and that is the job of the inferential resources of the explanation. Since it is easy to take a representational understanding of laws and causal relations, the distinction between inferential dependence (type 1–3) versus worldly dependence (type 4) seems less important in simple causal explanations. Nonetheless, once we get to more complicated explanations the distinction is crucial.

In particular, if we want to relax the putative truism on explanation, it is reasonable to do so for inferential dependence but not for worldly dependence (on these accounts of explanation). Non-representational aspects of modelling a target system (or, perhaps better, aspects that are not considered adequate or accurate enough representations of the target system) can play a role in mediating our reasoning about worldly dependence, but, since they cannot figure as relata in such relations, they cannot occur in the initial input into the explanans. For example, the intra-model inference might proceed via several steps where there is no demand of type 1 dependence at each step (of course, other more complicated relationships are possible too) (Figure 8.2).

Now we have a criterion internal to explanation that puts a systematic restriction on the explanatory roles of non-representational (or not adequately representing) aspects: the initial input into the explanans is ontologically committing and cannot be of this kind.

So far I have motivated a representational restriction on explanatory roles that nonetheless is compatible with many of the goals of Healey's pragmatist approach to quantum theory. The goal is to allow for the satisfaction of three desiderata. As a reminder, we would like to

1. retain an objective account of explanation (and in particular, allow for objective grounds to rule out the failed explanations of Section 8.2).
2. weaken the putative truism in Section 8.2.
3. allow non-causal dependencies to be explanatory.

Of course, whether the above criterion will stand up to scrutiny from examples is yet to be seen. Let me first turn to the examples in Section 8.2 to see if they—as we would hope—run afoul of the weakened version of the putative truism.[23]

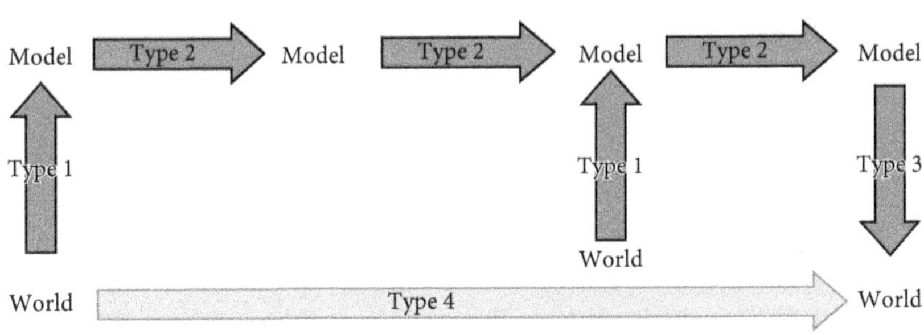

Figure 8.2 Intra-model dependence in several steps

[23] I am not arguing that the way that I have suggested in this section is the only way to reconcile the three desiderata. In particular, I am grateful to an anonymous referee for raising the suggestion that

8.4.3 *Non-explanations and explanatory distortions*

Consider again the cases 1–4 in Section 8.2 that the putative truism seemed to handle nicely by ruling them out as explanations since they do not accurately represent the systems of interest. Intuitively, the answer to why epicycles in the orbit of Mars cannot explain retrograde motion, or why the casting of malicious spells cannot explain the strange behaviour of children in Salem, etc., is simply that there are no epicycles and there are no (efficacious) malicious spells. The explanantia in question have failed to accurately represent what in the world the explananda depend on.

To illustrate how this is captured on the account presented in Section 8.4.2, let us reconsider the case of the geocentric model and the putative explanation of the retrograde motion of Mars. Our target explanandum is why Mars periodically seems to 'change' direction when observed from Earth. The view that I sketched above suggested that aspects of the explanans that do not (accurately enough) represent the target system can only be harmless in a model if they are *mediating* inferences within the model but are not playing the role of the initial input into the explanans.[24] In the case of the explanation of the retrograde motion, this phenomenon is supposedly explained by the motion of Mars around the Earth, including the epicycles in the model. However, here the very aspects that have to be taken as the initial input into the explanans in question—the epicycles in the orbit of Mars around the Earth—are not accurate representations of the target system. Moreover, the epicycles invoked in this model are not merely mediating inferences within the model. Notice that we cannot redescribe the case in order to simply put the epicycles into the inferential resources. We are trying to explain why it looks like Mars reverses its direction when viewed from Earth. The Ptolemaic answer is: Mars does reverse its direction of motion! We cannot remove the motion of Mars from the initial input into the explanans without simply destroying the Ptolemaic answer.[25]

Notice that we cannot recover the explanation by simply relying on the *relative* motion of Earth and Mars (that the Ptolemaic system might be taken to accurately enough describe). At least, this is ruled out by a fairly common and minimal restriction on scientific explanations. Namely, that we cannot explain the explanandum phenomenon simply by taking it to depend on itself. Here we are interested in explaining why the relative motion of Earth and Mars has a certain qualitative feature (retrograde motion). Simply citing the specific relative motion is just citing the specific instantiation of the general feature that we are interested in.[26]

inferential stability or power might be a pragmatist friendly way of reconciling the desiderata. The reason for not pursuing this option is that I do not see how inferential reliability will avoid judging seemingly non-explanations or bad explanations as superior to seemingly good (but inferentially limited) explanations.

[24] Of course, they do not have to be harmless even then. They also have to be justified. That is, we have to have reasons to believe that they are reliable for reasoning about the worldly dependence.

[25] This is not to say that the Ptolemaic system is incapable of providing any explanations. For some explananda the relative motion might be appropriate as input into the explanantia. For any such cases, the account in Section 8.4.2 would allow that the Ptolemaic model could provide (at least a minimal) explanation.

[26] Of course, this might be an explanation in terms of, say, the metaphysical dependence of the general on the particular. However, this is not how the request for a scientific explanation of retrograde motion is generally understood.

A similar account can be given for the other cases above. If we remove the capability of air to absorb phlogiston from the story above, we simply have no answer left to the question posed. We have not just removed details about how exactly the dependence goes; rather, we have destroyed the foundation for any such answer. If we remove the casting of malicious spells, we destroy the witchcraft explanation of the behaviour of the children in Salem. If we remove my desire to buy durian fruits, there is no explanation left as to why I went to the fruit stall.

The above are all examples where the failure to accurately represent the target system seems to destroy the explanation. However, it was part of the goal of the account in Section 8.4.2 to allow that the failure to accurately represent the target system is not always fatal to explanation. Another example from Section 8.2 can illustrate this possibility.

Consider, for example, the explanation of the escape velocity of an object at the surface of the Earth. We could account for the escape velocity at the surface of the earth in Newtonian terms by relying on $v_{escape} = \sqrt{2GM_{earth}/r_{earth}}$. In the derivation of this equation we rely on idealizations, distortions, and the use of the Newtonian law of universal gravitation, that we now think is strictly speaking false. However, Newtonian theory of gravity does an accurate enough job of describing how the escape velocity depends on the mass of the earth, etc. Here the failures to accurately represent the target system—for example, by introducing a gravitational force—do not prevent the explanation from providing accurate information about how the target explanandum, the escape velocity, depends on some aspect of the system that is accurately represented—for example, the mass of the earth. The force of gravity does not enter into the initial input in the explanation but is rather playing the role of allowing us to reason about how the explanandum depends on this input.

It is a further (and difficult) question to characterize precisely what makes an explanatory input accurate enough. For the general case, I think that the most promising way forward is to specify what the relevant range of variation is for the explanandum and to let that dictate the right level of accuracy for the explanans. On this general approach, it is clear that the mass of the earth is accurately enough characterized for the purposes of a coarse-grained estimate of escape velocities. The details about how, for example, Newtonian mass differs from relativistic mass do not matter here. However, for the rest of this chapter we can sidestep these questions since the distinctively quantum theoretic aspects of quantum theory are supposed to not represent the target systems at all in the cases that we are interested in (thereby avoiding having to answer the question of whether the representation is accurate enough).

8.5 Explanations on the Pragmatic Approach to Quantum Theory

So far we have tried to rule out the problematic cases by showing that the inputs into the putative explanations are not accurately represented. The benefits of the account developed in Section 8.4.2 is that it allows us to rule these cases out while still retaining an objective and non-pragmatic account of explanation. In this section

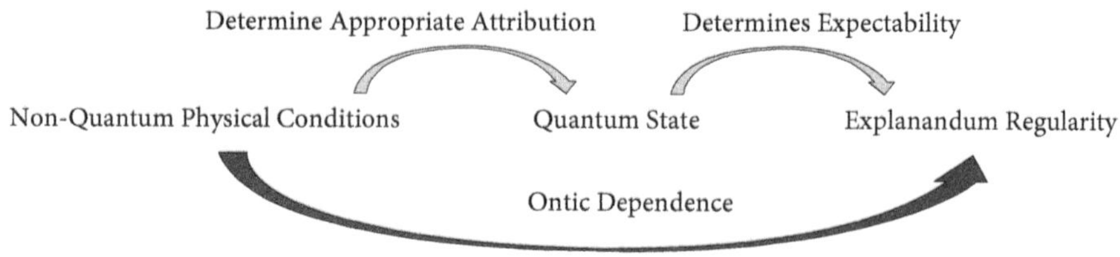

Figure 8.3 Types of dependence on the pragmatic approach to quantum theory

I will evaluate how far such an account can vindicate the explanatory claims of the pragmatic approach to quantum theory. After all, here the quantum state does not play the role of representing the target system (accurately or otherwise!). The idea that the quantum state cannot serve as an initial input into explanation fits well with the claim that what makes the dependence physical is "...that the correct quantum state ascription supervenes on the (non-quantum) physical conditions that ground it" (Healey, 2015, 38).

The worldly, or ontic, dependence seems to be between these physical conditions that make the ascription of a quantum state appropriate and the target explanandum. However, our reasoning about this dependence goes via the ascription of the quantum state (Figure 8.3).

If the account in Sections 8.4.1 and 8.4.2 is correct, we would expect that quantum states cannot feature as initial input into explanations on a pragmatic approach to quantum theory. On the one hand, this seems correct. The input into the explanatory model are the 'assumables', and the quantum state is not among them. On the other hand, we can worry that quantum states seem to be among the crucial initial input into many explanations. If the quantum state is explanatory only in the derivative sense of providing reliable resources for reasoning about worldly (type 4) dependence, then it seems as if the theory is not doing the real explanatory work after all. Healey considers an objection of this kind (but not, of course, following the exact proposal given here) and offers reasons to resist the conclusion that it is not the quantum state that is doing the explanatory work, and, in particular, that the quantum state *alone* is not capable of explaining.

It is often easier to demonstrate the correctness of a quantum state ascription than to make explicit the non-quantum claims that ground it. The subsequent explanatory application of that state then proceeds independently of the physical grounds of its ascription. One could choose to regard an explanation solely in terms of a quantum state as what Hempel called an explanation sketch; the explanation of which it is a sketch would then go on to specify the non-quantum claims that ground the ascription of that quantum state. But anyone making this choice would have to admit that we have yet to explain many or most of the phenomena we consider successfully explained using quantum theory just by reference to the quantum state.

There is a more important reason to accept an explanation in terms of a quantum state as satisfactory in its own right. Depending on the circumstances, one of a variety of different sets of non-quantum conditions may ground assignment of a given quantum state. We may already know a large number of different ways of preparing that quantum state while confidently expecting to find new ones. The supervenience base for a quantum state ascription is typically large, diverse, and open-ended. An explanation in terms of a quantum state acquires unifying power by abstracting from the details of the particular physical conditions in which that state is grounded. (Healey, 2015, 19–20)

For example, we might know several techniques for preparing entangled states where we do not have access to a statement of the physical grounds that make this quantum state ascription appropriate.[27] On the account in Section 8.4.2, this looks like a situation where we only have a partial handle on the explanation in question. There is a role for the quantum state, and the unification might be epistemically valuable, but it has not yet been given any unified ontic backing. If information about worldly dependencies is the goal of explanation, then this defence of taking the quantum state alone to be able to explain is only partially successful. On the one hand, the view that I have put forward allows that the quantum state might be indispensable *to us* in offering explanations. On the other hand, explanations on the pragmatic approach to quantum theory do black-box an important part of the explanatory project. In particular, we have only a partial understanding of what it is that makes the assignment of a given quantum state appropriate and so only a partial understanding of the worldly unificatory power offered by the quantum state.

Moreover, the quantum state is not, on this view, a mere abstract description of the physical conditions that ground the appropriate attribution of such a state. The quantum state is, as we have already seen, not even playing the role of describing some aspect of the target system. The type of unification that it achieves is, therefore, not the type of unification that comes from abstracting away from irrelevant details of the target system as we can understand other abstract explanations to do.[28]

Mere unification cannot be a good criterion for explanatory power on the account in Section 8.4.2. Only unification where there is an underlying unity in terms of the dependencies picked out should be expected to matter. The benefit of the account in Section 8.4.2 is that it allows us to merely weaken, rather than jettison, the truism about explanation and keep an objective worldly restriction on successful explanation. This also makes it possible to retain a weakened form of common realist arguments such as the explanatory indispensability argument and inference to the best explanation. However, the type of explanatory power that comes purely from unification is best suited to notions of explanation without such ontic constraints. If the power to unify is only of importance in terms of reasoning about the worldly dependence relations, then, no matter how great the unifying power of the explanations in terms of quantum states, the latter cannot explain on their own. On the suggestion made in this chapter, quantum states can play a genuinely explanatory role on the pragmatic approach to quantum theory. However, they cannot do so *alone*.

Since the Born rule is also treated as not representing some feature of, or law-like constraint on, the target system, the same worry applies to explanations that seem to take this as an initial input. For example, it seems natural to take the Born rule as such an initial input into the explanation of why, in cases of interference of more than two paths, the interference term is the sum of the interference terms taken pairwise. That

[27] For examples, see Healey (2015, footnote 14).
[28] See, for example, Jansson and Saatsi (2019).

is, the interference term from three paths A, B, and C is just a sum of I_{AB}, I_{AC}, and I_{BC}, with no additional term from I_{ABC}. This seems to depend on the form of the Born rule.

> The nonzero interference term I_{AB} is expected in all wave theories, including quantum mechanics (3,4). The next higher-order (i.e. three-path) interference term I_{ABC} will be zero in all wave theories, with a square-law relation between the field energy (or probability density) and field amplitude, which is the case in quantum mechanics with Born's rule.
>
> (Sinha et al., 2010, 418)

On this way of putting things, that the three-path interference term is zero depends on the Born rule being a square-law relation. If the Born rule is akin to a law then this looks like a law-based explanation. However, what makes the application of the Born rule appropriate on the pragmatic approach to quantum theory is not guaranteed to be uniform or law-like. The way that Healey approaches the explanation above highlights these features.

> [T]he Born rule is not a law of quantum theory: it emerges only piecemeal in its applications. When applying models of quantum theory to a variety of different physical systems, one can use different instances of the Born rule to explain probabilistic phenomena each manifests in basically the same way. This exhibits a wider pattern to which they all conform.
>
> (Healey, 2017, 163)

Here, the same worries as above apply. Generally, it is not enough to show that some phenomena all exhibit a wider pattern to have a genuine explanation. The ontic constraint that would remedy this problem would be to take the physical grounding that makes the application of the Born rule appropriate to be part of the explanation. However, these various circumstances are black-boxed if we understand this explanation from a pragmatic approach to quantum theory.

8.6 Conclusion: Explanation, Realism, and the Cost of the Pragmatic Approach

The pragmatic approach to quantum theory promises a middle road between realism and instrumentalism. However, balancing the demands of weakening the putative truism, retaining the objectivity of explanation, and allowing non-causal dependencies to do explanatory work without running afoul of seeming counterexamples, is a difficult balancing act. In Section 8.4.2, I have suggested a way in which all three desiderata could be fulfilled. It is, of course, open to the pragmatic approach to find another way of balancing the desiderata on explanation, but I hope to have illustrated that it is not easy to see how to do so. The way forward that I have suggested also retains the possibility of weakened versions of familiar arguments from *explanatory* considerations, such as explanatory indispensability or inferences to the best explanations, to scientific realism. On the proposal put forward here, it is not enough that some entity or fact plays an indispensable role in an explanation for us to be committed to it. Rather, the entity or fact must play an indispensable *explanatory role of a specific type.*[29] Similarly, we cannot infer to the best explanation

[29] See, for example, Baker (2005; 2012) and Colyvan (2001; 2012).

wholesale, but this account of explanation does allow for the possibility that inferences to the inputs of our best explanations are legitimate. The solution put forward in this chapter thus allows for weakened versions of common realist inferences from explanatory commitment to ontological commitment; it occupies a middle ground between instrumentalism and realism.

There are, therefore, several ways in which the suggestion made in this chapter fits nicely with the goals of the pragmatic approach to quantum theory. However, this account of explanation is not without costs. Many explanations according to this approach to quantum theory seem to at least partially black-box crucial information about the physical ground for the appropriate assignment of quantum states or applications of the Born rule. The restriction on explanation that I have suggested has the consequence that neither quantum states nor the Born rule can act as initial explanatory input. While this is a serious cost, it is not clear that a pragmatist approach to quantum theory has to resist this conclusion. It can be allowed that it is partially an open question what the worldly unification underlying the appropriateness of assigning a certain quantum state might be. The pragmatist approach does not entail taking quantum theory to provide a (metaphysically) fundamental and complete description of reality such that we could demand that an answer to this question should come from within quantum theory itself.

Acknowledgements

Thank you to the editors and to an anonymous referee for helpful comments that greatly improved this chapter. Thank you also to the feedback from everyone at the symposium on Scientific Realism and Quantum Theory, EPSA, 2017, and the workshop on Realism, Underdetermination, and Quantum Physics, Leeds, 2016, where earlier versions of the work in this chapter were presented. Many thanks also for the discussions (too many to list individually) about related issues concerning explanation and representation that have taken place at other conferences during the time I was working on this chapter.

References

Baker, A. (2005). Are there genuine mathematical explanations of physical phenomena? *Mind 114*(454), 223–38.

Baker, A. (2012). Science-driven mathematical explanation. *Mind 121*(482), 243–67.

Baron, S. (2016). The explanatory dispensability of idealizations. *Synthese 193*(2), 365–386.

Berenstain, N. and J. Ladyman (2012). Ontic structural realism and modality. In E. Landry and D. Rickles (Eds), *Structural Realism: Structure, Object, and Causality*, The Western Ontario Series in Philosophy of Science, pp. 149–68. Dordrecht: Springer.

Bokulich, A. (2012). Distinguishing explanatory from nonexplanatory fictions. *Philosophy of Science 79*(5), 725–37.

Bokulich, A. (2016). Fiction as a vehicle for truth: moving beyond the ontic conception. *The Monist 99*(3), 260–279.

Bueno, O. and M. Colyvan (2011). An inferential conception of the application of mathematics. *Noûs 45*(2), 345–74.

Chakravartty, A. (2017). Scientific realism. In E. Zalta (Ed.), *The Stanford Encyclopedia of Philosophy* (Summer 2017 Edition). https://plato.stanford.edu/archives/sum2017/entries/scientific-realism/.

Colyvan, M. (2001). *The Indispensability of Mathematics*. New York: Oxford University Press.
Colyvan, M. (2012). Road work ahead: heavy machinery on the easy road. *Mind 121*(484), 1031–46.
Craver, C. F. (2006). When mechanistic models explain. *Synthese 153*(3), 355–76.
Faye, J. (2014). *The Nature of Scientific Thinking: On Interpretation, Explanation, and Understanding*. Basingstoke: Palgrave Macmillan.
Healey, R. (2015). How quantum theory helps us explain. *British Journal for the Philosophy of Science 66*(1), 1–43.
Healey, R. (2017). *The Quantum Revolution in Philosophy*. Oxford: Oxford University Press.
Hempel, C. (1965). Aspects of scientific explanations. In *Aspects of Scientific Explanations and other Essays in the Philosophy of Science*, pp. 331–496. New York: The Free Press.
Hempel, C. (2001). Deductive nomological versus statistical explanation. In J. H. Fetzer (Ed.), *The Philosophy of Carl G. Hempel: Studies in Science, Explanation, and Rationality*. New York: Oxford University Press.
Hempel, C. and P. Oppenheim (1965). Studies in the logic of explanation. In *Aspects of Scientific Explanations and other Essays in the Philosophy of Science*, pp. 245–95. New York: The Free Press.
Hitchcock, C. and J. Woodward (2003). Explanatory generalizations, Part II: plumbing explanatory depth. *Noûs 37*(2), 181–99.
Jansson, L. (2015). Explanatory asymmetries: laws of nature rehabilitated. *The Journal of Philosophy 112*(111), 577–99.
Jansson, L. and J. Saatsi (2019). Explanatory abstractions. *The British Journal for the Philosophy of Science 70* (3), 817–44.
Jenkins, C. (2008). Romeo, René, and the reasons why: what explanation is. *Proceedings of the Aristotelian Society 108*(1), 61–84.
Kitcher, P. (1993). *The Advancement of Science: Science Without Legend, Objectivity Without Illusions*. New York: Oxford University Press.
Leng, M. (2005). Mathematical explanation. In C. Cellucci and D. Gillies (Eds), *Mathematical Reasoning and Heuristics*, pp. 167–89. London: King's College Publications.
Maddy, P. (1992). Indispensability and practice. *The Journal of Philosophy 89*(6), 275–89.
Maddy, P. (1995). Naturalism and ontology. *Philosophia Mathematica 3*(3), 248–70.
Morrison, M. (1999). Models as autonomous agents. In M. S. Morgan and M. Morrison (Eds), *Models as Mediators: Perspectives on Natural and Social Science*, pp. 38–65. Cambridge: Cambridge University Press.
Psillos, S. (1999). *Scientific Realism: How Science Tracks Truth*. London: Routledge.
Saatsi, J. (2016). On the 'indispensable explanatory role' of mathematics. *Mind 125*(500), 1045–70.
Saatsi, J. (2018). Realism and the limits of explanatory reasoning. In J. Saatsi and A. Reutlinger (Eds), *The Routledge Handbook of Scientific Realism*, pp. 200–11. London: Routledge.
Salmon, W. C. (1978). Why ask, "why?"? An inquiry concerning scientific explanation. *Proceedings and Addresses of the American Philosophical Association 51*(6), 683–705.
Sinha, U., C. Couteau, T. Jennewein, R. Laflamme, and G. Weihs (2010). Ruling out multi-order interference in quantum mechanics. *Science 329*(5990), 418–21.
Strevens, M. (2008). *Depth: An Account Of Scientific Explanation*. Cambridge, MA: Harvard University Press.
Timpson, C. (2008). Quantum Bayesianism: a study. *Studies in History and Philosophy of Modern Physics 39*(3), 579–609.
Van Fraassen, B. (1977). The pragmatics of explanation. *American Philosophical Quarterly 14*(2), 143–50.
Van Fraassen, B. (1980). *The Scientific Image*. Oxford: Clarendon Press.

Woodward, J. (2003). *Making Things Happen: A Theory of Causal Explanation*. New York: Oxford University Press.
Ylikoski, P. and J. Kuorikoski (2010). Dissecting explanatory power. *Philosophical Studies 148*(2), 201–19.

9

Quantum Mechanics and its (Dis)Contents

Peter J. Lewis

9.1 The Pragmatist Project

One standard way of illustrating the problematic nature of quantum mechanics goes like this. The quantum state plays the same role in quantum mechanics as the particle distribution does in classical physics—as a representation of physical state of the system under study. For example, the spin of an electron might be represented by the state $2^{-1/2}(|\uparrow\rangle + |\downarrow\rangle)$, where $|\uparrow\rangle$ represents a spin-up electron and $|\downarrow\rangle$ represents a spin-down electron (relative to some axis). If a spin measurement along this axis is performed on the electron, the Born rule says there is a 50% chance of getting the result 'spin-up' and a 50% chance of getting 'spin-down.' But the standard dynamical law of quantum mechanics—the Schrödinger equation—entails that the post-measurement state is $2^{-1/2}(|U\rangle|\uparrow\rangle + |D\rangle|\downarrow\rangle)$, where $|U\rangle$ represents a measuring device reading 'spin-up' and $|D\rangle$ represents a measuring device reading 'spin-down.' This state is entirely symmetric between the spin-up result and the spin-down result. So when the result is spin-up, what in the physical state of the system makes it the case that the result is spin-up? Quantum mechanics doesn't say. So it looks like quantum mechanics is either not right, or it is incomplete.

Various strategies have been proposed for solving this problem—the *measurement problem*. Bohmians propose to complete quantum mechanics by adding *particle locations* to the representation, where the particles are either associated with the $|U\rangle|\uparrow\rangle$ term or the $|D\rangle|\downarrow\rangle$ term in the state (Bohm, 1952). Spontaneous collapse theorists propose to change the dynamics by which the state evolves, so that the measurement process precipitates a 'collapse' to one term in the state, where the other term (essentially) disappears (Ghirardi, Rimini, and Weber, 1986). Everettians propose that both terms remain, and that an observer should be represented like the measuring device above, with one component in each term of the sum (Everett, 1957). That is, the observer splits into two, with one successor seeing 'spin-up' and one successor seeing 'spin-down' (Wallace, 2003).

All these approaches have their difficulties. The dynamics governing the Bohmian particles and the dynamics of the spontaneous collapse mechanism are both non-local, in prima facie conflict with special relativity. The Everettian approach

Peter J. Lewis, *Quantum Mechanics and its (Dis)Contents* In: *Scientific Realism and the Quantum*. Edited by: Steven French and Juha Saatsi, Oxford University Press (2020). © Peter J. Lewis.
DOI: 10.1093/oso/9780198814979.003.0009

doesn't add new dynamics, and the dynamics of standard quantum mechanics can arguably be made consistent with special relativity, so this difficulty doesn't arise. Nevertheless, the branching observers of the Everettian approach stretch our credulity, and also present difficulties for understanding probability, since every outcome of a measurement is observed by one or other of my successors. These problems (and others) have been addressed at great length, but the foundations of quantum mechanics remain hotly contested territory.

Despite this, quantum mechanics is arguably the most empirically successful theory ever devised. How can it be so successful, given that we don't understand what it says? A reasonable suspicion is that the success of the theory is evidence that we *do* understand what it says. That is, if we focus on how quantum mechanics is actually *used* by physicists, we will find that there are no genuine foundational problems.

This is the promise of a pragmatist dissolution of the measurement problem pursued separately by Richard Healey (2012; 2015; 2017) and Simon Friederich (2015). Their philosophical inspirations are somewhat different—Healey takes his motivation from the American pragmatists and Robert Brandom, whereas Friederich sees his approach as Wittgensteinian. But the details of their dissolution of the measurement problem are very similar. They each suggest that if we understand the meaning of quantum mechanical claims in terms of their use, then we can see that quantum mechanical claims function very differently from 'ordinary' non-quantum mechanical claims. Further, they argue that this difference is such that the measurement problem outlined above cannot even be formulated.

I shall not take issue with the pragmatist account of meaning, in either its Brandomian or Wittgensteinian variant. Nor do I insist on a realist understanding of quantum mechanics in a way that would beg the question against pragmatism. Rather, my argument here is that a pragmatist understanding of *quantum mechanics* raises its own special difficulties, ones that do not arise in other pragmatist dissolutions of ontological problems.

9.2 The Pragmatist Framework

What are the distinctive features of a pragmatist understanding of quantum mechanics? A key ingredient is Healey's distinction between two kinds of claim: quantum claims and non-quantum magnitude claims. Quantum claims "concern a quantum state, quantum probability (or expectation value), or other model element introduced by quantum theory" (Healey, 2015, 11). Non-quantum magnitude claims (or canonical magnitude claims) are characterized negatively—as claims concerning the magnitude of some physical quantity that do *not* involve quantum states, quantum probabilities etc. So, for example, 'The total squared-amplitude of quantum state $|\psi\rangle$ in region R is 0.9' is a quantum claim, but 'The particle is located in region R' is a non-quantum magnitude claim.

Healey bases this distinction on a difference in the way the two kinds of claim are used: quantum claims are not used to describe or represent physical systems, but non-quantum magnitude claims are so used. Instead of describing, the role of a quantum claim is to license a user of quantum theory to express non-quantum magnitude claims and to warrant the user to adopt appropriate epistemic attitudes toward these claims (Healey, 2015, 11; 2017, 131). Returning to the example, in circumstances in which it is appropriate to use state $|\psi\rangle$ to guide our beliefs about a particular particle, the fact that its squared amplitude in region R is 0.9 licenses an appropriately situated observer in believing to degree 0.9 that the particle is in region R.

Friederich endorses Healey's distinction between quantum and non-quantum claims (2015, 75). Friederich, too, takes non-quantum claims to be descriptive and quantum claims to be non-descriptive (2015, 114). In particular, he takes the role of quantum claims to be epistemic, in the sense that they prescribe the rational degree of credence in non-quantum claims (2015, 84). Hence there is broad agreement between Healey and Friederich concerning the central framework at the heart of a pragmatist understanding of quantum mechanics.

Healey also introduces a helpful example to illustrate the pragmatist approach: the demonstration of single-particle interference for C_{60} molecules by Juffmann et al. (2009). Thc apparatus is shown schematically in Figure 9.1. A beam of C_{60} molecules with well-defined velocity is produced (top left) and passes through two gratings of a Talbot–Lau interferometer in a high vacuum (top right). The molecules are deposited on a silicon surface, which is later moved into a second high-vacuum chamber and scanned with a scanning tunneling electron microscope (bottom right).

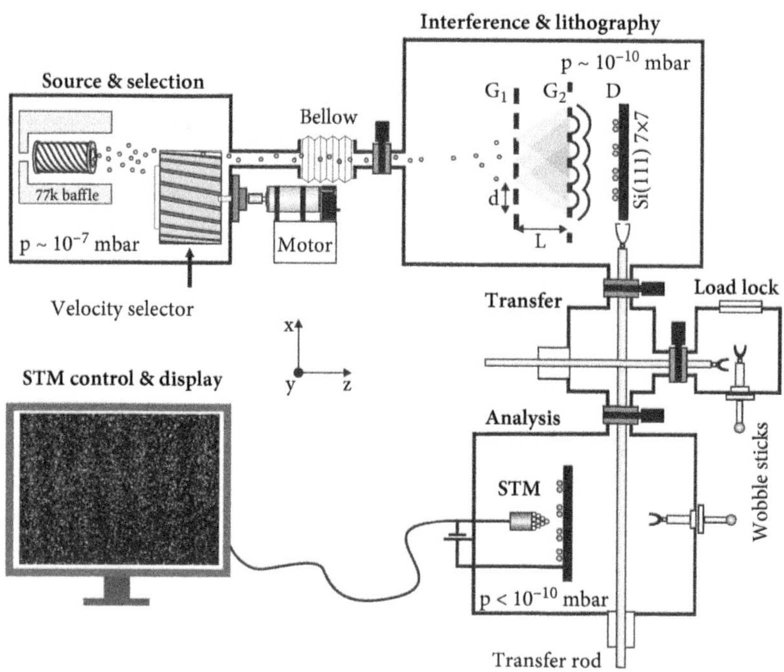

Figure 9.1 The apparatus of Juffmann et al. (2009). Reprinted figure with permission from Thomas Juffmann et al., 'Wave and Particle in Molecular Interference Lithography', *Physical Review Letters*, Volume 103, Issue 26, page 2, Figure 1, http://dx.doi.org/10.1103/PhysRevLett.103.263601, Copyright ©2009 by the American Physical Society.

The result is an image of a thousand or so individual C_{60} molecules forming an interference pattern (bottom left).

Juffmann et al. use a plane wave as the quantum state of an incoming C_{60} molecule. They then apply the Schrödinger equation to obtain the time evolution of the quantum state through the apparatus. In particular, the interaction between the state and the gratings produces a state with multiple terms, and these terms interfere to the right of the gratings to produce a state at the silicon surface exhibiting the characteristic pattern of high-amplitude and low-amplitude bands.

A suitably situated agent—one who is in a position to observe the image generated by the electron microscope—can use this final state as a guide to what to believe. In particular, she can use the Born rule to generate the probability that a C_{60} molecule will be located in a particular region of the silicon surface. That is, the quantum state prescribes how this agent should ascribe credences to non-quantum magnitude claims of the form 'The C_{60} molecule is located in region R.' The Born rule thus entails that her credences in claims of this form should be higher for some regions than for others. Hence quantum mechanics explains interference, not in the sense that the quantum state describes a real entity, but in the sense that it tells the agent to *expect* interference effects.

Decoherence plays an important role in this explanatory story (Healey, 2015, 13; 2017, 215). When the C_{60} molecule adheres to the silicon surface, it interacts with the surface in complicated ways. This has the result that the state of the C_{60} molecule undergoes environmental decoherence: the state diagonalizes when written in the position basis, meaning that off-diagonal terms in the state—interference terms—become very small. That is, although there is interference prior to the molecule adhering to the surface, after it has adhered, further interference effects are essentially ruled out. And this in turn warrants the applicability of the Born rule to the quantum state: when the state is approximately diagonal in the position basis, then the Born rule can be used to ascribe probabilities to non-quantum magnitude claims concerning the molecule's position. Friederich concurs with this role for decoherence (2015, 77).

The key feature of the Healey–Friederich approach is that quantum claims do not *describe* physical systems, but instead *prescribe* degrees of belief in non-quantum claims (taken as descriptive). Since it is a presupposition of the measurement problem that the quantum state is descriptive, adopting the pragmatist framework *dissolves* the problem; there is no need for a solution (Healey, 2017, 93). If the quantum state is not a description at all, it cannot be an *incomplete* description. And there is no conflict between a *prescriptive* quantum state that is spread out over a number of different experimental outcomes and a *description* that one of these outcomes actually occurs.

The Healey–Friederich approach naturally invites accusations of instrumentalism: at first glance, it looks like quantum mechanics is treated purely as a predictive instrument. They are sensitive to these accusations, each noting that their approach, unlike instrumentalist or constructive empiricist approaches, does not restrict either meaning or belief to claims about the observable world (Friederich, 2015, 114; Healey, 2017, 253). Note, for example, that the electron microscope plays no essential role in Healey's analysis of C_{60} interference: the position of the microscopic C_{60} molecule can be described by a non-quantum magnitude claim whether it is imaged by a microscope or not. As long as the state decoheres to a suitable extent, then the Born rule can be used to ascribe a probability to a claim about the position of the C_{60} molecule, and a suitably situated agent should believe the claim to the corresponding extent.

However, there is an important sense in which Healey–Friederich pragmatism is similar to instrumentalism, and distinct from constructive empiricism: Healey and Friederich seek to dissolve our foundational worries about quantum mechanics by denying that quantum states have descriptive content. We shouldn't worry that the quantum state of the C_{60} molecule at the point of measurement is a superposition of terms representing distinct positions, because careful attention to the way quantum mechanics is used shows us that the quantum state isn't used to *represent* the C_{60} molecule at all. Constructive empiricism, on the other hand, enjoins us to take quantum claims (and all other theoretical claims) as literally descriptive, even if we shouldn't (all things considered) actually *believe* those claims (van Fraassen, 1980, 11). Hence constructive empiricism, even if successful, does not dissolve the measurement problem; this is why van Fraassen (1991) felt compelled to offer a *solution*.

Healey and Friederich also both insist that quantum state ascriptions, and the degrees of belief they prescribe, should be regarded as *relative* to a physical situation. This feature of quantum claims is designed to dissolve the apparent non-locality of quantum mechanics. To see why non-locality is a potential problem, consider a pair of electrons in the spin singlet state $2^{-1/2}(|\uparrow\rangle_1|\downarrow\rangle_2 + |\downarrow\rangle_1|\uparrow\rangle_2)$. A spin measurement on either electron produces either a spin-up outcome or a spin-down outcome with probabilities of 1/2 each. But a spin measurement on one electron allows us to predict with certainty the outcome of a spin measurement on the other, no matter how far apart the electrons are. This might lead you to say that the spin measurement on the first electron affects the physical state of the other—and in fact this is just what Bohm's theory and spontaneous collapse theories do say. But any such influence, to be effective, would have to be instantaneous, and instantaneous space-like influences are apparently ruled out by special relativity.

Healey and Friederich each respond that since quantum states aren't used to describe the physical world, there is no need for agents physically situated by the two electrons to use the same quantum state to prescribe their degrees of belief (Healey 2012, 754; 2017, 175; Friederich 2015, 60). When a spin measurement is performed on one electron, the state decoheres, and the Born rule entails that the agent performing the measurement should ascribe equal degrees of belief to the two outcomes. When this agent learns that the outcome was spin-up, it is appropriate for her to also fully believe that the other electron is spin-down. That is, if she were to travel to the location of the other electron, she should fully expect a spin measurement on it to yield spin-down. But the agent who is already stationed by the other electron should continue to use the singlet state to prescribe her degrees of belief: if her electron is measured, she should expect spin-up and spin-down with credences of 1/2 each.

9.3 Content

There are at least two questions one can raise concerning this pragmatist understanding of quantum mechanics. The first concerns motivation: *Why* should we conceive of quantum states as Healey and Friederich suggest, rather than as descriptive of physical systems? The second concerns adequacy: Does the pragmatist conception of quantum states *succeed* in dissolving our worries about the foundations of quantum mechanics? Obviously these two questions are related: success in dissolving the foundational problems of quantum mechanics would be a powerful motivation for accepting a pragmatist approach! But nevertheless, they can be addressed somewhat separately, insofar as there are general reasons to favor a pragmatist understanding of the scientific enterprise. That is, one can ask whether pragmatism does in fact yield the understanding of quantum mechanics that Healey and Friederich endorse, and then one can go on to consider whether this understanding helps us with the foundational problems of quantum mechanics. That is the order I will follow here.

At one level the pragmatist position is quite easy to motivate. In any application of quantum mechanics to a system, empirical predictions are generated by the Born rule. That is, the role of the quantum state is ultimately to ascribe probabilities to observed outcomes, and it is quite plausible to regard these probabilities as prescribing the credences a suitably situated agent should have in each of these outcomes. The outcomes themselves are not expressed in terms of quantum states at all, but in terms that, in general, conform well to Healey's characterization of non-quantum magnitude claims. So quantum states are used to prescribe the proper credences in non-quantum magnitude claims, and non-quantum magnitude claims are used to describe the world.

By itself, though, this kind of motivation doesn't get us particularly far. Advocates of realist views of the quantum state—Bohmians, Everettians, and collapse-theorists—accept that quantum mechanics generates its predictions in probabilistic terms, but deny that this tells us that quantum state ascriptions are *purely* prescriptive. After all, a descriptive claim can generate a probabilistic inference: from 'Atmospheric pressure over New England is high' I can infer that the probability of rain in Vermont is low. So similarly, a Bohmian will claim that from a quantum state we can infer the probabilities of various measurement outcomes, in this case because the quantum state, though descriptive, is an *incomplete* description of the system under study. Furthermore, it would be naive to expect a direct motivation for the pragmatist framework in terms of the actual use of quantum and non-quantum claims by physicists. A general feature of pragmatism is the foregrounding of the variety of roles of human discourse, and the rejection of the assumption that the only role of a declarative statement is description or representation (Friederich, 2015, 51). It is typical of pragmatist accounts that superficially descriptive claims are argued to have some other function. So, for example, expressivism about moral values can be seen as a kind of pragmatism (Price, 2011, 9): claims about moral values look superficially descriptive, but instead, it is argued, they function to express approval or disapproval. So the fact that physicists use quantum claims in apparently descriptive ways does not provide direct evidence against the pragmatist assertion that these claims in fact function to prescribe our degrees of belief in non-quantum claims.

Juffmann et al. (2009), for example, use quantum and non-quantum claims in prima facie descriptive ways prior to decoherence. They describe the "quantum wave features" of the C_{60} molecules that account for the interference phenomena, and also the "composite particle nature of individual molecules" that accounts for the deposition of the molecules imaged on the silicon surface (2009, 1). That is, at times they use quantum claims in ways that might be taken as describing the passage of a wave through the apparatus, and at times they use non-quantum claims in ways that might be taken as describing the passage of a particle through the apparatus. But a pragmatist will warn against the automatic assumption that such claims really are descriptive.

Hence neither Healey nor Friederich motivate their accounts directly from the actual usage of physicists. Indeed, Friederich provides no direct motivation for his account, relying on its ability to dissolve quantum paradoxes as motivation enough (2015, 6). Healey, on the other hand, bases his account on a general pragmatist theory of propositional content. He thinks we can interpret the usage of physicists in a way that is consistent with this theory of content. That is, the theory of content tells us what we should say and what we should refrain from saying, and physicists' actual usage conforms to a large extent to these norms.

So let us turn to Healey's account of propositional content. Healey adopts an inferentialist account of meaning, in the spirit of Brandom (1994; 2000). That is, the meaning of a claim lies in the material inferences it supports, rather than in direct representation of reality. The claim 'There is a bat in the barn' derives its meaning from the material inferences I can draw from it, for example 'There are droppings on the barn floor,' or 'Something flies around the barn when I turn on the light.' It does not derive its meaning from correspondence between words and world.

How does that apply to quantum mechanics? Consider a non-quantum magnitude claim ascribing a location to a particular C_{60} molecule. When the molecule has adhered to the silicon surface, a claim of this kind licenses plenty of inferences, for example 'The electron microscope forms an image of the molecule at this location.' Hence it has empirical content (or empirical significance), and it is worth asserting.

But what about earlier on? Suppose I assert that a C_{60} molecule is close to the first diffraction grating. Healey thinks that such an assertion licenses erroneous inferences. In particular, if the molecule has a precise location close to the diffraction grating, then it passes through exactly one slit in the grating, and if it passes though exactly one slit, then no interference pattern is possible (2012, 745).

Of course, material inference is not deductive inference. For one thing, it is non-monotonic: addition of extra premises can undercut the inference. I can't infer 'Something flies around the barn when I turn on the light' from 'There is a bat in the barn' given the additional premise that the bat is dead. So similarly, perhaps the moral is that I can't infer 'There is no interference pattern' from 'The molecule is close to the diffraction grating' given *quantum mechanics* as an extra premise.

Healey's response is that this is correct, but then in the pre-decoherence context the claim about the location of the molecule has so little content that maybe one shouldn't assert it at all (2012, 747). It would be a mistake to infer the absence of interference, given quantum mechanics. But by the same token, it would be a mistake to infer that it passes through the first slit, or the second slit, or any other particular slit. It would be a mistake to infer that it passes through no slit at all, because then we might further infer that it doesn't arrive at the silicon surface. It would be a mistake to infer that it passes through all the slits, because then we might further infer that it arrives at the silicon surface at multiple locations at once. Since pretty much any inference one might draw from the non-quantum magnitude claim would be erroneous, the claim is almost contentless, and hence not worth asserting. If I assert that a molecule is close to the first diffraction grating, then those who do not recognize this lack of content might be misled into making one of the above inferences, and those who do recognize the lack of content will not have learned anything much.

What about quantum claims? Claims about the quantum state of a system license different inferences from non-quantum claims. In particular, from a quantum claim one can only infer the *probability* of some non-quantum magnitude claim, via the Born rule. This is why quantum claims are prescriptive rather than descriptive.

Beyond this difference, though, it looks like similar considerations should apply to quantum claims. Quantum claims prescribe degrees of belief in non-quantum claims via the Born rule, and the Born rule is only applicable in decoherent contexts. This is arguably a consequence of no-go theorems like Kochen and Specker (1967): since not all quantum observables can consistently be assigned simultaneous values distributed in accordance with the Born rule, and given that there is no way to privilege some observables over others, we should refrain from drawing any probabilistic inferences from a quantum claim prior to decoherence (Healey, 2017, 80). Decoherence picks out a preferred basis: it provides a way to privilege one set of observables, and guarantees that the application of the Born rule for those observables does not generate any inconsistency (Healey, 2012, 749; 2017, 81).

It apparently follows from this that the quantum state at the silicon surface has prescriptive content, but the earlier quantum state, say just before the diffraction grating, has very little prescriptive content concerning the system at that time, because the Born rule is inapplicable at that time. This is consistent with the lack of descriptive content of non-quantum claims about the C_{60} molecule at that time.

As noted earlier, Friederich does not adopt Healey's inferential account of propositional content. Instead, he wishes to leave open the possibility that non-quantum magnitude claims have well-defined descriptive content prior to decoherence (2015, 79). Nevertheless, he does agree with Healey that decoherence is a precondition for applying the Born rule, not because the non-quantum claims to which the Born rule ascribes probabilities have no content prior to decoherence, but simply because the Born rule is unreliable when applied to such claims (2015, 79). Hence it looks like Friederich, too, must say that quantum claims lack prescriptive content prior to decoherence.

9.4 Concerns about Content

Healey's motivation for the pragmatist framework lies in his inferentialist account of content. How successful is this motivation? I will raise two initial worries, neither of which I think is ultimately fatal to the project. I will then consider whether Friederich's position on content is an improvement over Healey's.

The first worry, and perhaps the most obvious, concerns the distinction between the prescriptive content of quantum claims and the descriptive content of non-quantum magnitude claims. Consider again the claim that there is a bat in the barn. This claim is descriptive, if anything is. It licenses an inference to 'There are droppings on the barn floor.' But that doesn't mean you should be *certain* that there are droppings on the barn floor. Presumably the background assumptions operative in this material inference suggest a certain (roughly specified) *degree* of belief. That is, the claim that there is a bat in the barn *prescribes* degrees of belief in various further claims; this is what its inferential content consists in.

The worry, then, is that the distinction between the prescriptive content of quantum claims and the descriptive content of non-quantum claims is not supported by the inferentialist account of content. According to inferentialism, no claims are descriptive if description requires representation. The content of every claim lies in its licensing of inferences, where an inference is typically to the prescription of a degree of belief in some further claim. All that distinguishes quantum claims from non-quantum claims, then, is that the prescription of degrees of belief is mediated by quantum mechanics, and hence calculated via the Born rule.

Perhaps this is all to the good, though. The main point that Healey seeks to establish is that quantum claims are prescriptive rather than descriptive, so if no claims are descriptive, Healey's point goes without saying. But on the other hand, I don't think Healey takes his point about quantum claims to be trivial in this way; his point is to *contrast* the way quantum claims are used with the way non-quantum magnitude claims are used (2017, 134). And there may well be various pragmatist tools to do just that. After all, one of the key tenets of pragmatism is to pay attention to the diversity of functions that our claims fulfill, not to reduce those functions to a single function, such as prescribing credences. Contemporary pragmatists typically acknowledge a sense in which many claims can be said to represent or describe, although not in the sense of standing in a traditional word–world relation (Brandom, 1994, 76). For example, Healey might appeal to Price's (2011, 20) distinction between i-representation and e-representation: our assertions are i-representational, in that they have inferential content, but a subset of them are also e-representational, in that the content is answerable to the environment via notions such as tracking and covariance. Using such a distinction, one might make the case that non-quantum claims are used to represent in ways that quantum claims are not, while denying that any claim represents the world in a straightforward *picturing* sense.

A second worry about content concerns the lack of content of claims about non-decoherent quantum systems. The general form of the worry is that Healey's position on content fails to adequately take into account the role of conditional or counterfactual inferences. Much of the content of our ordinary 'descriptive' claims has this conditional or counterfactual nature. For example, the content of 'There is a bat in the barn' includes a great deal of content that could be thought of as counterfactual: from 'There is a bat in the barn' I can infer 'If I were to turn on the light, I would see something flying around.'

Now consider the prescriptive content of the quantum state of the C_{60} molecule as it passes through the Juffmann apparatus. Prior to decoherence, the state prescribes no credences to non-quantum claims, and to this extent is devoid of content. But nevertheless, it *would* ascribe credences to various non-quantum claims given a suitable intervention on the system. If the diffraction grating were replaced by a screen, decoherence at the screen would allow the prescription of a probability (close to 1) to 'The molecule hits the screen.' If detectors were placed behind each slit, then decoherence at the detectors would allow the prescription of a (low) probability to 'The particle is located behind the leftmost slit.' Hence if counterfactuals contribute to content (and it is hard to see why they should not), then quantum claims have a good deal of content even prior to decoherence.

The same goes for non-quantum magnitude claims, and for the same reason: if from a quantum claim I can infer a particular credence in a non-quantum claim, then that non-quantum claim thereby acquires content. So even prior to decoherence, the claim that the C_{60} molecule approaches the diffraction grating has content, in that were the grating replaced by a detector, the Born rule would prescribe a credence in the claim. Similarly, the claim that the C_{60} molecule passes through the leftmost slit has content, in that were there a detector behind each slit, the Born rule would prescribe a credence in the claim.

It is worth noting, though, that the pragmatist project doesn't stand or fall with the denial of content to claims about systems prior to decoherence. It is possible to accept Healey's assertion that quantum claims function differently from non-quantum claims, while denying what he says about the content of such claims prior to decoherence. That is, one might accept that 'The C_{60} molecule passes through the leftmost slit' has content, and that a quantum claim about the state at the grating has content, where the latter prescribes the appropriate degree of belief in the former. The role of quantum mechanics, on such a view, is to make sure that our material inferences based on the content of the non-quantum claims do not land us in trouble. That is, while one might be tempted, based on classical intuitions, to conclude that if the particle passes through a determinate slit, then there is no interference, quantum mechanics blocks such an inference.

Understood in this way, the quantum state prior to decoherence has rich counterfactual content, prescribing credences in a wide variety of non-quantum claims. By the same token, those non-quantum claims have content. But because of the counterfactual nature of the relevant inferences, no contradictions result. If one were to put a detector behind the leftmost slit, there is a (small) probability a C_{60} molecule would be found there, but if there is no detector, the deposition of molecules on the silicon surface generates an interference pattern. However, even though the quantum state has rich content prior to decoherence, it would be a mistake to regard the quantum state as giving us a *picture* of the system in any sense.

On this view, the role of decoherence is not, as Healey argues, to delimit the range over which our claims have content, but to delimit the range over which our material inferences can draw unproblematically on our classical intuitions. The basic structure of the pragmatist approach remains intact: quantum claims are prescriptive rather than descriptive, and since the function of quantum claims isn't to describe physical systems, the measurement problem does not arise.

Friederich, recall, is not committed to the lack of content of non-quantum claims prior to decoherence, and indeed explicitly explores the possibility that all observables have sharp values at all times, via which the content of the corresponding non-quantum claims can be specified (2015, 162). However, Friederich notes that this proposal "goes beyond the boundaries of the therapeutic approach" (2015, 157). If the goal of the therapeutic (or pragmatist) approach is to dissolve the worry that quantum mechanics is either not right or incomplete, then Friederich's proposal threatens to reintroduce the latter worry: quantum mechanics says nothing about these sharp values, and hence is incomplete. His therapeutic ends would be better served, perhaps, by adopting Healey's inferentialism, modified along the lines just sketched.

9.5 Time and Explanation

If the above is correct, then despite worries about the details of Healey's and Friederich's respective accounts of content, the pragmatist can give a good account of the predictive success of quantum mechanics, and one that is not at odds with the foundational problems of the theory, since those problems cannot even be formulated under a pragmatist understanding of the quantum state.

Nevertheless, some may feel that even if the pragmatist account gives us a good understanding of how quantum mechanics is used to *predict*, it does not thereby give us a good understanding of how quantum mechanics is used to *explain*. That is, some may feel that the predictive success of quantum mechanics was never in question, and that the pragmatist account really just dodges the real problems of quantum mechanics, which lie in its inability to provide adequate explanations of the phenomena it so accurately predicts.

Such objections to the pragmatist project have to be couched carefully though. Realist objections against the pragmatist approach are liable to beg the question. The realist may object that the pragmatist fails to give an explanation for interference, for example, because the pragmatist doesn't give a mechanistic account of the formation of the interference pattern analogous to the classical explanation of interference in light or water waves. But the pragmatist will reply that since the role of the quantum state is to prescribe, not describe, the call for a mechanism amounts to nothing more than a flat denial of the pragmatist account of the function of quantum claims. Furthermore, Healey argues at length that pragmatist accounts of phenomena like interference fulfill two requirements of good scientific explanation: "(i) they show that the phenomenon to be explained was to be expected, and (ii) they say what it depends on" (2015, 4). Friederich concurs with this assessment (2015, 116). Nevertheless, I think there is something to the suggestion that pragmatism about quantum mechanics renders explanation problematic. In this section I will try to diagnose a sense in which the pragmatist approach seems to fall short in terms of explanation, in a way that does not simply amount to a flat denial of the pragmatist approach.

To do so, it will be helpful to contrast the pragmatist approach to quantum mechanics with pragmatist approaches to other issues, such as the status of the mental, the mathematical, or the moral. Here, too, the pragmatist tries to dissolve foundational problems by arguing that apparently descriptive claims in fact have some other function. When we say 'Kicking your sister is bad', we appear to be ascribing a property, badness, to physical acts of a certain sort. But we can examine the physical act as closely as we like and fail to find the *badness*. Here the pragmatist responds that to say that kicking your sister is bad is not to *describe* the act as having a certain property, but to *express* our disapproval of such acts.

Note, though, that the pragmatist account of the function of moral claims does nothing to restrict the ordinary physical claims we make about bodily actions—what a kick is, physiologically speaking, and how it is caused. It is certainly not the case that in applying a moral evaluation to a kicking action we are thereby precluded from explaining it in physiological terms. Indeed, since the function of the moral evaluation is distinct from that of physiological description, it is hard to see how the former could restrict the latter.

This is where the pragmatist account of quantum discourse deviates from standard pragmatist analyses—because applying a quantum claim to a physical system *does* preclude certain sorts of ordinary physical explanation. Consider the Juffmann interference experiment again. We explain interference via a three-step temporal process. First, we describe the preparation of the system. Given this preparation, a certain quantum state is appropriate—in this case, a plane wave that evolves into a set of overlapping arc-shaped wave fronts. This quantum state in turn prescribes degrees of belief for the various locations at which the C_{60} molecule might be located on the silicon surface. But note here that in the middle stage—the stage at which one uses a non-decoherent quantum state—non-quantum descriptions of the location of the molecule are to some extent blocked. According to Healey, such claims have almost no content, and so are not worth asserting. In the previous section I urged that we should regard such claims as having content—but nevertheless, many of the standard inferences from these location claims are blocked, so on an inferentialist account of meaning, they have significantly different content from ordinary location claims. So the applicability of a non-decoherent quantum state to a system precludes, or at least limits, our ability to describe that system in ordinary non-quantum terms.

There are two senses in which this distinctive feature of the application of pragmatist analysis to quantum mechanics seems prima facie problematic. The first has to do with explanation. As noted above, it would beg the question against the pragmatist to demand explanations in which quantum states describe physical systems. But if quantum claims do not have the function of describing the physical world, it does not seem out of place to expect such claims to be consistent with our ordinary non-quantum descriptions of physical systems, and hence to expect explanations in non-quantum terms. In an interference experiment, we can describe the preparation of the system in terms of the locations of objects, and the results in terms of the locations of objects, so why not the processes during the intervening time?

Can decoherence provide a satisfactory answer to this question? Certainly decoherence provides a mathematical demarcation between situations in which non-quantum descriptions are appropriate and those in which they are not. But how does decoherence perform this feat? Physicists are happy to provide descriptive explanations: in Juffmann-type experiments, decoherence occurs when a C_{60} molecule exchanges energy with its environment in complicated ways (Hackermüller et al., 2004). Collision of a C_{60} molecule with the silicon surface satisfies this condition. According to Healey, however, since it is decoherence that provides the precondition for non-quantum claims to be assertable, no explanation of decoherence in terms of molecular collision is possible. Decoherence is central to the pragmatist framework, but according to that framework, descriptive explanations of decoherence are ruled out. Decoherence tells us when non-quantum descriptions are possible and when they are not, but it cannot tell us *why*.

So decoherence doesn't explain why non-quantum descriptions are unavailable prior to decoherence. Can we appeal to the no-go theorems to provide this explanation? The no-go theorems show that the prescriptive content of the quantum state cannot be realized by a descriptive model of a certain sort, a model in which every observable has a determinate value. That is, the no-go theorems preclude one kind of model; but others are available, notably Bohm, spontaneous collapse, and Everett. In addition, Friederich (2015, 161) suggests that a model in which every observable has a sharp value is not ruled out by the no-go theorems, although he doesn't explicitly construct such a model. Of course, this would take us back to the project of *solving* the measurement problem that the pragmatist project is designed to avoid. If the pragmatist approach is to succeed in *dissolving* the measurement problem, it should show us why trying to construct a descriptive account of quantum processes is not just problematic, but conceptually misguided. That is, it should give us a principled explanation of the inapplicability of non-quantum descriptions to systems between preparation and measurement. Analogy with other pragmatist projects, for example concerning the status of moral claims, provides little guidance as to why that should be the case.

A second sense in which the distinctive features of quantum pragmatism might be problematic is that the content of descriptive claims can change radically over time. Consider the time period during which a C_{60} molecule approaches and adheres to the silicon surface. After it has adhered, the claim that the molecule has a particular location has straightforward descriptive content. But just prior to this point, the quantum state has not yet decohered, and the claim that the molecule has a particular location lacks content (according to Healey), or at least has significantly modified content. That is, the content of a claim is highly sensitive to the *physical environment* of the system concerned (Healey, 2017, 137).

This is again a radical departure from other applications of pragmatism. The pragmatist can certainly account for contextuality of content: the give and take of reasons is a social process, so the meaning of a claim can be sensitive to the *social* environment. But it is hard to see why the content of a claim should depend on the *physical* environment. Sometimes we *make* the physical environment relevant, for example when we say 'Gold is whatever is of the same kind as this,' pointing at a sample. But there is no indexicality of this kind when I ascribe a location to a molecule. Why should the *meaning* (as opposed to the truth) of my utterance 'The molecule is in region *R*' depend on the physical environment of the target system? According to the pragmatist approach, whether you should take my utterance as meaningful or (practically) meaningless depends on whether decoherence has occurred: it can be meaningless at one moment, and meaningful a fraction of a second later. This is very strange semantic behavior. Of course, strange consequences are typical of interpretations of quantum mechanics, and perhaps this is a consequence we can live with. But it does seem that the pragmatist approach *relocates* the problematic aspects of quantum mechanics rather than dissolving them.

Again, Friederich avoids this consequence, but at a cost. Friederich is open to the possibility that claims about the molecule's position always have content, based on his proposal of sharp values for all observables (2015, 79). However, again this speculation undermines the advantages of the pragmatist approach, in that it suggests that quantum mechanics is incomplete as it stands.

So there are at least two ways in which the application of pragmatism to dissolve the foundational problems of quantum mechanics differs from other applications.

When I say 'Kicking your sister is bad,' I don't preclude physiological descriptions of kicking, yet a quantum claim can preclude non-quantum descriptions of a system. And the content of a physiological claim about kicking doesn't change radically over time depending on the physical environment of the person concerned, but the content of a non-quantum claim about a molecule does change radically over time depending on the physical environment of the system concerned. Perhaps this means that there is something problematic in the way Healey (in particular) employs pragmatism to help us understand the way quantum mechanics works. Or perhaps it just means that the application of pragmatism to quantum mechanics is a unique case, and further philosophical therapy can dissolve these remaining concerns too.

9.6 Conclusion

I have done nothing here to evaluate the general pragmatist approach to meaning. But if it is otherwise defensible, it certainly seems like a good place to look for a dissolution of the foundational problems of quantum mechanics. The notable successes of pragmatism lie in areas where a straightforward realism lands us in deep philosophical difficulties—about mathematical entities, for example, or moral properties. And that is not a bad initial diagnosis of the situation regarding quantum mechanics: it is an excellent instrumental recipe, but taking quantum claims as literally representational quickly leads to problems, the measurement problem among them.

So the pragmatist approaches of Healey and Friederich are welcome. And their proposal for understanding quantum claims—as prescriptive of our degrees of belief in non-quantum claims—has a good deal of initial plausibility. But it is also important to note the ways in which the application of pragmatism to quantum mechanics differs from previous pragmatist projects. Generally in such projects, a domain of statements is singled out as having a function other than representation, but the representational work of other statements in the vicinity goes on unchanged. Things are less straightforward in quantum mechanics, where the identification of quantum claims as performing a prescriptive function apparently also sharply circumscribes the applicability of representational non-quantum claims to physical systems. Hence a certain representational kind of explanation is also circumscribed: it is only available after decoherence, so representational explanation prior to decoherence, or of decoherence itself, is ruled out. Furthermore, the content of non-quantum claims exhibits an unusual kind of dependence on the physical environment, so that a claim about the location of a molecule can be meaningless at one moment and meaningful the next. Again, this has no analog in standard applications of pragmatism.

For these reasons, I think that further work is needed to ascertain whether pragmatist therapy can succeed in dissolving the problems of quantum mechanics. In particular, it looks like the pragmatist approach favored by Healey (and Friederich, insofar as he doesn't go beyond the pragmatist approach) requires fairly radical changes to our understanding of propositional content and of explanation, even when compared to other pragmatist projects. The surrounding issue is whether the pragmatist approach, all things considered, is less problematic than realist approaches to the foundations of quantum mechanics. If so, then this could provide a powerful independent motivation to adopt the pragmatist program. But at present, I think this remains unresolved.

References

Bohm, D. (1952). A suggested interpretation of the quantum theory in terms of "hidden" variables, parts i and ii. *Physical Review* (85), 166–93.

Brandom, R. (1994). *Making It Explicit*. Cambridge, MA: Harvard University Press.

Brandom, R. (2000). *Articulating Reasons: An Introduction to Inferentialism*. Cambridge, MA: Harvard University Press.

Everett III, H. (1957). "Relative state" formulation of quantum mechanics. *Reviews of Modern Physics* 29, 454–62.

Friederich, S. (2015). *Interpreting Quantum Theory: A Therapeutic Approach*. New York: Palgrave Macmillan.

Ghirardi, G. C., A. Rimini, and T. Weber (1986). Unified dynamics for microscopic and macroscopic systems. *Physical Review D* 34, 470–91.

Hackermüller, L. et al. (2004). Decoherence of matter waves by thermal emission of radiation. *Nature* 427, 711–14.

Healey, R. (2012). Quantum theory: a pragmatist approach. *British Journal for the Philosophy of Science* 63, 729–71.

Healey, R. (2015). How quantum theory helps us explain. *British Journal for the Philosophy of Science* 66, 1–43.

Healey, R. (2017). *The Quantum Revolution in Philosophy*. Oxford: Oxford University Press.

Juffmann, T. et al. (2009). Wave and particle in molecular interference lithography. *Physical Review Letters* 103, 263601.

Kochen, S. and E. P. Specker (1967). The problem of hidden variables in quantum mechanics. *Journal of Mathematics and Mechanics* 17, 59–81.

Price, H. (2011). *Naturalism Without Mirrors*. Oxford: Oxford University Press.

Van Fraassen, B. (1980). *The Scientific Image*. Oxford: Oxford University Press.

Van Fraassen, B. (1991). *Quantum Mechanics: An Empiricist View*. Oxford: Oxford University Press.

Wallace, D. (2003). Everett and structure. *Studies in History and Philosophy of Modern Physics* 34, 87–105.

PART IV

Wavefunction and Quantum State Realism

10

Losing Sight of the Forest for the Ψ

Beyond the Wavefunction Hegemony

Alisa Bokulich

10.1 Introduction

Second only to experiment, the formalism of a theory is a central avenue through which to explore the ontological implications of a scientific theory. When engaging in such a realist project, it is essential to keep in mind the distinction between the representational vehicle (the mathematics) and the thing in the world it represents. In the context of quantum mechanics, however, this obvious distinction has often been ignored. In particular, there has been an elision between the wavefunction (ψ) and the quantum state (properties of the system in the world) it represents. This practice of using ψ to stand in for both the mathematical representational vehicle and the thing in the world is problematic for a number of reasons: first, it encourages a reification of the mathematics; and, second, there may be properties of the representational vehicle that are not properties of the thing in the world. We see these issues particularly vividly in the case of wavefunction realism, which takes the wavefunction to be a concrete physical object, and takes the $3n$-dimensional configuration space on which the wavefunction is defined, to be the space that we (despite appearances to the contrary) actually live in.

A useful antidote to this habit of eliding the mathematics with the thing in the world is a consideration of alternative mathematical formalisms. A careful consideration of alternative formalisms is particularly important in the context of realist projects that try to read the ontology of a theory off of the mathematics. My aim in this paper is to highlight a little-known alternative formalism for quantum theory that does not make use of the wavefunction, ψ, in representing the quantum state. Instead, it represents the quantum state by means of a displacement function of a continuum of interacting 'particles' following spacetime trajectories, $q_i(a, t)$. The Schrödinger equation is recast as a second-order Newtonian-type law for this congruence of spacetime trajectories. The congruence of trajectories can be computed independently of the wavefunction, thus providing a 'quantum mechanics without wavefunctions.' As will become clear, this is a full formulation or representation of quantum mechanics—not an approximation scheme—and also not merely an 'interpretation' of quantum mechanics. It is a

Alisa Bokulich, *Losing Sight of the Forest for the ψ: Beyond the Wavefunction Hegemony* In: *Scientific Realism and the Quantum*. Edited by: Steven French and Juha Saatsi, Oxford University Press (2020).
 DOI: 10.1093/oso/9780198814979.003.0010

full alternative formalism for representing quantum theory that is equivalent to the standard formalism. I will refer to this alternative formalism for quantum theory, with its 'displacement' or 'trajectory' model of the quantum state, as Lagrangian quantum hydrodynamics (LQH). It should be emphasized, however, that 'hydrodynamics' here simply refers to the analogy used in the development of the mathematical formalism, and in no way requires a commitment to anything like a quantum fluid.

LQH is a representation—not interpretation—of quantum mechanics, and all questions related to the measurement problem or interpretations of quantum mechanics will be bracketed for the purposes of this chapter. It is not my aim here to offer a realist interpretation of quantum mechanics, that is, to answer the question of what quantum mechanics (on any formalism) tells us is physically real. Rather my aim is to argue that a more careful consideration of alternative possible formalisms is a critical prerequisite to any such realist project that attempts to read the ontological implications of a theory off the formalism.

Exploring alternative formalisms for a theory, such as LQH, is a fruitful endeavor for at least four reasons. First, as has already been emphasized, it undercuts a facile identification of the formalism with the world, bringing to the fore the substantive questions about what sorts of ontological inferences can legitimately be made. Second, it may be that some phenomena or questions are more readily treated in one formalism rather than another (indeed this is typically why such alternative formalisms are developed and widely taught). Third, some formalisms may be more fertile in suggesting future lines of theory development (e.g., think of how the Hamilton–Jacobi formulation of classical mechanics was seminal in the development of quantum mechanics).[1] And, fourth, alternative formalisms can challenge common presuppositions about what features are supposedly demanded by the theory (e.g., the presumption that the quantum state must live in a $3n$-dimensional configuration space, as will be discussed later on).

The structure of this chapter is as follows: I begin in Section 10.2 by reviewing the claims of wavefunction realism, and then turn to some cautionary tales that are worth recalling when undertaking such a realist project. In order to motivate Lagrangian quantum hydrodynamics and introduce the central conceptual elements needed to understand this formulation, I review in Section 10.3 *classical* hydrodynamics and the fertility of continuum hydrodynamic representations at the classical scale. This review of classical hydrodynamics is helpful both because quantum hydrodynamics is built in close formal analogy with it, and because it provides a broader context in which to think about the ontological implications of LQH. Section 10.4 turns to hydrodynamic representations at the quantum scale, beginning with the early 'Eulerian' hydrodynamic representation of Erwin Madelung (1927), and then turning to the more recent 'Lagrangian' quantum hydrodynamics (LQH) developed by Peter Holland (2005a).

Section 10.5 explores some of the broader questions about LQH. In particular, this formulation of quantum mechanics will be clearly distinguished from the de Broglie–Bohm interpretation of quantum mechanics, which also emphasizes

[1] See, e.g., Butterfield (2005).

trajectories. As we will see, LQH exhibits a novel quantum symmetry, which is obscured on the standard formulation, and this symmetry (by Noether's theorem) corresponds to a set of conservation laws. This section also explores whether or not, in analogy with arguments that have been made in the classical case, there is some sense in which the Lagrangian formulation of quantum mechanics is more fundamental.

Finally, in Section 10.6, I draw out some of the implications of LQH for realist projects in quantum theory. In connection with wavefunction realism, the existence of LQH—a formulation of quantum mechanics without wavefunctions—challenges the hegemony of the wavefunction in representing the quantum state. Moreover, it undermines the supposed necessity of identifying the $3n$-dimensional configuration space of ψ as the space we live in. I conclude by briefly identifying three interpretive attitudes one can take toward LQH, situating my preferred approach within an alternative tradition in scientific realism that understands realism in terms of the development of fruitful metaphors, rather than literal construals.

10.2 Quantum Physics Realism Debate

10.2.1 ψ Realism

Traditionally, the realist project in quantum theory has been understood as one of trying to read the mathematical formalism of the theory as a *literal depiction* of the world. Interestingly, Bas van Fraassen takes this literalist commitment to be required not just for the realist, but also for his own (antirealist) constructive empiricism. He writes, "The agreement between scientific realism and constructive empiricism is considerable and includes the literal interpretation of the language of science" (1991, 4).[2] A prominent thread in this quantum realism project, which has recently experienced a resurgence of interest, is known as 'wavefunction realism.'

Alyssa Ney concisely defines this view as follows: "the view that the wave function is a fundamental object and a real, physical field on configuration space is today referred to as 'wave function realism'" (2013, 37). Wavefunction realism has been defended by David Albert since the mid-1990s, and he claims it is a necessary part of the realist project in quantum theory on any interpretation.[3] He writes,

> [I]t has been essential (that is) to the project of quantum-mechanical *realism* (in *whatever* particular form it takes—Bohm's theory, or modal theories, or Everettish theories, or theories of spontaneous localization), to learn to think of wave functions as physical objects *in and of themselves*. (Albert 1996, 277; emphasis original)

According to Albert, this reifying of the wavefunction (which we should keep in mind, that as a *function*, is strictly speaking a mathematical object) is both a necessary and obvious consequence of taking quantum mechanics seriously under any

[2] As will be discussed in Section 10.6, it is precisely this commitment to a literal construal that I urge the realist should abandon.

[3] J. S. Bell defended a precursor of this view in connection with Bohm's theory when he wrote, "No one can understand this theory until he is willing to think of Ψ as a real objective field rather than just a 'probability amplitude'. Even though it propagates not in 3-space but in $3n$-space." (1987a, 128).

interpretation. Of course, what he has in mind here is not a mere Platonism about mathematical objects, but rather a realism about a physical field, which the wavefunction is supposed to represent (e.g., Albert 2013, 53).

Furthermore, since the wavefunction (the mathematical object) is defined on a $3n$-dimensional configuration space (where n is the number of particles in the universe) Albert concludes that this astronomically large configuration space must also be interpreted literally as our actual physical space. He writes:

> And of course the space those sorts of objects live in, and (therefore) the space we live in, the space in which any realistic understanding of quantum mechanics is necessarily going to depict the history of the world as playing itself out . . . is configuration space. And whatever impression we have to the contrary (whatever impression we have, say, of living in a three-dimensional space, or in a four-dimensional space-time) is somehow flatly illusory. (Albert 1996, 277)

Here again, we see the realist project in quantum theory being understood as taking certain features of the mathematical formalism of the theory at face value as a *literal depiction* of what our world is really like.

Despite the enormous amount of ink and intellectual resources that have been poured into wavefunction realism and the interpretation of quantum mechanics, neither has managed to produce a broad consensus, despite the former project going on for over two decades and the latter project nearing a century. When a field has remained in such a holding pattern for an extended period of time, it is reasonable to ask whether the time has come to reframe the realist project in quantum physics and perhaps start asking a different set of questions. Before engaging in such a reframing, however, it is important to recall a few cautionary tales.

10.2.2 A few cautionary tales

The first cautionary tale recalls that, despite its fundamental role in quantum mechanics, the wavefunction is both enigmatic and unobservable. As a letter to Nature recently lamented:

> [The wavefunction] is typically introduced as an abstract element of the theory with no explicit definition. Rather physicists come to a working understanding of the wavefunction through its use to calculate measurement outcome probabilities by way of the Born rule. At present, the wavefunction is determined through tomographic methods, which estimate the wavefunction most consistent with a diverse collection of measurements. The indirectness of these methods compounds the problem of defining the wavefunction. (Lundeen et al. 2011, 188)[4]

In practice, the wavefunction must be estimated and inferred from a collection of various measurements (of quantities that are observable) made on an ensemble of identically prepared systems. Even someone like John S. Bell, who interprets the wavefunction in accordance with Bohmian mechanics as a real, physical field, notes its inaccessibility:

[4] There have been various attempts to measure the wavefunction of a single system, using weak or protective measurements, including the paper cited here. For a recent review, see Gao (2017), and for criticisms see, for example, Combes et al. (2017). An alternative way to understand such measurements will be briefly discussed in Section 10.6.

Although ψ is a real field it does not show up immediately in the results of a single 'measurement,' but only in the statistics of many such results. It is the de Broglie-Bohm variable x that shows up immediately each time. That x rather than ψ is historically called a 'hidden' variables is a piece of historical silliness. (Bell 1987b, 162–3)

The point here, of course, is not that one should never be a realist about unobservables. Rather the point is to recognize that there is an additional level of difficulty involved in such cases, so the evidential bar must be set higher. Extra caution should be exercised when trying to read ontological implications off a formalism when the entity or state in question is unobservable and only indirectly accessible.

The second cautionary tale is that quantum mechanics is not the final, most fundamental, theory of everything, but rather is only an effective theory. The non-fundamental status of quantum mechanics has been emphasized in connection with wavefunction realism by Wayne Myrvold who argues that we should

bear in mind that quantum mechanics—that is, the nonrelativistic quantum theory of systems of a fixed, finite number of degrees of freedom—is not a fundamental theory, but arises, in a certain approximation, valid in a limited regime, from a relativistic quantum field theory. (2015, 3247)

He goes on to note that configuration spaces are not fundamental and that wavefunctions are not really like classical fields.

An effective theory can be understood as a framework for capturing the essential physics within some circumscribed domain, without claiming to describe the one true fundamental ontology. In his book on effective theories, James Wells describes this recent shift in the physics community away from thinking about the "Theory of Everything," toward thinking of theories in physics as effective. As Wells emphasizes, one of the important lessons to take away from this shift is the recognition of "the power that explicitly agreeing to the Effective Theory mindset can have in developing richer theories of nature and achieving a deeper understanding" (2012, *v*). Although the implications of this shift for the philosophy of physics have yet to be fully explored, the suggestion here, which I think is worth exploring, is the idea that a collection of effective theories provides us with a deeper understanding of nature than a single theory of everything does. The cautionary lesson is that it is important to recall the effective nature of quantum mechanics and view the theory within the broader context of other physical theories, rather than trying to draw ontological conclusions from it as if it were the final true theory of everything.

The third cautionary tale, noted at the outset, is that one should always clearly distinguish the mathematical representation from the thing being represented. Buried in a footnote to his discussion of the reality of the wavefunction, Peter Lewis notes,

strictly speaking the wave function itself is a mathematical representation of an (unnamed) physical state of affairs.... However, [...] it is traditional to use the term 'wave function' for both the mathematical representation and the physical object represented. (2016, 192)

This habit of the field is problematic for several related reasons: First, failing to distinguish the mathematical representation from the target can lead one to unreflectively reify the mathematics. Second, it can lead to confusion in that there may be properties of the representational vehicle that are not properties of the thing in the world. Third, and perhaps most importantly, there are ways to represent the state of a quantum system other than by means of the wavefunction, as we will discuss in detail. By clearly distinguishing the representation from the thing represented, the conceptual space is opened up to explore alternative representations, which may provide further insights.

The fourth cautionary tale is that mathematical representations do not wear their ontological interpretations on their sleeves. Many equations in physics today are so familiar to us that their physical meaning appears indistinguishable from their formal expression. That this identification is nontrivial, however, is evident to anyone who has studied the history of physics.[5] Moreover, there is often more than one formalism to choose from in representing a particular physical system. As Tim Maudlin rightly emphasizes,

> If one intends to try to read the physical ontology of a theory off of the mathematical structure used to present the theory, then one should give a great deal of consideration to alternative mathematical structures and the reasons for choosing one or another. (2013, 136)

It is precisely such a case, where we have more than one formalism for representing a given physical system and a literal reading-off of the ontology is ill advised, that I want to examine here.

As will be discussed in detail in the coming sections, there is a hydrodynamic formulation of quantum mechanics that does not make use of the ψ in representing the quantum state. This alternative quantum formalism undercuts a facile identification of the ψ with the quantum state, and highlights the importance of these four cautionary lessons when trying to read realist implications off the formalism of quantum theory. It is my contention that debates about realism in quantum physics have lost sight of the big-picture forest through an excessive focus on the ψ. My aim is not to replace the ψ conception of state with the hydrodynamic one, but rather to underscore the importance of recognizing the legitimacy of both. These two conceptions or models of the quantum state, though empirically equivalent, paint very different pictures of the unobservable world. However, neither one should be read as a literal depiction. In order to properly understand the implications of the hydrodynamic formulation of quantum mechanics, it is helpful to contextualize it within the class of hydrodynamic representations in physics more broadly. Hence, the next step on our path to regain sight of the proverbial forest is to take a scenic tour of hydrodynamic representations through a variety of length scales.

10.3 Hydrodynamic Representations at the Classical Scale

Classical hydrodynamics is the study of fluid flow, where the fluid is modeled as a continuum (that is, as a continuous distribution of mass) rather than as being composed of molecules or atoms. On this approach, the macroscopic properties of the

[5] See Bokulich (2015) for a discussion of this point, using the example of how Helmholtz and Maxwell, though agreeing on the same formal equation, disagreed about the right way to hook up the physical quantities with the elements of the equation.

fluid, such as density and pressure, are taken to be well defined down to infinitesimal volume elements. Although strictly false as a depiction of real fluids, for many domains the continuum assumption is an example of what I have elsewhere called a "credentialed fiction" (e.g., Bokulich 2016). More generally, hydrodynamics can be understood as an 'effective theory,' which captures the essential physics at certain length scales (viz., greater than inter-atomic), and it is the appropriate way to represent fluids in fields such as oceanography and meteorology (or what is more generally referred to as geophysical fluid dynamics).[6] The continuum model of hydrodynamics allows one to take advantage of the powerful mathematical framework of continuum mechanics, which provides insights that would be difficult (if not impossible) to achieve on an atomistic approach.

There are two different pictures or formulations of (continuum) hydrodynamics: the Eulerian formulation and the Lagrangian formulation.[7] On the Eulerian formulation, one records the evolution of the fluid at each point in space, $\boldsymbol{x}$, and time, t, in a fixed inertial coordinate system. It is a field description, where the fluid properties such as density, velocity, and pressure are thought of as fields $\rho_p(\boldsymbol{x},t)$, $\boldsymbol{v}_p(\boldsymbol{x},t)$, $p_p(\boldsymbol{x},t)$ defined at a particular point. Perhaps counterintuitively, the properties are ascribed to the points in space, and not to some substance moving through that space. An example of a Eulerian measuring device is a probe fixed in space. On the Eulerian picture, to say that a fluid flows, is to say that properties defined at various spatial locations are changing over time.

On the Lagrangian picture, by contrast, one considers the fluid as a dense set of fluid particles or parcels,[8] each of which carries its own properties and maintains its identity as it follows a classical trajectory. Each fluid parcel has its own unique particle label, $\boldsymbol{a} = (a,b,c)$, which can be taken to be the position of the parcel at some initial time, and the trajectories do not cross. The trajectories are expressed by the function $\boldsymbol{x}(\boldsymbol{a},t)$, which follows in time the position of the fluid parcel initially at $\boldsymbol{a}$. Conceptually, there are two ways to think about the fluid motion on this picture:

> We can think of a label space with coordinates (a,b,c) and a location space with coordinates (x,y,z). Then the fluid motion ... is a time-dependent mapping between these two spaces. Alternatively, we can think of the label variables (a,b,c) as curvilinear coordinates in location space. Then the fluid motion drags these curvilinear coordinates through location space.
>
> (Salmon 2014, 5)

The flow, described by $\boldsymbol{x}(\boldsymbol{a},t)$, can be considered as a continuous differentiable mapping of the three-dimensional Euclidean space onto itself $\boldsymbol{x}(\boldsymbol{a},t) = \Phi_t(\boldsymbol{a})$, which allows one to construct the Jacobian matrix of the mapping:

$$J_{ik}(t) \equiv \left(\frac{\partial x_i}{\partial a_k}\right)(t) = \alpha$$

[6] There are, of course, certain situations where one must take care in using continuum representations, such as when it comes to shocks and certain boundary conditions.

[7] Despite its name, the Lagrangian picture was first introduced by Euler in the context of acoustics (see, e.g., Darrigol 2005, 29 and references therein).

[8] These parcels or 'particles' are understood as pieces of the continuum.

The familiar continuity equation, which in the Eulerian picture is expressed as

$$\frac{d\rho}{dt}|_x + \nabla \cdot (\rho \boldsymbol{v}) = 0$$

can be understood as arising from the Lagrangian-picture requirement that fixed volumes in particle-label space always contain the same mass (Roulstone 2015, 369).

One can write down the Lagrangian for a perfect fluid as

$$\mathcal{L} = \int d^3\boldsymbol{a}\left[\frac{1}{2}\dot{x}^2 - E(\alpha, S(\boldsymbol{a})) - \Phi(\boldsymbol{x})\right],$$

where $E(\alpha, S(\boldsymbol{a}))$ is the internal energy, which is a function of the volume, α, and entropy, $S(\boldsymbol{a})$, and $\Phi(\boldsymbol{x})$ is the potential for external forces, with the integration being a measure over particle label space.

Hamilton's principle, which is a variational principle equivalent to Newton's second law, states that the action is stationary for arbitrary variations in $\delta\boldsymbol{x}(\boldsymbol{a}, t)$:

$$\delta \int \mathcal{L}dt = 0.$$

Because the particle labels $\boldsymbol{a}(\boldsymbol{x}, t)$ enter the Lagrangian only through the density $\partial(\boldsymbol{a})/\partial(\boldsymbol{x})$ and entropy $S(\boldsymbol{a})$, the potential energy terms in the Lagrangian are unaffected by particle-label variations $\delta\boldsymbol{a}(\boldsymbol{x}, t)$ that leave the density and entropy unchanged. The physical oceanographer Rick Salmon was the first to recognize in 1982 that this relabeling symmetry corresponds by Noether's theorem to a conservation law, namely, the most general statement of vorticity conservation (Salmon 1988, 238):

$$\frac{\partial}{\partial t}(\nabla_a \times \mathbf{A}) = 0$$

where ∇_a is the gradient operator in particle-label space and

$$\mathbf{A} = u\nabla_a x + v\nabla_a y + w\nabla_a z$$

is the projection of the velocity components on the basis of the gradient operator in particle-label space.

From this general statement of vorticity conservation, one can derive the well-known Ertel's theorem of potential vorticity conservation,

$$\frac{\partial}{\partial t}[(\nabla_a \times \mathbf{A}) \cdot \nabla_a \theta] = 0,$$

(where θ is any conserved quantity on the flow), Helmholtz's various theorems, and Kelvin's circulation theorem

$$\frac{d}{dt}\oint \boldsymbol{v} \cdot d\boldsymbol{x} = 0.$$

This vorticity conservation law, which results from the particle relabeling symmetry of the Lagrangian picture, is scientifically one of the most important results in geophysical fluid dynamics. The meteorologist Peter Névir, for example, writes that the "conservation of potential vorticity is a cornerstone in dynamic meteorology" (2004, 486). Similarly, the oceanographer Peter Müller writes, "most aspects of large-scale oceanography can be understood in terms of potential vorticity and its evolution, as is stressed in textbooks" (1995, 68). Intuitively potential vorticity conservation is the idea that a rotating column of fluid, despite changing shape, will conserve volume and angular momentum.

Although these conservation of vorticity and circulation theorems were known long before the particle-relabeling symmetry was recognized, their derivation from the symmetry property provides a unification and simple explanation that is missing on the standard approaches. Moreover, as Salmon points out, "the symmetry approach shows that vorticity conservation is a consequence of the continuum approximation. It has no analogue in particle mechanics, where the particle labels cannot be varied continuously" (Salmon 1988, 241).

The rich representational machinery of classical hydrodynamics is not just useful for understanding ordinary fluids, but can also be used to describe granular materials, which are conglomerations of discrete macroscopic particles, such as piles of sand or grain in a silo. Although granular materials sometimes behave as a solid, they can also behave as a fluid and flow (even though no water is involved) as is familiar in an hourglass. Granular hydrodynamics is the use of classical hydrodynamics to model the flow of granular materials.[9] Granular hydrodynamics is particularly useful for helping geoscientists understand earthquake-induced landslides, for example.

Similarly, classical hydrodynamics representations are extremely fruitful at the cosmological scale and are regularly employed in cosmology and astrophysics. The hydrodynamic equations—in both the Eulerian and Lagrangian formulations—are productively used to understand the formations of stars, galaxies, and large-scale structures in the universe.[10] As with ordinary hydrodynamics, each of these pictures is useful for bringing out certain sorts of insights. As Shy Genel and colleagues write:

"The equations of hydrodynamics are usually solved in astrophysical applications using either... particle-based Lagrangian-like schemes such as smoothed particle hydrodynamics (SPH) or mesh-based Eulerian-like schemes such as adaptive mesh refinement (AMR). There are various advantages and shortcoming of each approach" (Genel et al. 2013, 1426)

As an example, Genel and colleagues note that one of the most debated questions in galaxy formation is how galaxies get their gas, for which the ability to track the mass flow in a Lagrangian manner is crucial. They further note, however, that using the Lagrangian formulation comes at a price, and hence they explore how the strengths of each representation can be exploited, collectively giving deeper insights into galaxy formation than would be obtained by using one representation alone.

In all of these applications of classical hydrodynamics, from ordinary geophysical fluids in oceanography and meteorology to hydrodynamic representations in astrophysics and cosmology, the interesting question is not whether the continuum representation is a literally true depiction of the world. Rather, the relevant question

[9] For a classic introduction, see Jaeger, Nagel, and Behringer (1996), and for a more recent review, see Trujillo, Sigalotti, and Klapp (2013).

[10] For textbook reviews of hydrodynamic representations at the cosmological scale see, for example, Murkhanov's (2005) *Physical Foundations of Cosmology* or Regev, Umurhan, and Yecko's (2016) *Modern Fluid Dynamics for Physics and Astrophysics*.

is what true physical insights and correct inferences does this representation allow us to draw about the world? I urge that we should keep this lesson in mind as we turn to hydrodynamic representations at the quantum scale.

10.4 Hydrodynamic Representations at the Quantum Scale

10.4.1 *Eulerian quantum hydrodynamics*

Hydrodynamic representations of quantum mechanics are almost as old as quantum mechanics itself (though as we will see, a fuller elaboration of the formal analogy was only carried out quite recently). Within months of Erwin Schrödinger's seminal papers introducing the wave equation, Erwin Madelung published a paper titled "Quantum Theory in Hydrodynamical Form" (*Quantentheorie in hydrodynamischer form*). Madelung describes the purpose of this paper to "show that far-reaching analogies with hydrodynamics exist" (Madelung 1927, 322). He begins by writing the wavefunction in polar form[11]

$$\psi = Re^{iS/\hbar}, \tag{10.1}$$

which when substituted into the single-particle Schrödinger equation

$$-\frac{\hbar^2}{2m}\nabla^2\psi + V\psi = i\hbar\frac{\partial\psi}{\partial t} \tag{10.2}$$

and separated into real and imaginary parts yields the following two real, coupled partial differential equations:[12]

$$\frac{\partial S}{\partial t} + \frac{(\nabla S)^2}{2m} - \frac{\hbar^2}{2m}\frac{\nabla^2 R}{R} + V = 0 \tag{10.3}$$

and

$$\frac{\partial R^2}{\partial t} + \nabla \cdot \left(\frac{R^2 \nabla S}{m}\right) = 0. \tag{10.4}$$

The heart of this hydrodynamic analogy is to identify $\rho = R^2$ as the fluid density and $v = \nabla S/m$ as the velocity field of this fluid; then equation (10.4) becomes

$$\nabla \cdot (\rho v) + \frac{\partial \rho}{\partial t} = 0, \tag{10.5}$$

which is a continuity equation expressing the conservation of mass of the fluid. The other part of equation (10.4), after applying the operator ∇ and using $v = \nabla S/m$ becomes[13]

$$\frac{\partial \boldsymbol{v}}{\partial t} + (\boldsymbol{v} \cdot \nabla)\boldsymbol{v} = -(1/m)\nabla(V + Q), \tag{10.6}$$

[11] I am using a slightly different notation here from Madelung's original in order to be consistent with the notation used later in connection with Holland's extensions of the hydrodynamic analogy.

[12] For further details on the derivation in this same notation, see, for example, Cushing (1994, ch. 8, Appendix 1).

[13] Again, for the explicit intervening steps, see Cushing (1994, ch. 8, Appendix 1) or Holland (1993, ch. 3).

where

$$Q \equiv -\left(\frac{\hbar^2}{2m}\frac{\nabla^2 R}{R}\right). \tag{10.7}$$

Equation (10.6) is analogous in form to the classical Euler equation:

$$\frac{\partial \boldsymbol{v}}{\partial t} + (\boldsymbol{v}\cdot\nabla)\boldsymbol{v} = -\frac{1}{\rho}\nabla p, \tag{10.8}$$

where p is the pressure of the fluid, and ρ, recall, is the density.

After deriving the two key equations, Madelung concludes (with our notation in brackets),

> [This equation] has the character of a hydrodynamical equation of continuity when one regards α^2 [i.e., ρ] as a density and ϕ [i.e., S/m] as the velocity potential of a flow $u = \text{grad}\phi$ [i.e., $\boldsymbol{v}$].... [The other equation] also corresponds precisely to a hydrodynamical one, namely that of an irrotational flow moving under the action of conservative forces.... We therefore see that [the equation from Schrödinger] is completely explainable in terms of hydrodynamics, and that a peculiarity appears only in one term, which represents the internal mechanism of the continuum. (Madelung 1927, 323).

The 'peculiarity' Madelung notes in one term is what would later, in the context of David Bohm's (1952) interpretation of quantum mechanics, be called the 'quantum potential.'

Madelung's hydrodynamical model of quantum theory was further developed by Takehiko Takabayasi (1952), who elaborated this 'peculiar' term as an expression of the internal stress in the fluid, which he represents as a stress tensor, σ_{ij}. If we rewrite the classical Euler equation as[14]

$$\frac{\partial v_i}{\partial t} + (\boldsymbol{v}\cdot\nabla)v_i = -\frac{1}{m}\partial_i V - \frac{1}{m\rho}\partial_j(p\delta_{ij}) \tag{10.9}$$

we can see the formal analogy with the quantum version of the Euler equation, which (from equation (10.6)) can be written

$$\frac{\partial v_i}{\partial t} + (\boldsymbol{v}\cdot\nabla)v_i = -\frac{1}{m}\partial_i V - \frac{1}{m\rho}\partial_j(p\sigma_{ij}), \tag{10.10}$$

where the stress tensor is

$$\sigma_{ij} = -\left(\frac{\hbar^2\rho}{4m}\right)\partial_{ij}\log\rho. \tag{10.11}$$

Takabayasi concludes: "This showes [*sic*] that the motion can just be pictured as that of [a] fluid with (symmetric) stress tensor σ_{ik} [i.e, our σ_{ij}] and that the influence of quantum mechanics is regarded as introducing this 'quantum-theoretical' stress" (Takabayasi 1952, 180). As can be seen above (in equations (10.9) and (10.10)), the classical Euler and 'quantum Euler' equations are formally identical, apart from the classical pressure tensor, $p\delta_{ij}$, being replaced with the quantum-mechanical stress tensor (equation (10.11)).[15]

[14] Here I am following the notation of Holland (1993, 121) for consistency.

[15] As Bohm and Vigier (1954, 209) point out, the quantum stress is unlike the classical stress in that it depends on derivatives of the fluid density.

To complete the hydrodynamic analogy, the appropriate boundary and subsidiary conditions need to be imposed on the density and velocity fields ρ and $\boldsymbol{v}$ (Holland 1993, 121). In ordinary quantum mechanics one imposes the condition that the wavefunction is single-valued, which means that at every instant, each point of space can be assigned a unique value of the function ψ. In the context of the hydrodynamic representation, the velocity field is irrotational (the fluid parcels are not rotating), except at the nodes where it is undefined, and the single-valuedness condition means

$$\oint \boldsymbol{v} \cdot d\boldsymbol{x} = nh/m, \tag{10.12}$$

which is interpreted as the fluid possessing 'quantized vortices.'[16]

Takabayasi takes pains to distinguish this hydrodynamical model of quantum mechanics from David Bohm's (1952) interpretation. While on the hydrodynamic analogy the quantum potential term is an internal stress in the fluid, on Bohm's interpretation it is understood as an external force, $-\nabla Q$. Further differentiating his approach, Takabayasi writes: "Instead of doing this Bohm has reintroduced the quantity ψ not as a mere mathematical tool but as an objectively real field" (Takabayasi 1952, 155). Takabayasi further emphasizes the distinction between a *formulation* and an *interpretation*. He writes:

> The hydrodynamic analogy... though sometimes useful for the analysis of the Schrödinger equation and rather appropriate to make one visualize the presence of internal force, does not necessarily prove the *reality* of the hydrodynamic picture. (Takabayasi 1952, 150)

While Madelung seems somewhat agnostic about how literally it should be interpreted, describing the use of hydrodynamics as an 'analogy,' Takabayasi is quite explicit in describing the use of hydrodynamics as just a 'model' or a 'formulation' of quantum mechanics, analogous to Feynman's path integral formulation. By contrast, Bohm and Vigier (1954), for example, go further in describing it as a theory of a real fluid. They write: "Since the Madelung fluid is being assumed to be some kind of physically real fluid, it is therefore quite natural to suppose that it too undergoes more or less random fluctuations in its motion" (Bohm and Vigier 1954, 209). As will be discussed later, these same questions about whether to understand the hydrodynamic analogy in quantum mechanics as an interpretation or a formulation persist to this day.

The hydrodynamic analogy in quantum theory, as developed by Madelung, Takabayasi, and others in the twentieth century, has been exclusively in the Eulerian picture. As discussed in the previous section, hydrodynamics (like quantum mechanics) admits of two different formulations: the Eulerian picture and the Lagrangian picture. Although Takabayasi, in a footnote to his 1953 paper, briefly alludes to the possibility of a self-contained Lagrangian formulation, surprisingly, it would be another fifty years before such a full Lagrangian formulation of quantum hydrodynamics would be developed.

[16] This is reminiscent of the quantum condition of the old quantum theory, though of course is given a different physical interpretation in that context (Holland 1993, 72).

10.4.2 *Lagrangian quantum hydrodynamics*

The systematic development of a Lagrangian quantum hydrodynamics, as a full alternative picture or formulation of quantum mechanics, was only carried out recently in a series of papers by Peter Holland beginning in 2005.[17] Recall that the Lagrangian picture treats the fluid as a continuum of fluid particles (or parcels), each of which maintains its identity throughout the evolution of the fluid. Unlike Madelung's Eulerian formulation of quantum hydrodynamics, the Lagrangian formulation introduces new variables (the particle labels a) into the quantum formalism. One begins by introducing a vector (e.g., in three-dimensional Euclidean space) label a_i for each particle, which can be taken to be the position of that fluid particle at $t = 0$. As Holland notes,

> This step is not merely of mathematical significance for the labeling allows us to conceive of fluid functions such as density and pressure in terms of notions not available in the Eulerian picture, namely, interparticle interactions described by the deformation matrix $\partial q_i/\partial a_l$.
>
> (Holland 2017, 340)

The full congruence of all the particle trajectories is described by the displacement function $q_i(a,t)$. This displacement function provides a new, alternative conception of the quantum state, in place of the wavefunction ψ. In order to have a complete Lagrangian hydrodynamic representation of QM, one needs an independent way of calculating the congruence of trajectories. Following Holland (2017), if one substitutes $x_i = q_i(a,t)$ into the quantum analog of Euler's force law

$$\frac{\partial v_i}{\partial t} + v_j \frac{\partial v_i}{\partial x_j} = -\frac{1}{m}\frac{\partial}{\partial x_i}(V + V_Q), \tag{10.13}$$

then one obtains

$$m\frac{\partial^2 q_i(a)}{\partial t^2} = -\frac{\partial}{\partial q_i}\big(V(x) + V_Q(x)\big)\,\big|_{x = q(a,t)} \tag{10.14}$$

where the derivatives with respect to q_i are defined as

$$\frac{\partial}{\partial q_i} = J^{-1} J_{ij} \frac{\partial}{\partial a_j}$$

and J_{ij} is the adjoint of the deformation matrix $\partial q_i/\partial a_j$. Holland concludes that equation (10.14) should be understood as Schrödinger's equation in the form of Newton's second law. To have a flow representative of Schrödinger evolution, one must restrict the admissible solutions to what Holland calls a "quasi-potential" flow.[18] He concludes:

> The quantum state is now represented by the 'displacement amplitude' $q_i(a,t)$ encoding the history of an infinite ensemble of particles whose interaction is described by the derivatives of

[17] There were also efforts toward a Lagrangian quantum hydrodynamics developed as an *approximation scheme* (not full formulation) in the quantum chemistry community, most notably by Robert Wyatt (2005).

[18] The restriction to quasi-potential flow means the initial velocity field is of the form $\frac{\partial q(a)}{\partial t} = \frac{1}{m}\frac{\partial S(a)}{\partial a}$, but the flow is not irrotational everywhere since the quantum phase obeys the quantization condition. Vortices occur only in the nodal regions, where the density vanishes.

q_i with respect to a_i .... With the appropriate initial conditions the vector $q_i(a,t)$ determines the motion completely, without reference to $\psi(x)$. (Holland 2017, 342)

The Lagrangian model of quantum hydrodynamics provides a complete formulation of quantum mechanics and an alternative representation of the quantum state. It is thus a formulation of *quantum mechanics without wavefunctions.*[19] Furthermore, beginning with the congruence of trajectories it is possible to deduce the time-dependence of the wavefunction.[20] In other words, the congruence of trajectories can be understood as the fundamental quantum state entity through which to understand the time evolution of the system.

Although the hydrodynamic formulation of quantum mechanics was initially developed for a single-body, spin-0 system (Holland 2005a,b), Holland shows how it can be readily extended to an n-body system by allowing the indices $i, j, \ldots$ to range over $3n$ values and the congruence of trajectories in configuration space can be mapped into ensembles of interlacing trajectories in 3-space (Holland, 2017, 343). Holland (2006) further extends the Lagrangian hydrodynamic model to spin-1/2 systems by allowing the fluid particles to have an internal rotational freedom. The standard angular momentum approach, where the rotational degrees of freedom appear as discrete indices in the wavefunction, does not provide sufficient information (since the fluid quantities are 'averaged' over these indices), so Holland instead uses the angular *coordinate* representation, where the spin degrees of freedom are represented by the continuous parameter Euler angles, α. To construct a fully general Lagrangian hydrodynamic model, Holland generalizes to an arbitrary-dimensional Riemannian manifold, $\mathcal{M}$, with a static metric $g_{\mu\nu}(x)$, $\mu, \nu, \ldots = 1, \ldots, N$, where the history of the fluid is encoded in the positions $\xi(\xi_0, t)$ of the distinct fluid elements at time t. The result is a simpler and physically clearer hydrodynamic model embracing both fermions and bosons, that is a straightforward generalization of the spin-0 theory (Holland 2006, 371).[21]

Holland similarly uses this method to construct a full (Eulerian and Lagrangian) hydrodynamic 'fluid' model of the electromagnetic field (a feat which famously eluded the nineteenth-century ether theorists, though of course here without the literal

[19] Indeed this is the title of the paper Schiff and Poirier (2012) where they further emphasize that what we are here calling Lagrangian quantum hydrodynamics is a "standalone reformulation of quantum mechanics, that neither relies on the TDSE [time-dependent Schrödinger equation] nor makes any mention of any external constructs such as ψ" (p. 031102–1).

[20] Although the Eulerian variables ρ, v are not canonical, Holland also shows how one can set up a canonical formulation of Eulerian QH by introducing potentials for the velocity, which result in a Clebsch-like representation. One can define new dependent variables Q, P_i which are canonically conjugate (defining three pairs position and momentum variables for each space point, whose temporal development is governed by Hamilton's equations) which become the new descriptors of the state. Hamilton's equations plus initial conditions imply the Schrödinger equation via the QH equations (continuity and Euler-type force law). Thus, the trajectory formulation of QM can be obtained by a canonical transformation of wave mechanics when the latter is formulated in terms of the hydrodynamic phase space variables or gauge potentials $Q_i(x), P_i(x)$. See Holland (2017, §4) for details.

[21] Holland also notes that this provides an alternate method of quantization: from the single particle case one can pass to a continuum of particles, introduce the interparticle interaction, generalize to Riemann space with external scalar and vector potentials, and finally pass to the Eulerian description (Holland 2011, 85).

physical interpretation). More specifically he shows how "the relativistic spin-1 field obeying the source-free Maxwell equations can be computed from the Lagrangian trajectories" (Holland 2005b, 3660). As in the spin-1/2 case, one endows the fluid particles with a rotational degree of freedom and makes use of the generalized Riemannian manifold construction, from which electromagnetic theory can be extracted as a special case.[22] Thus he provides a

> continuum mechanics model from which we may deduce Maxwell's equations as the Eulerian counterpart to the equation of motion of the Lagrangian trajectories, and in particular with an algorithm to compute the electromagnetic field from the latter. (Holland 2005b, 3671)

He further argues that this method of construction can be extended to produce a Lagrangian trajectory formulation of any generic field theory (Holland 2012, 2013). This work thus provides researchers with new conceptual and computational resources for exploring field theories.

10.5 Exploring Lagrangian Quantum Hydrodynamics

With the details of this hydrodynamic formulation of quantum mechanics in hand, let us now turn to exploring some of the broader questions about this formalism. In particular, we first distinguish the LQH formulation from the de Broglie–Bohm interpretation, which similarly uses trajectories. Second, we examine the novel particle-relabeling symmetry and conservation laws that arise in the LQH formulation. And third, we explore whether there is any sense in which the LQH formulation of quantum mechanics should be thought of as more fundamental than the ψ representation.

10.5.1 LQH is not the de Broglie–Bohm interpretation

It is important to clearly distinguish between Lagrangian quantum hydrodynamics (LQH) and the de Broglie–Bohm (dBB) interpretation of quantum mechanics. The dBB interpretation differs from LQH in five respects. First, dBB attributes ontological status to the wavefunction, while LQH is a formulation of quantum mechanics without wavefunctions. Second, dBB depends on the wavefunction to obtain the trajectories; that is, one first solves the time-dependent Schrödinger equation to obtain the wavefunction, and then uses this to derive the quantum trajectories. On LQH, by contrast, one obtains the trajectories directly by solving the second-order Newtonian-type equation (equation (10.14)). Third, unlike LQH, dBB postulates the existence of a corpuscle of mass m following one of the trajectories. As one researcher notes, Bohm was interested in "corpuscle propagation using information gleaned from precomputed wave functions" (Wyatt 2005, 2), while LQH dispenses with both the corpuscle and the wavefunction, and instead determines the fluid trajectories directly.

The similarity to note between LQH and dBB is that one of the paths of the fluid parcels coincides with that of the dBB corpuscle. However, as Holland notes, "The corpuscle is ... to be distinguished from a fluid particle both in its mass and in its

[22] This method makes use of the fact that Maxwell's 'curl' equations are a form of Schrödinger's equation (with a constraint from the divergence equations) and that the corresponding wave equation has a continuous representation in the Euler angles (Holland personal communication).

dynamical behavior" (Holland 2005a, 525). This points to a fourth difference between LQH and dBB: On dBB, the motion of the corpuscle is determined by the external quantum potential, while on the LQH approach the trajectories are determined by the interactions between the fluid parcels themselves. This arises because the LQH introduces new variables (the particle labels) into the quantum formalism, while the dBB interpretation does not. This means that the LQH fluid parcels do not suffer from the dBB problem of no back-reaction (i.e., the asymmetry of the wavefunction, via the quantum potential, acting on the particle, but the particle not back-reacting on the wavefunction).[23]

Finally, the fifth and perhaps most salient difference between dBB and LQH is that the former is an *interpretation* of quantum mechanics, and the latter is best understood as a *formulation* of quantum mechanics. While the dBB interpretation seeks to solve the measurement problem by underpinning the indeterministic and statistical story told by standard quantum mechanics with a completely deterministic, causal story of individual quantum systems and processes, the LQH formulation, on its own, has no such interpretive ambitions.[24]

10.5.2 Symmetries in quantum hydrodynamics

The Lagrangian formulation of quantum hydrodynamics provides new insights into symmetries and conservation laws in quantum mechanics. Most strikingly, it allows for the discovery of a novel quantum symmetry that is obscured on the standard formulation. Recall that the Lagrangian formulation of quantum hydrodynamics introduces new variables into the quantum formalism, namely the particle labels, a_i. Moreover, no particular choice of particle label is preferred. This freedom corresponds to a new quantum symmetry or gauge freedom, namely the infinite-parameter particle relabeling group, with respect to which the Eulerian variables of position, density, and velocity, are invariant. As Holland explains,

> The origin of the relabelling symmetry is that the deformation coefficients (derivatives of the current position with respect to the label) appear in the field equations only through the Jacobian .... [This is] a characteristic feature of fluid mechanics that is not displayed in other continuum theories. (Holland 2013, 57)

By Emmy Noether's (first) theorem, we know that this relabeling symmetry implies a conservation law. Indeed this symmetry is the quantum analog of the classical symmetry in ordinary hydrodynamics, examined earlier, which is responsible for the vorticity and circulation conservation theorems of Helmholtz, Kelvin, and Ertel. In the Lagrangian picture one can, for example, derive the *quantum* version of Kelvin's conservation of circulation theorem (Holland 2017, 342):

$$\frac{\partial}{\partial t} \oint \dot{q}_l dq_i = 0, \tag{10.15}$$

[23] See, for example, Holland 1993, §3.3.2 for a discussion of this feature of de Broglie–Bohm.

[24] Not surprisingly, some have tried to turn the LQH approach into an interpretation of quantum mechanics, such as in what is known as the "many interacting worlds" (MIW) interpretation (Hall, Deckert, and Wiseman 2014). I will return to briefly discuss this interpretation in Section 10.6.

which states that the closed loop of particles remains closed during the flow due to the continuity of the function $q_i(a)$. There is also a conserved density and current associated with the relabeling symmetry (Holland 2013, 71):

$$P(a,t) = -m\rho_0 \frac{\partial q_i}{\partial t}\frac{\partial q_i}{\partial a_j}\xi_j, \quad J_i(a,t) = \rho_0 \xi_i \left(\frac{1}{2} m \frac{\partial q_i}{\partial t}\frac{\partial q_i}{\partial t} - V(q(a,t)) - V_Q \right),$$

which also have analogs in classical hydrodynamics. Although the importance of these conservation theorems is indisputable in the classical context, their significance in non-relativistic quantum mechanics is less clear. Moving to the relativistic context, however, Poirier (2017) has argued that the conservation law associated with the particle relabeling symmetry is what allows one to define global simultaneity manifolds, restoring absolute simultaneity, and allowing a foliation into a 3+1 space and time.

The Lagrangian hydrodynamic formulation of electromagnetism, mentioned earlier, also provides an alternative perspective on the Lorentz covariance of the theory. The fluid paths in the Lagrangian hydrodynamic formulation do not form a Lorentz covariant structure. In that respect they are similar to the electric and magnetic field lines, or the energy flow lines derived from Poynting's vector, which, though not a covariant structure themselves, are derived from the Lorentz covariant electromagnetic theory. Holland argues that the fact that the

> non-covariant hydrodynamic ones [trajectories] may be employed as a basis from which to derive the relativistic theory... suggests that [the Eulerian field variables] $\rho(x)$ and $S(x)$ (and hence $E(x)$ and $B(x)$) may be regarded as 'collective coordinates'—functions that describe the bulk properties of the system without depending on the complex details of the particulate substructure. Features peculiar to the Eulerian picture, such as Lorentz covariance, may therefore be viewed as collective rather than fundamental properties. (Holland 2005b, 3678)

This question of whether the Lagrangian formulation should be considered as, in some sense, more fundamental than the Eulerian formulation is an interesting issue that comes up in the context of classical hydrodynamics as well, and will be briefly examined next.

10.5.3 Is the Lagrangian picture more fundamental?

The Lagrangian and Eulerian formulations of hydrodynamics are taken to be—in some not yet philosophically precise sense—equivalent.[25] It is important to note, however, that they are not merely coordinate transformations of each other, but rather are two different ways of specifying the fluid flow (i.e., one can use the Eulerian and Lagrangian formulations in any frame of reference and using any coordinate system). A recent textbook describes the relation between the Eulerian and Lagrangian formulations as follows: "[The] two approaches ultimately describe physical objects that are equivalent to one another, as well-defined mathematical manipulations transform one perspective into the other. In certain practical situations, however, one approach may be superior to the other... both in terms of the mathematical formulation of a given problem and its interpretation" (Regev, Umurhan, and Yecko 2016, 2)

[25] There is actually a substantive philosophical debate about what it means for two theories, models, or formalisms to be 'theoretically equivalent' (see, for example, Nguyen 2017 and references therein).

In other words, although they are two ways of describing the same physical system, one formulation may be superior to another in a given situation in terms of the calculational tractability or the physical insight it provides.

Despite the empirical equivalence and inter-transformability of the Lagrangian and Eulerian descriptions, there are two (related) arguments one finds in the classical fluid dynamics literature for why the Lagrangian formulation should be thought of as the more fundamental description.[26] The first argument comes from the perspective of geometric mechanics and reduction theory. Reduction is a powerful tool in the study of mechanical systems, whereby one can exploit conserved quantities and symmetry groups to reduce the dimensions of a phase space.[27] We can think of a Hamiltonian system as consisting of a phase space and two geometrical objects: the Poisson bracket $\{,\}$ and the Hamiltonian, $\mathcal{H}$. When the Poisson bracket is *singular* there exist a set of what are called Casimir functions $\{C\}$ for which $\{C, A\} = 0$ for every function A, including, $\mathcal{H}$, meaning C is conserved by the dynamics. As Salmon points out,

> Singular Poisson brackets typically arise from a transformation from canonical coordinates (in which the bracket is nonsingular) to a reduced set of (fewer) coordinates in which the dynamics comprises a fewer number of equations but is nevertheless closed. Then . . . the Casimirs are the conserved quantities corresponding to the symmetries that permit the reduction.
> (Salmon 2014, 339)

In fluid mechanics, the reduction is precisely the transformation from the (canonical) Lagrangian variables to the non-canonical Eulerian variables. There is thus a sense in which the Lagrangian formulation is a more complete description of the fluid. As Salmon notes,

> If we know [the Eulerian variables of velocity, density, and entropy at fixed locations], then we know everything we need to compute the Hamiltonian . . . but our knowledge of the fluid motion is incomplete; we cannot say which fluid particle went where. (Salmon 2014, 337)

The symmetry that permits the reduction from Lagrangian variables (the velocities and locations of marked fluid particles) to the Eulerian variables—and thus what is responsible for there being a closed Eulerian formulation of fluid mechanics—is the relabeling symmetry.

This last point brings us to the second, related, argument for why the Lagrangian formulation can be thought of as more fundamental: The general law of vorticity conservation is a consequence of the relabeling symmetry, which can only be articulated on the Lagrangian formulation. As one researcher remarks, despite the central importance of the Eulerian description,

> the general vorticity law cannot be formulated without referring to the positions of marked fluid particles; an example proving that the continuum model of the fluid particles and the Lagrangian description are in a sense more fundamental than the Eulerian description.
> (Sieniutycz 1994, 55)

[26] As we will see, the appropriate notion of 'fundamentality' still needs to be elaborated.

[27] The inverse of reduction is known as reconstruction, whereby one can get back the dynamics of the full Hamiltonian system from the reduced system; holonomies, related to phenomena such as Berry's phase and A–B effect, can be viewed as an instance of reconstruction (Marsden and Ratiu 1999, 256).

The physical oceanographer Peter Müller likewise notes,

> The [usual analysis of vorticity conservation] is unsatisfactory since it neither reveals the underlying cause for the material conservation of potential vorticity nor offers any explicit expressions for homentropic and homogeneous fluids. These issues become resolved in a Lagrangian description of the fluid motion. (Müller 1995, 72)

And Salmon similarly writes,

> [T]he general vorticity law cannot be stated without referring to the locations of marked fluid particles. This is but one of several important examples in which the greatest simplicity and generality are achieved only by considering the complete set of Lagrangian fluid variables. These examples suggest that the primitive picture of a fluid as a continuous distribution of massive particles is in some sense the more fundamental, and that the simplicity of the conventional Eulerian description has been purchased at a definite price. (Salmon 1988, 226)

Interestingly the relevant notion of 'fundamentality' being used here is not that of being less idealized or closer to the truth. Both the Eulerian and Lagrangian formulations are continuum approximations for fluids that are ultimately understood to be composed of discrete molecules. Yet, as Salmon notes, "the particle-relabeling symmetry property is unique to fluid mechanics. It has no analogue in discrete-particle mechanics, where the particle labels cannot be varied continuously" (Salmon 2014, 304).

Returning to the quantum context, one might similarly argue that the Lagrangian formulation of quantum hydrodynamics is more fundamental, where again 'fundamental' need not be construed as 'closer to the truth.' Recall that on the Lagrangian quantum hydrodynamics approach, the quantum state is represented by the full congruence of fluid particle (parcel) trajectories described by the displacement function $q_i(a,t)$. As discussed earlier, these trajectories can be calculated directly (by means of equation (10.14)) and provide a self-contained formulation of quantum mechanics, without reference to the wavefunction. Moreover, as Holland shows, if one wishes, one can then derive the time-dependent wavefunction from these independently calculated trajectories. Hence, "one may make the displacement function of the collective the basis of the quantum description with the wavefunction being regarded as a derived quantity" (Holland 2017, 334). In other words, rather than viewing the wavefunction as the fundamental entity and the trajectories as a derived or interpretive overlay, on the Lagrangian approach it is natural to view the trajectories as the more fundamental description. In analogy with the classical hydrodynamics case, on the Eulerian formulation of quantum hydrodynamics (e.g., of Madelung), one may view the Eulerian functions $\rho(x)$ and $S(x)$ as the "'collective coordinates'—functions that describe the bulk properties of the system without depending on the complex details of the particulate substructure" (Holland 2006, 384).

Given the equally idealized and effective nature of both the Eulerian and Lagrangian formulations (of either classical hydrodynamics or quantum mechanics), their physical equivalence, and their inter-transformability, I am not sure these arguments about fundamentality are likely to convince anyone not already sympathetic to the view. However, what I think these arguments do succeed in doing is shifting the burden of proof: there are no longer grounds for automatically assuming that the Eulerian field picture and the wavefunction representation of the quantum state (or relatedly the $\rho(x)$ and $S(x)$ representation of the quantum state) are privileged.[28]

10.6 Beyond the Wavefunction Hegemony

With this alternative formulation of quantum mechanics in hand, let us return to drawing out some of the preliminary implications of LQH for various realist projects in quantum theory. While a full exploration of these implications is not possible here, some promising avenues for future work are indicated. Most immediately, the LQH formulation, with its displacement representation of the quantum state, $q_i(a,t)$ challenges the hegemony of the wavefunction. Unlike the de Broglie–Bohm interpretation of quantum mechanics, which has a dual ontology of trajectories and wavefunctions (or pilot waves), the LQH dispenses with the wavefunction entirely; the quantum evolution is borne by the congruence of trajectories alone. Hence it shows how it is possible to do 'quantum mechanics without wavefunctions.' The two pictures (ψ and q_i) are, of course, inter-transformable, and share the common mathematical data of the initial conditions. That is, the initial density and velocity of the congruence is equal to the initial squared magnitude and phase gradient of ψ_0. Not assuming one picture to be more fundamental than the other, one could say that certain mathematical functions (initial data) may be ascribed different interpretations.[29] Although the trajectory formulation does not disprove wavefunction realism, it does shift the burden of proof, raising substantive questions about what ontological conclusions can legitimately be read from the formalism, given the non-necessity of the wavefunction concept of state.

Not only does LQH show that one can formulate quantum mechanics without wavefunctions, but it also provides a different perspective on experiments that claim to measure the wavefunction of a single system as an extended object. In a well-known paper, Yakir Aharonov and colleagues argue that so called 'protective measurements,' which use a suitable adiabatic interaction to measure the expectation values of operators without appreciably disturbing a quantum state, provides evidence that the wavefunction is ontologically real (Aharonov, Anandan, and Vaidman 1993). Holland (2017) shows, however, that what is directly measured in such cases are the hydrodynamic variables (ρ and v), and not the wavefunction itself. Since these (Eulerian) hydrodynamic variables can be used to construct either the $\psi(x,t)$ or $q_i(a,t)$, the protective measurement scheme cannot be used to privilege the wavefunction conception of the quantum state as being more ontologically real than the displacement conception.

As noted in Section 10.1, a consideration of alternative formalisms is not only helpful in avoiding an elision between the mathematical formalism and what it represents, but can also challenge the presumption that some feature of the mathematical

[28] That, of course, doesn't mean that one formulation might not be pragmatically preferred in a given situation, just like one may find Feynman's path integral formulation of quantum mechanics more useful in some situations.

[29] I owe this way of expressing it to Holland (personal communication).

representation is also necessarily a feature of the world. A striking example of this is the claim by wavefunction realists that because the wavefunction ψ lives in a $3n$-dimensional configuration space (where n is the number of particles in the universe), we too must live in this high-dimensional abstract space, and the three-dimensional physical space of our experience is in fact an illusion (e.g., Albert 1996, 277). This view, sometimes referred to as 'configuration-space realism,' arises from a literalist approach to scientific realism, as others have noted (e.g., Dorato and Laudisa 2015, 120).

Against configuration-space realism, the LQH formalism shows that it is possible to represent the quantum state of a many-body system as a set of states in ordinary, three-dimensional physical space. As is well known, the many body wavefunction for a system of n particles with masses m_r with $r = 1, \ldots, n$ is $\psi(x_1, \ldots, x_n)$, which is defined in a $3n$-dimensional configuration space. In the equivalent LQH representation, such an n-body quantum state is represented as a single-valued congruence of curves $q_{ri}(a_1, \ldots, a_n)$ in the $3n$-dimensional configuration space, where the indices r, i collectively range over the $3n$ values. The $a_1, \ldots, a_n$, recall, uniquely label the initial positions $q_{roi} = a_{ri}$. As Holland emphasizes in a recent paper,

From the grouping of the indices, we see immediately that in this picture each configuration space trajectory is composed of n trajectories in three-dimensional physical space, the rth trajectory being given by the position vector q_{ri}. The whole nondenumerable configuration space congruence is therefore composed of n families of trajectories in 3-space. *The n-body quantum state may be represented as a collection of n states in 3-space.*

(Holland 2018, 269–3; emphasis original)

In other words, on the LQH formulation, the many-body quantum state can be represented in ordinary physical 3-space, consistent with our experience—no grand illusions required.

In the generalization of the previously discussed one-body case to a many-body system, the Schrödinger equation can be cast as a set of n Newtonian-type equations describing the coupled evolution of the set of n displacement 3-vectors:

$$m_r \frac{\partial^2 q_{ri}(a_i, \ldots, a_n)}{\partial t^2} = -\frac{\partial}{\partial q_{ri}} \left[V(x_i, \ldots, x_n) + V_Q(x_i, \ldots, x_n) \right] \big|_{x_r = q_r(a_1, \ldots, a_n, t)}$$

As Holland explains, from this equation we see that the trajectory q_{ri} is generally coupled with all the other current locations, such that if the rth family of trajectories is acted upon by an external force, the whole congruence will respond. In this way, nonlocality is still captured in this trajectory formulation, as one would expect. In sum, the LQH formulation undermines configuration-space realism by showing how one can represent the full many-body quantum state as living in ordinary 3-space, while still capturing the essential feature of nonlocality.

As emphasized at the outset, LQH is a full mathematical representation or formulation of quantum mechanics—not an interpretation. Nonetheless in the context of realist explorations of quantum theory, one can ask what the further project of interpreting this LQH formalism might look like. In other words, what hints might this formalism give us about the way the world really is? So far, three different philosophical approaches towards LQH have emerged, which I will refer to as the interpretations approach, the duality approach, and the inferential realist approach.

The first option is to turn LQH into an interpretation of quantum mechanics. Examples of this approach include the "Many Interacting Worlds (MIW)" interpretation of Bill Poirier (2010) and Michael Hall and colleagues (Hall, Deckert, and Wiseman 2014) or the "Newtonian QM" of Chip Sebens (2015). In contrast with Bohmian mechanics, which takes only one trajectory to be real, Poirier notes that on this alternative interpretation,

> one might prefer to regard all trajectories in the quantum ensemble as equally valid and real. It is hard to imagine how this could be achieved without positing that each trajectory inhabits a separate world.... this version of the many worlds interpretation would be very different from the standard form. (Poirier 2010, 14)

Similarly Sebens describes this interpretation as a novel no-collapse interpretation that

> combines elements of Bohmian mechanics and the many-worlds interpretation to form a theory in which there is no wave function.... Unlike the many worlds of the many-worlds interpretation, these worlds are fundamental, not emergent; they are interacting, not causally isolated; and they never branch. (Sebens 2015, 267)

To make this interpretation fly, however, one must assume a finite number of worlds. As Sebens notes,

> The meaning of ρ becomes unclear if we move to a continuous infinity of worlds since we can no longer understand ρ as yielding the proportion of all worlds in a given volume of configuration space upon integration over that volume. (Sebens 2015, 283)

Michael Hall and colleagues similarly try to avoid the "ontological difficulty" of a continuum of worlds by "replacing the continuum of fluid elements in the Holland–Poirier approach by a huge but finite number of interacting 'worlds'" (Hall, Deckert, and Wiseman 2014, 1).[30] By moving to a finite number of trajectories or worlds, this is strictly speaking a break from the full LQH formulation of standard QM. Recall that LQH is based on a continuum mechanics approach which, like classical hydrodynamics, rests on an infinite continuum of 'fluid' parcel trajectories. So the MIW interpretation is only equivalent to standard quantum dynamics in the limit where the number of worlds becomes uncountably infinite.

Instead of turning this formulation of quantum mechanics into an interpretation, the second approach takes a step back and asks what realist implications might follow from the fact that quantum mechanics admits of both a wavefunction and a trajectory concept of state. Holland himself sees this as indicative of a new kind of wave-particle duality. He writes,

> The full hydrodynamic model of quantum mechanics therefore provides an interpretation of two pictures—the wave-mechanical (Eulerian) and the particle (Lagrangian), and the latter is just as valid a representation of quantum processes as the former.... This mapping therefore gives a new and mathematically precise meaning to the notion of "wave-particle duality."
> (Holland 2005a, 508)

[30] Hall et al. note their approach is "broadly similar" to that of Sebens (2015).

These two pictures each emphasize and bring out different aspects of the quantum state. For example, the second order Newtonian equation of the trajectory picture, discussed above, highlights the force behind quantum propagation. Holland moreover argues that this wave-particle duality is not unique to quantum theory, but rather is a generic feature of field theories which can admit of Lagrangian trajectory formulations, such as in the case for electromagnetism (Holland 2005b) and relativity (Holland 2012). Whether this duality is to be understood simply as a feature of our mathematical representations, or whether it is to be read as revealing an ontological duality inherent in the world remains unclear.

A third approach might be along the lines of what I call inferential realism. Similar to the duality approach, this approach emphasizes the importance of a plurality of representations, though it does not read these two pictures of the quantum state as a joint depiction of some more fundamental ontological duality. Inferential realism shifts the focus of the realism question from 'what there is,' to 'what true things can we learn.'[31] Inferential realism rejects the literalist approach to scientific realism characteristic of van Fraassen, and instead traces its roots back to Ernan McMullin's (1984) understanding of realism as the development of scientifically fruitful metaphors.

In the heyday of the realism–antirealism debate in the 1980s there were two diametrically opposed conceptions of scientific realism. One conception, articulated by the antirealist van Fraassen, construed scientific realism as follows:

> *Science aims to give us, in its theories, a literally true story of what the world is like, and acceptance of a scientific theory involves the belief that it is true.* This is the correct statement of scientific realism. (van Fraassen 1980, 8; emphasis original)

Although van Fraassen claimed that this is a minimal construal that any realist would assent to, McMullin, one of the chief defenders of scientific realism at the time, flatly rejected this way of conceiving the realist project. In his paper "A Case for Scientific Realism," McMullin outlines a very different conception of realism:

> Science aims at fruitful metaphor and at ever more detailed structure. . . . The realist would not use the term 'true' to describe a good theory. He [or she] would suppose that the structures of the theory give some insight into the structures of the world. (McMullin 1984, 35)

Here we see McMullin rejecting the view that realism is committed to a literal construal of scientific theories, and instead conceiving of the realist project as one of developing fruitful metaphors, analogies, and models. On this alternative view, the realism comes in the new discoveries, insights, and deeper understanding that these metaphors, analogies, and models enable, rather than in their interpretation as a literal depiction of world.[32] In this tradition, inferential realism is not about finding the one true depiction of the world, but rather about developing a plurality of fertile representations. Representations, like depictions, are aimed at and purport to tell us about the world, but their connection is often looser involving pragmatic

[31] Although these questions are often linked, they can also come apart in significant ways, as discussed in Bokulich (2016).

[32] For a discussion of how metaphors, analogies, and models, despite being not literally true, can nonetheless give true insights into the physical systems they represent, again see "Fiction as a Vehicle for Truth" (Bokulich 2016).

elements; hence, a greater caution is required in drawing ontological conclusions. Moreover, different representations can be more or less useful in different contexts and domains. This view is nonetheless committed to realism because it holds that these representations yield genuine knowledge and advance our understanding of the actual world.

When trying to decide what philosophical attitude to take towards LQH, it is helpful to contextualize it within the broader class of Lagrangian hydrodynamic approaches in the physical sciences (e.g., in geophysical fluid dynamics and astrophysics), as I have done here. Within this broader context, it would be odd to say that the Eulerian formulation, for example, is the only legitimate mathematical formulation of classical hydrodynamics (as one might with the standard ψ formulation of QM); or at the other extreme, it would be odd to allow the Lagrangian formulation only if it is taken as a literal depiction of real fluids (as for example in the MIW interpretation). In the classical case, we clearly see the enormous fertility and explanatory power of the hydrodynamic analogy (and formalism of continuum mechanics more broadly) as a non-literal representation of our world. The credentials of classical hydrodynamics come not from its status as a literal depiction of the world, but rather from the many correct inferences it licenses.

In the context of quantum theory, the hydrodynamic analogy has given rise to the discovery of a new way of representing and time-evolving the quantum state. Although it is an empirically equivalent formulation of quantum mechanics, it nonetheless has profound implications for the quantum realism debate, as we have seen. In particular, the LQH formulation, with its displacement or trajectory representation of the quantum state, challenges the hegemony of the wavefunction. By showing that the wavefunction is neither a necessary—nor even the most fundamental—representation of quantum systems, LQH undercuts the central argument for wavefunction realism. Just as strikingly, it falsifies the claim that configuration space realism is a necessary consequence of any realist understanding of quantum mechanics, by showing how one can represent an n-body quantum system as a set of n states in ordinary three-dimensional space. To reiterate, it has not been my contention that we should replace the ψ conception of state with the trajectory one—we should admit both. Regardless of what philosophical attitude one takes towards the LQH formulation, regaining sight of the proverbial forest of quantum representations beyond the ψ is an essential step in exploring the realist implications of quantum theory.

Acknowledgments

In writing this chapter I have benefited greatly from many helpful discussions with Peter Holland; I am grateful to him both for his stimulating work on this topic and for his helpful comments on an earlier draft. I would also like to thank Rick Salmon for helpful discussions about classical hydrodynamics. Thanks are also owed to Tim Maudlin, as well as the editors and referee, for suffering through an earlier draft of this paper. None of them, of course, are to be blamed for the final product.

References

Aharonov, Y., J. Anandan, and L. Vaidman (1993). Meaning of the wave function. *Physical Review A 47*, 4616–26.

Albert, D. (1996). Elementary quantum metaphysics. In J. Cushing, A. Fine, and S. Goldstein (Eds), *Bohmian Mechanics and Quantum Theory: An Appraisal*. Boston Studies in the Philosophy of Science, pp. 277–84. Dordrecht: Kluwer.

Albert, D. (2013). Wave function realism. In A. Ney and D. Albert (Eds), *The Wave Function: Essays on the Metaphysics of Quantum Mechanics*, pp. 52–7. Oxford: Oxford University Press.

Bell, J. (1987a). Quantum Mechanics for Cosmologists. In *Speakable and Unspeakable in Quantum Mechanics*, pp. 117–138. Cambridge: Cambridge University Press.

Bell, J. (1987b). On the impossible pilot wave. In *Speakable and Unspeakable in Quantum Mechanics*, pp. 159–68. Cambridge: Cambridge University Press.

Bohm, D. (1952). A suggested interpretation of the quantum theory in terms of 'hidden' variables. I. *Physical Review 85*, 166–93.

Bohm, D. and J. P. Vigier (1954). Model of the causal interpretation of quantum theory in terms of a fluid with irregular fluctuations. *Physical Review 96*(1), 208–16.

Bokulich, A. (2015). Maxwell, Helmholtz, and the unreasonable effectiveness of the method of physical analogy. *Studies in History and Philosophy of Science 50*, 28–37.

Bokulich, A. (2016). Fiction as a vehicle for truth: moving beyond the ontic conception. *The Monist 99*(3), 260–79.

Butterfield, J. (2005). On Hamilton–Jacobi theory as a classical root of quantum theory. In S. D. A. Elitzur and N. Kolenda (Eds), *Quo Vadis Quantum Mechanics?*, pp. 239–73. Berlin: Springer-Verlag.

Combes, J., C. Ferrie, M. Leifer, and M. Pusey (2017). *Why Protective Measurement Does Not Establish the Reality of the Quantum State*. https://arxiv.org/abs/1509.08893.

Cushing, J. (1994). *Quantum Mechanics: Historical Contingency, and the Copenhagen Hegemony*. Chicago: University of Chicago Press.

Darrigol, O. (2005). *Worlds of Flow: A History of Hydrodynamics from the Bernoullis to Prandtl*. Oxford: Oxford University Press.

Dorato, M. and F. Laudisa (2015). Realism and instrumentalism about the wavefunction: how should we choose? In S. Gao (Ed.), *Protective Measurements and Quantum Reality: Toward a New Understanding of Quantum Mechanics*, pp. 119–34. Cambridge: Cambridge University Press.

Gao, S. (2017). *The Meaning of the Wavefunction: In Search of the Ontology of Quantum Mechanics*. Cambridge: Cambridge Univeristy Press.

Genel, S., M. Vogelsberger, D. Nelson, D. Sijacki, V. Springel, and L. Hernquist (2013). Following the flow: tracer particles in astrophysical fluid simulations. *Monthly Notices of the Royal Astronomical Society 435*, 1426–42.

Hall, M., D. Deckert, and H. Wiseman (2014). Quantum phenomena modeled by interactions between many classical worlds. *Physical Review X 4*, 41013–17.

Holland, P. (1993). *The Quantum Theory of Motion: An Account of the de Broglie–Bohm Causal Interpretation of Quantum Mechanics*. Cambridge: Cambridge University Press.

Holland, P. (2005a). Computing the Wavefunction from Trajectories: Particle and Wave Pictures in Quantum Mechanics and their Relation. *Annals of Physics 315*, 505–531.

Holland, P. (2005b). Hydrodynamic construction of the electromagnetic field. *Proceedings of the Royal Society A 461*, 3659–79.

Holland, P. (2006). Hidden variables as computational tools: the construction of a relativistic spinor field. *Foundations of Physics 36*(3), 369–84.

Holland, P. (2011). Quantum field dynamics from trajectories. In P. Chattaraj (Ed.), *Quantum Trajectories*, pp. 73–86. Boca Raton: CRC Press, Taylor & Francis Group.
Holland, P. (2012). Hydrodynamics, particle relabelling and relativity. *International Journal of Theoretical Physics* 51, 667–83.
Holland, P. (2013). Symmetries and conservation laws in the Lagrangian picture of quantum hydrodynamics. In S. Ghosh and P. Chattaraj (Eds), *Concepts and Methods in Modern Theoretical Chemistry: Statistical Mechanics*, pp. 55–78. Boca Raton: CRC/Taylor Francis.
Holland, P. (2017). The quantum state as spatial displacement. In R. Kastner, J. Jeknić-Dugić, and G. Jaroszkiewicz (Eds), *Quantum Structural Studies: Classical Emergence from the Quantum Level*, pp. 333–72. London: World Scientific Publishing.
Holland, P. (2018). Three-dimensional representation of the many-body quantum state. *Journal of Molecular Modeling* 24(269), 1–5.
Jaeger, H., S. Nagel, and R. Behringer (1996). Granular solids, liquids, and gases. *Reviews of Modern Physics* 68(4), 1259–73.
Lewis, P. (2016). *Quantum Ontology: A Guide to the Metaphysics of Quantum Mechanics*. Oxford: Oxford University Press.
Lundeen, J., B. Sutherland, A. Patel, C. Stewart, and C. Bamber (2011). Direct measurement of the quantum wavefunction. *Nature* 474, 188–91.
Madelung, E. (1927). Quantentheorie in hydrodynamischer form. *Zeitschrift für Physik* 40, 322–6.
Marsden, J. and T. Ratiu (1999). *Introduction to Mechanics and Symmetry: A Basic Exposition of Classical Mechanical Systems*, 2nd edn. New York: Springer-Verlag.
Maudlin, T. (2013). The nature of the quantum state. In A. Ney and D. Albert (Eds), *The Wave Function: Essays on the Metaphysics of Quantum Mechanics*, pp. 126–53. Oxford: Oxford University Press.
McMullin, E. (1984). A case for scientific realism. In J. Leplin (Ed.), *Scientific Realism*, pp. 8–40. Berkeley: University of California Press.
Müller, P. (1995). Ertel's potential vorticity theorem in physical oceanography. *Reviews of Geophysics* 33(1), 67–97.
Murkhanov, V. (2005). *Physical Foundations of Cosmology*. Cambridge: Cambridge University Press.
Myrvold, W. (2015). What is a wavefunction? *Synthese* 192(10), 3247–74.
Névir, P. (2004). Ertel's vorticity theorems, the particle relabelling symmetry and the energy-vorticity theory of fluid mechanics. *Meteorologische Zeitschrift* 13(6), 485–98.
Ney, A. (2013). Introduction. In A. Ney and D. Albert (Eds), *The Wave Function: Essays on the Metaphysics of Quantum Mechanics*, pp. 1–51. Oxford: Oxford University Press.
Nguyen, J. (2017). Scientific representation and theoretical equivalence. *Philosophy of Science* 84(5), 982–95.
Poirier, B. (2010). Bohmian mechanics without pilot waves. *Chemical Physics* 370, 4–14.
Poirier, B. (2017). Quantum mechanics without wavefunctions. *Talk* in Quantum Foundations Seminar, Perimeter Institute, October 24.
Regev, O., O. Umurhan, and P. Yecko (2016). *Modern Fluid Dynamics for Physics and Astrophysics*. New York: Springer.
Roulstone, I. (2015). Lagrangian dynamics. In J. P. G. North and F. Zhang (Eds), *Encyclopedia of Atmospheric Sciences*, 2nd edn, Vol. 2, pp. 369–74. London: Academic Press.
Salmon, R. (1988). Hamiltonian fluid mechanics. *Annual Review of Fluid Mechanics* 20, 225–56.
Salmon, R. (2014). *Lectures on Geophysical Fluid Dynamics*. New York: Oxford University Press.
Schiff, J. and B. Poirier (2012). Quantum mechanics without wavefunctions. *The Journal of Chemical Physics* 136, 31102–4.
Sebens, C. (2015). Quantum mechanics as classical physics. *Philosophy of Science* 82, 266–91.

Sieniutycz, S. (1994). *Conservation Laws in Variational Thermo-Hydrodynamics*. Dordrecht: Springer.

Takabayasi, T. (1952). On the formulation of quantum mechanics associated with classical pictures. *Progress of Theoretical Physics 8*(2), 143–82.

Takabayasi, T. (1953). Remarks on the formulation of quantum mechanics with classical pictures and on relations between linear scalar fields and hydrodynamical fields. *Progress of Theoretical Phyiscs 9*(3), 187–222.

Trujillo, L., D. Sigalotti, and J. Klapp (2013). Granular hydrodynamics. In *Fluid Dynamics in Physics, Engineering, and Environmental Applications*, pp. 169–83. Berlin: Springer-Verlag.

Van Fraassen, B. (1980). *The Scientific Image*. Oxford: Clarendon Press.

Van Fraassen, B. (1991). *Quantum Mechanics: An Empiricist View*. Oxford: Oxford University Press.

Wells, J. (2012). *Effective Theories in Physics: From Planetary Orbits to Elementary Particles Masses*. New York: Springer.

Wyatt, R. (2005). *Quantum Dynamics with Trajectories: Introduction to Quantum Hydrodynamics*. New York: Springer.

11

Scientific Realism without the Wave Function

Valia Allori

11.1 Introduction

Can we describe what the world is like through a fundamental physical theory? In the answer to this question lies the historic disagreement between scientific realists and antirealists. According to the antirealist, we should be content with physics to be, for example, just empirically adequate. In contrast, the realist believes that physics informs us about metaphysics. Quantum mechanics has always been taken to be devastating for the realist program, being so full of contradictions and mysteries. This situation started changing after the 1950s, since when several realist quantum theories have been proposed. However, until recently, scientific realists were convinced that in order to make quantum mechanics amenable to a realist interpretation, one would have to give a material interpretation of the fundamental mathematical object of the theory, namely the wave function. In this chapter, I explore an alternative realist proposal, dubbed the primitive ontology (PO) approach, which is not committed to this.[1]

I will show how the PO approach provides a distinctive account of theory construction in which the PO is chosen first, and then the rest of the theory is built around it to ensure empirical adequacy. Since many such theories are produced in this way, which cannot be ruled out by empirical means, I argue that coherence and parsimony considerations together with general reflections on the supervenience relation (or lack thereof) between the PO and the wave function, as well as on the type of scientific realism that this approach suggests, allow us to make a more informed decision about which theories are the best candidates for the scientific realist.

The following is the outline of the chapter. In Section 11.2, I present the PO approach for classical theories and then I extend it to the quantum domain. I contrast it with wave function realism, the view that the wave function is a material field, and I argue that the source of the tension between quantum mechanics and scientific realism is the idea of considering the wave function as representing physical objects. In Section 11.3, I outline the general theory-construction schema for the PO approach,

[1] This proposal has been introduced in Dürr, Goldstein, and Zanghì (1992; 1995; 1997), and Goldstein (1998), and further developed in Allori et al. (2008; 2011; 2014) and Allori (2013a; 2013b).

Valia Allori, *Scientific Realism without the Wave Function* In: *Scientific Realism and the Quantum*. Edited by: Steven French and Juha Saatsi, Oxford University Press (2020). © Valia Allori.
DOI: 10.1093/oso/9780198814979.003.0011

and in Section 11.4, I present some possible empirically adequate quantum theories in which the wave function is not physical. These theories are obtained by varying the type of PO (particles, fields, or spatio-temporal events, dubbed flashes), its evolution, and the evolution of the wave function. In Section 11.5, I move to the meaning of the wave function in these theories, which is taken to have a nomological character. These observations lead into Section 11.6, which discusses in more detail the type of realism this approach suggests, where one is realist about the PO but not about the wave function. I continue in Section 11.7, noting how the theories presented earlier could be better classified in terms of the PO being independent or dependent of the wave function. In Section 11.8, I discuss how this, among others, can be a consideration in theory selection.

11.2 Theory Architecture

The PO approach, like any realist framework, emphasizes that any theory should specify its scientific image: what the world is like according to the theory. Since theories are formulated mathematically, this implies the specification of the mathematical entities representing the physical ones. However, differently from other views, the emphasis is that scientific realism is better served when this is done at the very beginning of theory construction. That is, when the scientist proposing the theory has already a metaphysical hypothesis in mind, so that the correspondence rule is there from the beginning. For instance, Newton assumed the world is made of particles. Then, he put the theory in mathematical form using the mathematical object that naturally corresponds to his metaphysical hypothesis, namely points in three-dimensional space (as opposed to, for instance, vectors or scalar-valued functions from three-dimensional space to three-dimensional space that would naturally represent various types of fields). The other variables in the equations are interpreted accordingly: some represent matter, and some represent its properties. For instance $G\frac{mM}{r^2} + qE(r) = m\frac{d^2r}{dt^2}$ represents the temporal evolution of a material entity, namely a point-particle with mass m and charge q, whose trajectory in space-time is given by the solution $r(t)$ of this equation. In this way, some mathematical objects in the theory are privileged over the others since they capture the fundamental nature of reality. These variables are the *primitive ontology*, PO, of the theory.

Thus, theories are born with a hierarchical structure, determined by the role the various variables play in the theory. On the foundational level there is the primitive variable, which captures the metaphysical hypothesis. Then we have many other variables: some are constants, like G; some others may be taken as describing properties of matter, like m or q. These variables suitably 'dress up' the fundamental entities so that the theory accurately accounts for the phenomena. None of these variables represents matter: table and chairs are not made of charges, say; they are made of particles with charges.

What about other variables, like E above? Traditionally, they are taken to represent electromagnetic fields, and accordingly they have been dubbed 'local beables,' to use a terminology introduced in Bell (1987). They are mathematically specified by functions on three-dimensional space ('local'), as opposed to in a higher-dimensional space ('nonlocal'). The 'beable' part of the name comes in opposition to 'observable': they may represent something which exists rather than merely what is observed in an experiment. However, in this approach they do not represent matter: since they are needed to make the theoretical particle trajectories empirically adequate, they are best regarded as non-primitive variables. (See Allori 2015 for an elaboration on the possible choices of the PO of classical electrodynamics and its consequences.)

As a result of taking particles as its PO, classical mechanics is arguably able to account for observed phenomena. As long as we can neglect quantum effects, macroscopic bodies and their properties can arguably be accounted for in terms of the motion of point-like particles moving in three-dimensional space using the familiar notions of reduction and compositionality. There are two ingredients for this explanatory schema to be satisfactory: (1) the PO is *microscopic*; (2) the PO is in *space-time*. The first requirement ensures that the objects in the PO are the building blocks of everything else: particles clump together to form bigger objects, which behave independently of their initial composition (for instance, protons, neutrons, and electrons bind together to form atoms, which behave as prescribed by the laws of chemistry in a way which is independent of their subatomic composition; then atoms bind together to form more complicated molecules that obey the laws of biology, which similarly are opaque to their atomic composition, and so on). This hierarchy of objects straightforwardly allows the explanation of the macroscopic properties in terms, of the microscopic constituents.[2] This is where the second requirement comes in: this schema works if the building blocks live in the same space as the macroscopic entities, namely three-dimensional space. If the fundamental building blocks of nature live in a different space then there is an additional step to be made, namely to explain how we think we live in three-dimensional space while we actually do not (see Section 11.5).

If the PO represents matter, what about the other variables in the theory? In a very important sense, a theory has to give us an image of reality. In this approach, this is done through the spatio-temporal trajectories of the PO. They are like the output generated by a computer program simulating a system, while the other variables serve as means for generating this output: they are internal variables of the program, needed for the computation (Allori et al. 2008). Given this role, they may be dubbed nomological variables: they appear in the laws of nature which govern the behavior of matter. Be that as it may, setting aside for the moment the status of the non-primitive variables (see Section 11.5), let us see how to extend this framework to the quantum domain.

In contrast with classical theories, it was widely claimed that quantum mechanics is incompatible with scientific realism. The measurement problem, or the problem

[2] However, notice that theories with macroscopic fundamental entities have been put forward. For instance, Allori et al. (2008) propose that Bohr's quantum theory could be seen as an example of such a theory, where pointer positions, which reveal the experimental results, are fundamental. Nonetheless, as these authors point out, aside from being intrinsically vague (what counts as a macroscopic object?) the explanatory schema to recover the manifest image from the scientific image would fail to apply because in this theory microscopic entities such as atoms and molecules are 'made of' macroscopic ones, namely pointer positions.

of the Schrödinger cat, played a crucial role in this, and can be summarized as follows. Assume the fundamental object of quantum mechanics, the wave function ψ, is physically real, and assume it evolves in time as described by the solutions of the Schrödinger equation $\psi = \psi(t)$. As a mathematical fact, sums of solutions are also solutions, which therefore describe possible states of affairs of the system. If the description of matter provided by the wave function is complete, then these 'superpositions' may represent a cat which is dead and alive (i.e., non-dead) at the same time, or a particle being here and there (i.e., not-here) at the same time. Thus, in the 1930s this contradiction was taken to show that quantum mechanics could only have an instrumental value. However, it was realized in the 1950s that this instead only shows that at least one of the assumptions below is mistaken (Bell, 1987, Maudlin, 1995):

- The wave function completely describes every physical system;
- The wave function evolves according to Schrödinger's equation;
- Macroscopic objects have non-contradictory properties.

At least three theories have been proposed to solve this problem: the pilot-wave theory, the spontaneous collapse theory, and the many-worlds theory. The pilot-wave theory is taken to reject the first premise above in that it adds particles to the description of the wave function (de Broglie, 1927; Bohm, 1952; Bell, 1987; Dürr, Goldstein, and Zanghì, 1992). The theory has therefore another equation describing the motion of particles in terms of the wave function. The spontaneous localization theory is taken to reject the second assumption: here the wave function evolves according to the Schrödinger equation until a random time, at which it undergoes an instantaneous localization in a random point (Ghirardi, Rimini, and Weber, 1986). The many-worlds theory is taken to reject the third premise, since it accepts that physical objects may possess contradictory properties (Everett, 1957, de Witt, 1970, Wallace, 2002). That is, the two terms of the superposition that describe an alive cat and a dead cat both exist but they interact so little that they can be interpreted as living in distinct 'worlds' occupying the same region of space-time.

This is the traditional story: the problem of quantum mechanics is the presence of macroscopic superpositions. According to the proponents of the PO approach, instead, *the problem lies in the assumption that the wave function represents physical objects*. This is an implicit and seemingly undeniable assumption: all the theories just seen, as stated, take the wave function as representing matter. That is, it was assumed that the natural way of rescuing scientific realism in the quantum domain was to commit to some form of wave function realism. Notice that one is tempted to seriously consider the wave function as representing a physical field only in a theory like quantum mechanics, in which there is no starting metaphysical assumption. Quantum mechanics as developed by Heisenberg, Born, and Jordan in the 1920s merely provided a formalism. They did not care about providing an interpretation of it given their antirealist inclinations. In the years that followed, scientific realists, when facing the measurement problem, had to interpret this formalism *post hoc*, and it was natural for them to think of the wave function as providing (at least part of) the correspondence between physics and metaphysics.

In contrast, the proponent of the PO approach thinks that this is an unfortunate consequence of historical contingencies: the wave function is not the right kind of mathematical object that one would have naturally considered as representing a metaphysical assumption. If so, one should always deny the first premise, namely that the wave function completely describes physical systems, not because it is incomplete, but because it is not physical. In this sense, something in space-time representing matter is always needed. This goes back to the 1920s, when Lorentz, de Broglie, Heisenberg, and Einstein expressed perplexities about considering the wave function as a physical field.[3]

However, these worries have been long forgotten, receiving new attention only recently, as discussed in Albert and Ney (2013) and references therein. One problem, dubbed the *configuration space problem*, is that the link between the scientific and the manifest image provided by the wave function is not sufficiently explanatory. In fact, the wave function $\psi = \psi(r_1, \ldots, r_N)$ is a field that lives in configuration space, the space of possible particle configurations $q = (r_1, \ldots, r_N)$ (Albert, 1996; 2013; 2015, Lewis, 2004; 2005; 2006; 2013, Ney, 2012; 2013; 2015; 2017, North, 2013). The dimension of such space is $M = 3N$. If the wave function is a material field, then physical space would be configuration space. If so, one would have to explain why we think we live in a three-dimensional world instead, and it is controversial whether this can be successfully done. One would need to add a correspondence rule between configuration space and three-dimensional space (Albert, 1996) and whether the proposed maps are successful is up for debate (Monton, 2002, 2006; Allori, 2013a, 2013b). In any case, the explanatory schema developed in classical theories to derive the macroscopic properties in terms of the microscopic constituents has to be dropped in this framework, and a new one needs to be developed.[4]

However, it seems there is no need for re-thinking such schema, given that, as we will mention in Section 11.3, this is still available to the PO approach. Moreover, it is difficult to see how within wave function realism theories can have symmetry properties, in contrast with the PO approach (see Section 11.3). For instance, quantum mechanics turns out not to be Galilean invariant: the wave function is a scalar field in configuration space and as such will remain the same under a Galilean transformation, while invariance would require a more complicated transformation.[5]

11.3 Quantum Theory Construction Kit

Because of the reasons discussed in the previous section, the proponents of the PO account reject the assumption that the wave function represents matter. Consequently, the theories proposed as responses to the measurement problem are regarded as satisfactory only if they all postulate something in three-dimensional space to describe material objects. The role of the wave function is, similarly to that of electromagnetic fields in classical electrodynamics, to generate the space-time histories of the PO.

[3] See Bacciagaluppi and Valentini (2009) for a very interesting discussion of the various positions on this issue and others during the 1927 Solvay Conference.

[4] See Albert (2015) and Ney (2017) for two different proposals on how this may be accomplished.

[5] There are other ways to think of the wave function as material but avoiding some of these problems. See Forrest (1988), Norsen (2010), and Hubert and Romano (2018) for the proposals, and Belot (2012), Solé (2013), and Suárez (2015) for criticisms.

The pilot-wave theory can be naturally understood as a theory with a particle PO. However, the situation is trickier in the spontaneous localization theory and in many-worlds: one can add different, more or less natural, PO for these theories, as we will see in Section 11.4. In fact, the PO approach provides us with a set of rules for generating quantum theories:

- Make a metaphysical assumption and select a corresponding spatio-temporal PO, which therefore has an ontological role;
- Select an evolution law for the PO. This is likely implemented using some appropriate mathematical objects including the wave function (as it is clear in the pilot-wave theory: the particles evolve according to a law defined in terms of the wave function). In virtue of this, the wave function assumes a nomological role (i.e., its role is to help implementing the law of evolution of the PO);
- Select a law of evolution for the nomological object(s).

A variety of such theories have been proposed and analyzed. However, not all possible theories are good ones. A first constraint is empirical adequacy: the manifest image has to be successfully recovered. Quantum mechanics is empirically adequate, so all it takes is that the theory under consideration is empirically equivalent to quantum mechanics. One can distinguish between exact and effective empirical equivalence. Two theories are *exactly* empirically equivalent when there is no possible experiment that can in principle distinguish between the two. Theories are instead *effectively* empirically equivalent when they cannot be currently experimentally distinguished in practice.

Before discussing some of these theories, let me briefly explain how symmetries are implemented in this framework. Because the various solutions to the measurement problem are ultimately not about the wave function but about histories of a PO in space-time, the law of evolution of the wave function should no longer be regarded as playing a central role in determining the symmetries of the theory. Indeed, they are determined by the PO, not by the wave function. Roughly put, to say that a theory has a given symmetry is to say that the possible histories of the PO (those that are allowed by the theory), when transformed according to the symmetry, will again be possible histories for the theory (Allori et al., 2008). That means that the symmetries transform empirically adequate histories into other empirically adequate histories.[6] Changing PO could (and probably will) change the symmetry properties of the theory. This is particularly relevant in the context of developing relativistic invariant theories: without the PO one focuses on the relativistic invariance of the wave function, while in this framework one should look at a relativistic invariant evolution for the PO.

[6] Notice that the entities in the theory which do not represent matter, such as the wave function in quantum mechanics, can transform as needed to preserve the symmetry. There could be different ways of producing the same trajectories, for example using two wave functions that differ by a phase factor. Or one can generate the same trajectories by either a Schrödinger-evolving wave function and static operators (as in the Schrödinger picture) or a static wave function and evolving operators (as in the Heisenberg picture). If one assumes that the wave function does not represent matter, and wants to keep the symmetry, then one can allow the wave function to transform as to allow this. See Allori (2018a) for an argument based on Galilean symmetry that the wave function is best seen as a projective ray in Hilbert space rather than a physical field.

For instance, in reference to the theories that we will discuss later, GRWf has been modified to make it relativistic invariant in this sense (Tumulka, 2006), and relativistic extensions of Sm and GRWm have been proposed respectively in Allori et al. (2011) and Bedingham et al. (2014).

11.4 Primitive Ontology

Here are some examples of how different POs can be combined with different evolutions and different nomological variables in order to construct empirically adequate quantum theories according to the theory construction kit this approach provides. I consider here only three types of PO: particles, matter density fields, and flashes.

If matter is made of point-particles, the following is a list of some empirically adequate theories.

The pilot-wave theory: This theory can be naturally read as a theory of particles moving in three dimensions according to a suitable guidance equation that involves a Schrödinger-evolving wave function.

Sip: In this theory the PO is given by instantaneous randomly distributed configurations without any temporal correlation among them, whose probability distribution is governed by a Schrödinger-evolving wave function (Bell, 1987; Allori et al., 2008).[7]

GRWp3: This theory combines a particle PO, evolving according to the same guidance equation as in the pilot-wave theory, and a wave function that is stochastically evolving as in the original GRW theory. However, here each localization point is the actual position of the particle at the localization time which is 'displaced' at random (Bedingham, 2011; Allori et al., 2014).

GRWp6: The particles evolve according to the same guidance equation as in the pilot-wave theory between the localizations of the wave function, like in GRWp3. However, at the localization center all the particles jump at random.[8]

MBM: In this theory, called 'master equation Bohmian mechanics' and hence 'MBM', the wave function is completely absent. The particles evolve according to something similar to the pilot-wave's guidance equation, but instead of the wave function we have a density matrix which evolves according to the Limblad equation (Allori et al., 2014).

[7] The 'S' in the name comes from Schrödinger evolution of the wave function, 'i' stands for the fact that the particles are independent, and 'p' denotes the PO of particles. The logic is similar for the names of the other theories.

[8] Simpler theories with a particle PO and a GRW-like evolution for the wave function turn out not to be empirically adequate. The simplest attempt, GRWp1, would be a theory in which the motion of particles is the same as in the pilot-wave theory and the wave function is the same as the original GRW theory. The second simplest attempts, GRWp2, would allow the center of the localization of the wave function to be the actual particle position. Then GRWp4 would make the particles move according to how the average position in orthodox quantum mechanics would prescribe, rather than according to the guidance equation of the pilot-wave theory. Finally, in GRWp5 the particles move as in the pilot-wave theory between localizations; however, at that time also the configuration of the particle in the localization point would jump. See Allori et al. (2014) for more details about such theories.

Alternatively, the world could be continuous, with matter being represented by a scalar field m in three-dimensional space defined in terms of the wave function. Given this PO, here are some empirically adequate theories:

Sm: In this theory the matter density field evolves according to a law which involves a Schrödinger-evolving wave function (Allori et al., 2008).
GRWm: The matter density here evolves according to a law mediated by a stochastically evolving wave function (Ghirardi, Rimini, and Weber, 1995).
Mm: In analogy with MBM, the matter density field evolution is implemented by a Limblad-evolving density matrix rather than a wave function (Allori et al., 2014).

A particle theory usually provides space-time histories where configurations at different times are continuously connected. However, as shown in Sip, configurations may jump from one moment to the next without any connection. This suggests that matter could be thought as constituted by a set of spatiotemporal events, the 'flashes': $F = \{(X_1, T_1), \ldots, (X_k, T_k), \ldots\}$, k being a progressive natural number indicating the time progression of the flashes.[9] Below is a list of possible empirically adequate 'flashy' theories:

Sf: The flash distribution is determined by the wave function, like the localization point is determined by the stochastic evolution in the spontaneous localization theory. However, in Sf the wave function evolves according to Schrödinger's equation (Allori et al., 2008).
GRWf: In this theory the wave function evolves according to the GRW stochastic evolution and every flash corresponds to one of the spontaneous localizations of the wave function (Bell, 1987, Tumulka, 2006).
Mf: In analogy with MBM and Mm, Mf is a theory of flashes in which the rate of the flashes is not generated by the wave function but by a Limblad density matrix (Allori et al., 2008).

11.5 The Wave Function

As scientific realists we are interested in the metaphysics of the above theories. However, if the PO of the theory constitutes the building blocks of matter, what is the wave function? How is this approach compatible with scientific realism? Let us focus on the first question in this section, and move to the second one in the next. As briefly anticipated, the role of the wave function, just like electromagnetic fields in classical electrodynamics or more generally the potential or the Hamiltonian in classical mechanics, is to help implement the law governing the spatio-temporal trajectories of the PO. Hence, it is better understood as having a nomological character

[9] A note about all these theories: if the number N of 'particles' is large, as in the case of a macroscopic object, the number of flashes is large, too (if $\lambda = 10^{-15}\ s^{-1}$ and $N = 10^{23}$, we obtain a rate of 10^8 flashes/second). Therefore, for a reasonable choice of the parameters of the theory, a cubic centimeter of solid matter contains more than 10^8 flashes per second. That is to say, large numbers of flashes can form macroscopic shapes, such as tables and chairs. At almost every time, however, space is in fact empty, containing no flashes and thus no matter.

rather than representing matter: it is best regarded as akin to a law of nature (Dürr, Goldstein, and Zanghì, 1997, Goldstein and Teufel, 2000, Goldstein and Zanghì, 2013, Allori, 2018a). This view fits particularly well with a Humean account of laws, according to which laws are axioms and theorems of our 'best system' of the world. Since the wave function is part of the axioms, it can be naturally regarded as a Humean law (Callender, 2005, Esfeld, 2014, Miller, 2014, Bhogal and Perry, 2017).

Several objections have been raised against this view, the most compelling of which focuses on the disanalogies between the wave function and the general conception of laws (see, most notably, Belot, 2012; Brown and Wallace, 2005). First, one may argue that since the wave function interacts with the particles then it has to be material. However, in classical physics potentials play a similar theoretical role as well but no one considers them as real. More challenging is the observation that the wave function evolves in time, while laws do not. In response, one could notice that evidence suggests that in a future quantum cosmology the wave function would be static, eliminating the problem (Goldstein and Teufel, 2000). However, if so, one would have to wait until such a theory is developed to rightfully do metaphysics. Perhaps more convincingly, one could reply by noting that the Schrödinger evolution could be regarded as a constraint on a time-independent wave function of the universe rather than an evolution equation (Esfeld et al., 2014).

Another objection is that the wave function is contingent, since it varies with the subsystem, while laws are not. However, the wave function which is contingent is the wave function of the system. Instead, the wave function with a nomological status is the wave function of the universe, since it is the one for which the Schrödinger equation holds (Goldstein and Zanghì, 2013). One may counter-reply insisting that the universal wave function is also contingent in the sense that there could have been a physically distinct one. However, one could reply that there could have been other laws as well. I think that the best reply to this, as well as the previous objection, is to maintain that we should not force at any cost classical intuitions onto quantum mechanics especially about laws of nature, and we should be open to modify our nomic concepts accordingly, if needed (Callender, 2005). This may be a surprising reaction, given that the PO supporters have always emphasized their preference for traditional frameworks. However, it is by allowing a looser notion of laws of nature as entities 'guiding' the motion of the PO that one can still use the classical explanatory framework to recover macroscopic properties in terms of the microscopic PO.[10]

[10] There are other ways in which someone could think of the wave function, broadly speaking, as nomological. One can think of the wave function as a property which expresses some non-material aspect of the particles (Monton, 2013). Similarly, one can endorse a dispositional account where laws are understood in terms of dispositions, which in turn are described by the wave function (Esfeld et al., 2014, Suárez, 2015). Arguably, since dispositions can be time dependent, in this context the temporality objection seems less compelling. Having said that, I think these proposals are not very promising in that they rely on the notion of properties which are notoriously a tough nut to crack. As Esfeld (2014) has pointed out, there are several severe problems in trying to spell out what fundamental properties are, both in the classical and the quantum domain.

11.6 Explanationist Realism

The existence of the theories presented in Section 11.4 shows that scientific realism is alive and well even with respect to quantum mechanics, and even without considering the wave function as physically real. However, one may worry that this cannot be correct: doesn't scientific realism tell us to consider as real whatever our best theories postulate? Luckily, there is already a kind of scientific realism, dubbed selective or restricted realism, which gives up on such a complete correspondence between realist commitments and what our theories postulate. In particular, as we will see in this section, there is one kind of selective realism which is well suited to accommodate the PO approach.

The main argument for realism, the so-called no-miracle argument, states that the empirical success of a theory is evidence of its truth; otherwise this success would be a miracle. Nonetheless, the so-called pessimistic meta-induction argument aims to show that the empirical success of a theory is not a reliable indicator of its truth, given that past successful theories turned out to be false. One way to respond to this challenge is to restrict realism, and argue that one should not be realist about the whole theory, but about a restricted set of entities. If one can show that the entities that are retained in moving from one theory to the next are the ones that are responsible for the empirical success of the theory, the previous argument is blocked. One particular way of doing this is the so-called explanationist realism according to which one should be realist with respect to the *working posits* of the theory, the posits involved in explanations and predictions. If the working posits are preserved during theory change, one could argue that past theories were successful because they got the working posits right. At the same time, these theories on the whole are also false since they got something wrong too, namely the *presuppositional posits*. Thus, the realist is justified in believing in the physical reality of the working posits, without being committed to the existence of the presuppositional posits, which are somewhat 'idle' components of the theory (Kitcher, 1993, Psillos, 1999).[11]

This view has been articulated and advocated in the context of classical theories (e.g., Fresnel's theory of light). The case for explanationism is fundamentally incomplete, however, if one does not consider the theory change from classical to quantum mechanics. Interestingly, it bears striking similarities with the PO approach in that quantum theory's predictions, being encoded in pointer positions, are determined by the PO, not by the wave function. Similarly, the explanations of the phenomena are in terms of the PO, and only indirectly involve the wave function. That is, the PO is reminiscent of the working posits and the wave function is reminiscent of a presuppositional posit. If so, this type of realism finds a natural association to the wave function as conceived by the primitivists, and moreover the PO approach provides a very nice framework for the explanationist to extend her view into the quantum domain (Allori, 2018b).

[11] For a rediscovery of this approach and an application to effective quantum field theories, see Williams (2019).

11.7 Dependent and Independent Primitive Ontologies

The theories presented in Section 11.4 are mutually exclusive and, being all empirically adequate, cannot be distinguished on the basis of empirical constraints. Thus, they need to be assessed using some *super-empirical* virtues, as discussed in Section 11.8. Interestingly, one of such considerations is distinctive of the PO approach, as I discuss in this section. Notice that the matter field m and the flashes F are functionals of the wave function, in contrast with theories such as the pilot-wave theory in which PO and the wave function are independent. So, the construction kit presented in Section 11.3 could be spelled out more explicitly by emphasizing that the specification of the metaphysical assumption of the theory (the PO) may be in terms of some other mathematical object (the wave function). In philosophical jargon, dependence is sometimes translated in terms of supervenience: Y supervenes on X if no two possible situations are indiscernible with respect to X while differing in Y. For instance, chemical properties supervene on physical properties insofar as any two possible situations that are physically indistinguishable are chemically indistinguishable. The matter density and the distribution of the flashes supervene on the wave function: there cannot be a difference in the matter density or in the distribution of the flashes without a difference in the wave function. It has been argued that, because of this dependence, there is no need to add the matter density or the flashes to the description provided by the wave function (Lewis, 2006). However, by definition, in the PO approach the wave function does not represent matter and thus some other-spatiotemporal object in the theory should do so, and in these cases it is either the matter density or the flashes. The question remains, however, about what kind of dependence (or supervenience) holds between the PO and the wave function, and whether that can be used in theory selection.

An important distinction here is between *logical* (or conceptual) and *natural* (or nomic) supervenience. We have logical supervenience between X and Y when X entails or implicates Y, i.e., $Y = f(X)$. For instance, the description 'table' supervenes logically on the configuration of the particles composing the table. By contrast, there is natural supervenience between X and Y when Y is a function of X, i.e., $Y = f(X)$, *and* this function defines a law of nature. An example of natural supervenience is the relation between the pressure exerted by one mole of gas and its volume and temperature: $p = f(T, V) = KT/V$, where K is a constant. The function f defines Boyle's law and expresses an empirical truth, in contrast with logical supervenience (Chalmers, 1996). The distinction between logical and natural supervenience can be summarized as follows: Y logically supervenes on X when Y comes 'for free' once there is a certain X; Y naturally supervenes on X when one needs to specify a law-like relation to define Y in terms of X. Once the law is specified, X will automatically bring along the Y.

Is the dependence of the PO on the wave function logical or natural? There is a sense in which one would want it to be natural: there are other logically possible definitions of the matter density in terms of the wave function, but this one is the one that actually holds. The matter field is defined in that way as a matter of natural law, regardless of how many other possible definitions one could come up with. However, the fact that the PO is defined by a law of nature that involves the wave function specified by some function f as $PO = f(\psi)$, is puzzling: in these theories, there is a law of nature that defines the PO and a law of nature that defines its evolution, both of which involve the wave function. Doesn't that mean that the wave function is more primitive than the PO? Indeed, consider the natural dependence of the electric field on the charge density. This relation defines the field in terms of the charge density, but wouldn't we say that the charge density generates the field, which because of this turns out to be less primitive? This seems the exact opposite of what the primitivist needs. However, one could reverse the dialectic. That is, the dependence of PO and the wave function expressed in terms of the function f may be taken to define what the wave function is, rather than what the PO happens to be: $\psi = f^{-1}(PO)$. Notice that the matter field could have had many mathematical definitions, including some which do not involve the wave function. Theories like that are not among the ones we have presented in Section 11.4, but they seem possible, and they would be theories in which the PO and the wave function are independent.

Be that as it may, this dependence (or lack or thereof) suggests a novel distinction among the solutions of the measurement problem. Since in the PO approach this problem is taken to show the inadequacy of the wave function as representing matter, its solutions will be different in how they specify some entity in three-dimensional space to play the role of matter. That is, different solutions are characterized by whether the PO is independent of the wave function or it is not:

Theories of type-1: PO and ψ are independent; once the PO is specified, the wave function is introduced independently of it to make the theory empirically adequate as specified. This is the case for particle theories.

Theories of type-2: PO and ψ are dependent; once the PO is specified, the wave function is introduced so that the PO appears to be defined in terms of the wave function. This is the case of theories with flashes and matter density.

11.8 Theory Evaluation

As already mentioned, the dispute about which of the theories discussed in Section 11.4 is the best will have to be settled on grounds other than empirical adequacy. Here's a list of features that one could use during theory selection.

Lack of Many-worlds Character: Because of its linearity, in any theory with a Schrödinger-evolving wave function there are superpositions. In the pilot-wave theory this is not a problem: since configurations are continuously connected in time, it is not possible for the configuration to jump, in an instant, from the support of one term of the superposition to a macroscopically distinct one (that is, a dead cat will not become instantly alive). However, because of this, many other theories will show a many-worlds character.[12] For instance in Sip, since there is no connection between different configurations at different times, the configuration will likely visit distant regions at subsequent instances. That means, for instance, that if at time t there are

[12] Note that the concept of a 'world' is just a practical matter, relevant to comparing the matter density function provided by the theory to our observations. However, this is not a problem: there is no need for a precise definition of 'world,' just as we can get along without a precise definition of 'table.'

dinosaurs, at time $t + dt$ they have disappeared. Therefore, many worlds exist, not at the same time but one after another. A similar many-world character is shared by Sm and Sf, as well as Mf and Mm, in which the superpositions of the wave function are inherited by the flashes and the matter density field. By the linearity of the Schrödinger evolution, the flashes and the matter density form independent families of correlated flashes, or matter density, associated with the terms of the superposition, with no interaction between the families: the living cat and the dead cat do not interact with each other, as they correspond to alternative states of the cat. Thus, they can indeed be regarded as comprising many worlds, superimposed on a single space-time. Since the different worlds do not interact among themselves, they are, so to speak, reciprocally transparent. Notice that, since these worlds are undetectable, all other things being equal, none of these theories seems to be among the best alternatives. Moreover, note that in the theories in which the wave function localizes (like all GRW-type theories) these many-worlds exist, even if for a short moment.

Empirical Coherence: A theory is empirically incoherent when its truth undermines our empirical justification for believing it to be true (Barrett, 1999, Huggett and Wüthrich, 2013). Notice that Sip is empirically incoherent, given that it implies that our records of the past, including evidence to support the theory, may well be completely erroneous. In fact, its many-worlds character is so radical that it implies that our memories and records of the past are most likely to be false. Similar considerations lead us to rule out also GRWp6: since all configurations jump at the same time when the wave function gets localized, one could instantaneously move from one world in which there are dinosaurs to one in which there aren't any.

Simplicity: When considering ontology, particles are the simplest: they require just one parameter to be specified, their location in space. If so, then the pilot-wave theory stands out. It is not the only theory with a particle PO, but what is the point of complicating the theory with a non-linear evolving wave function, as in GRWp3? Density matrices also seem unnecessarily complicated. Sip and GRWp6, as we just saw, are ruled out because they are empirically incoherent.

Symmetries: Arguably, the PO of continuous fields is less developed and thus requires more work. However, one can argue that theories like GRWm and Sm have relativistic extensions which do not require a foliation, contrary to the pilot-wave theory, and thus they may be thought to be more compatible with relativity (Dürr, Goldstein, and Zanghì, 1992). Among the two, Sm is arguably better than GRWm, given that its wave function evolution is simpler. However, they both have a many-worlds character. While one may suggest that the relativistic invariance of Sm should be taken to be an independent justification for many-worlds, there are theories, like GRWf, which have relativistic extensions without a foliation and without a many-worlds character as severe as that of Sm. Flashes are a more exotic choice of PO, but they seem to be particularly well-suited for relativistic theories (see Bell, 1987; Tumulka, 2006). Among flash-based theories, therefore, GRWf seems to give the best balance of mathematical simplicity and metaphysical sensibility, since it does not possess a severe many-worlds character like that of Sf, and it is not as mathematically complicated as Mf.

Scientific Realism: General considerations about realism may lead us back to particles. Indeed, they are more familiar: well-developed and well-known theories like classical mechanics had a PO of particles. In this respect, one could argue that a particle ontology would help the scientific realist responding to the pessimistic meta-induction argument (Allori, 2018b). In fact, if the particle PO is preserved in the classical-to-quantum theory change, and the particle PO is responsible for the empirical success of both theories, then one could defeat the pessimistic meta-induction and therefore be justified in being realist about the PO.

Independent PO: Finally, and perhaps more interestingly, notice that in theories denoted as type-1 in Section 11.6 in which the PO is independent of the wave function, such as the pilot-wave theory or other particle theories, the theory architecture is straightforward: the PO evolves in time and as such represents the evolution of matter; the wave function appears as a suitable ingredient in this evolution. It may be odd that the wave function evolves in time, as we discussed in Section 11.5, but insisting that this is just a convenient representation of how the wave function may generate the spatio-temporal motion of the PO may make it less so. In contrast, in type-2 theories in which the wave function and the PO are dependent, the structure is more convoluted: the PO still represents matter but the role of the wave function seems more difficult to accommodate since not only does it appear in the definition of the PO but also defines its motion. One could argue that, because of this, independent POs should be preferred. This would lead directly to particle theories, but would leave the door open to yet-unexplored matter density theories which are defined independently of the wave function.

Summing up the results of these proposed criteria, which by any means are not the only ones possible, matter density theories (at least the ones presented so far) do not seem to win in any categories. The battle arguably remains between the pilot-wave theory and GRWf. The former seems to be leading, given that it wins in the categories of simplicity, realism and independence but loses in symmetries, while the latter wins in symmetries but loses in simplicity, independence, and realism. If anything, this explains why most proponents of the PO account prefer the pilot-wave theory over the alternatives.

Be that as it may, independently of theory evaluation criteria and considerations, if we follow the PO approach and we get the wave function out of the ontological picture as representing matter, quantum mechanics becomes a theory which, with the discussed qualifications, is compatible with scientific realism.

References

Albert, D. Z. (1996). Elementary quantum metaphysics. In J. Cushing, A. Fine, and S. Goldstein (Eds), *Bohmian Mechanics and Quantum Theory: An Appraisal*, pp. 277–84. Dordrecht: Kluwer.

Albert, D. Z. (2013). Wave function realism. In D. Z. Albert and A. Ney (Eds), *The Wave Function: Essays on the Metaphysics of Quantum Mechanics*, pp. 52–7. New York: Oxford University Press.

Albert, D. Z. (2015). *After Physics*. Cambridge, MA: Harvard University Press.

Albert, D. Z. and A. Ney (2013). *The Wave-Function: Essays on the Metaphysics of Quantum Mechanics*. New York: Oxford University Press.

Allori, V. (2013a). Primitive ontology and the structure of fundamental physical theories. In D. Z. Albert and A. Ney (Eds), *The Wave-Function: Essays on the Metaphysics of Quantum Mechanics*, pp. 58–75. New York: Oxford University Press.

Allori, V. (2013b). Primitive ontology and the structure of fundamental physical theories. In S. Lebihan (Ed.), *Precis de la Philosophie de la Physique*, pp. 116–51. Paris: Vuibert.
Allori, V. (2015). Maxwell's paradox: classical electrodynamics and its time reversal invariance. *Analytica 1*, 1–19.
Allori, V. (2018a). A new argument for the nomological interpretation of the wave function: the galilean group and the classical limit of nonrelativistic quantum mechanics. *International Studies in the Philosophy of Science 31*(2), 177–88.
Allori, V. (2018b). Primitive ontology and scientific realism. or: the pessimistic meta-induction and the nature of the wave-functionhugge. *Lato Sensu: Revue de la Societe de Philosophie des Sciences 5*(1), 69–76.
Allori, V., S. Goldstein, R. Tumulka, and N. Zanghì (2008). On the common structure of Bohmian mechanics and the Ghirardi–Rimini–Weber theory. *The British Journal for the Philosophy of Science 59*(3), 353–89.
Allori, V., S. Goldstein, R. Tumulka, and N. Zanghì (2011). Many-worlds and Schrödinger's first quantum theory. *The British Journal for the Philosophy of Science 62*(1), 1–27.
Allori, V., S. Goldstein, R. Tumulka, and N. Zanghì (2014). Predictions and primitive ontology in quantum foundations: a study of examples. *The British Journal for the Philosophy of Science 65*(2), 323–52.
Bacciagaluppi, G. and A. Valentini (2009). *Quantum Theory at the Crossroads: Reconsidering the 1927 Solvay Conference*. Cambridge: Cambridge University Press.
Barrett, J. A. (1999). *The Quantum Mechanics of Minds and Worlds*. New York: Oxford University Press.
Bedingham, D., D. Dürr, G. Ghirardi, S. Goldstein, R. Tumulka, and N. Zanghì (2014). Matter density and relativistic models of wave-function collapse. *Journal of Statistical Physics 154*(1), 623–31.
Bedingham, D. J. (2011). Hidden variable interpretation of spontaneous localization theory. *Journal of Physics A: Mathematical and Theoretical 44*, 275303.
Bell, J. S. (1987). *Speakable and Unspeakable in Quantum Mechanics*. Cambridge: Cambridge University Press.
Belot, G. (2012). Quantum states for primitive ontologists: a case study. *European Journal for Philosophy of Science 2*, 67–83.
Bhogal, H. and Z. Perry (2017). What the Humean should say about entanglement. *Noûs 51*(1), 74–94.
Bohm, D. (1952). A suggested interpretation of the quantum theory in terms of 'hidden' variables i and ii. *Physical Review 85*, 166–93.
Brown, H. and D. Wallace (2005). Solving the measurement problem: de Broglie–Bohm loses out to Everett. *Foundations of Physics 35*(4), 517–40.
Callender, C. (2005). One world, one beable. *Synthese 192*, 3153.
Chalmers, D. J. (1996). *The Conscious Mind: In Search of a Fundamental Theory*. Oxford: Oxford University Press.
de Broglie, L. (1927). La nouvelle dynamique des quanta. In *Solvay Conference, Electrons et Photons*, pp. 341–71. Cambridge: Cambridge University Press. Translated from French in: G. Bacciagaluppi and A. Valentini (2009), *Quantum Theory at the Crossroads: Reconsidering the 1927 Solvay Conference*.
de Witt, B. S. (1970). Quantum mechanics and reality. *Physics Today 23*(9), 30–40.
Dürr, D., S. Goldstein, and N. Zanghì (1992). Quantum equilibrium and the origin of absolute uncertainty. *Journal of Statistical Physics 67*, 843–907.
Dürr, D., S. Goldstein, and N. Zanghì (1995). Quantum physics without quantum philosophy. *Studies in History and Philosophy of Physics 26*, 137–49.

Dürr, D., S. Goldstein, and N. Zanghì (1997). Bohmian mechanics and the meaning of the wave function. In R. Cohen, M. Horne, and J. Stachel (Eds), *Experimental Metaphysics: Quantum Mechanical Studies for Abner Shimony Vol I*, pp. 25–38. Boston Studies in the Philosophy of Science. Dordrecht: Kluwer.

Dürr, D., K. Münch-Berndl, S. Goldstein, and N. Zanghì (1992). Hypersurface Bohm–Dirac models. *Journal of Statistical Physics 67*, 843–907.

Esfeld, M. A. (2014). Quantum humeanism, or: physicalism without properties. *The Philosophical Quarterly 64*, 453–70.

Esfeld, M. A., D. Lazarovici, M. Hubert, and D. Dürr (2014). The ontology of Bohmian mechanics. *The British Journal for the Philosophy of Science 65*(4), 773–96.

Everett, H. (1957). Relative state formulation of quantum mechanics. *Review of Modern Physics 29*, 454–62.

Forrest, P. (1988). *Quantum Metaphysics*. Oxford: Basil Blackwell.

Ghirardi, G., A. Rimini, and T. Weber (1986). Unified dynamics for microscopic and macroscopic systems. *Physical Review D 34*, 470.

Ghirardi, G. C., A. Rimini, and T. Weber (1995). Describing the macroscopic world: closing the circle within the dynamical reduction program. *Foundations of Physics 25*, 5–38.

Goldstein, S. (1998). Quantum theory without observers. *Physics Today I and II*, 38–46.

Goldstein, S. and S. Teufel (2000). Quantum spacetime without observer: ontological clarity and the conceptual foundations of quantum gravity. In C. Callender and N. Huggett (Eds), *Physics Meets Philosophy at the Planck Scale*. Cambridge: Cambridge University Press.

Goldstein, S. and N. Zanghì (2013). Reality and the role of the wave-function in quantum theory. In D. Z. Albert and A. Ney (Eds), *The Wave-Function: Essays on the Metaphysics of Quantum Mechanics*, pp. 91–109. New York: Oxford University Press.

Hubert, M. and D. Romano (2018). The wave-function as a multi-field. *European Journal for Philosophy of Science 8*(3), 521–37.

Huggett, N. and C. Wüthrich (2013). Emergent spacetime and empirical (in)coherence. *Studies in the History and Philosophy of Modern Physics 44*, 276–85.

Kitcher, P. (1993). *The Advancement of Science*. Oxford: Oxford University Press.

Lewis, P. J. (2004). Life in configuration space. *The British Journal for the Philosophy of Science 55*, 713–29.

Lewis, P. J. (2005). Interpreting spontaneous collapse theories. *Studies in History and Philosophy of Modern Physics 36*, 165–80.

Lewis, P. J. (2006). Grw: a case study in quantum ontology. *Philosophy Compass 1*, 224–44.

Lewis, P. J. (2013). Dimension and illusion. In D. Z. Albert and A. Ney (Eds), *The Wave-Function: Essays on the Metaphysics of Quantum Mechanics*, pp. 110–25. New York: Oxford University Press.

Maudlin, T. (1995). Three measurement problems. *Topoi 14*(7), 7–15.

Miller, E. (2014). Quantum entanglement, Bohmian mechanics, and Humean supervenience. *Australasian Journal of Philosophy 92*(3), 567–83.

Monton, B. (2002). Wave-function ontology. *Synthese 130*, 265–77.

Monton, B. (2006). Quantum mechanics and 3n-dimensional space. *Philosophy of Science 73*(5), 778–89.

Monton, B. (2013). Against 3-n-dimensional space. In D. Z. Albert and A. Ney (Eds), *The Wave-Function: Essays on the Metaphysics of Quantum Mechanics*, pp. 154–67. New York: Oxford University Press.

Ney, A. (2012). The status of our ordinary three-dimensions in a quantum universe. *Noûs 46*, 525–60.

Ney, A. (2013). Ontological reduction and the wave-function ontology. In D. Z. Albert and A. Ney (Eds), *The Wave-Function: Essays on the Metaphysics of Quantum Mechanics*, pp. 168–83. New York: Oxford University Press.

Ney, A. (2015). Fundamental physical ontologies and the constraint of empirical coherence. *Synthese* 192(10), 3105–24.

Ney, A. (2017). Finding the world in the wave-function: some strategies for solving the macro-object problem. *Synthese* 192, 1–23.

Norsen, T. (2010). A theory of (exclusively) local beables. *Foundations of Physics* 40(12), 1858–84.

North, J. (2013). The structure of the quantum world. In D. Z. Albert and A. Ney (Eds), *The Wave-Function: Essays on the Metaphysics of Quantum Mechanics*, pp. 184–202. New York: Oxford University Press.

Psillos, S. (1999). *Scientific Realism: How Science Tracks Truth*. London: Routledge.

Solé, A. (2013). Bohmian mechanics without wave-function ontology. *Studies in History and Philosophy of Modern Physics* 44, 365–78.

Suárez, M. (2015). Bohmian dispositions. *Synthese* 192(10), 3203–28.

Tumulka, R. (2006). A relativistic version of the Ghirardi–Rimini–Weber model. *Journal of Statistical Physics* 125, 821–40.

Wallace, D. (2002). Worlds in the Everett interpretation. *Studies in the History and Philosophy of Modern Physics* 33, 637–61.

Williams, P. (2019). Realism made effective. *The British Journal for the Philosophy of Science* 70 (1), 209–37.

12

On the Status of Quantum State Realism

Wayne C. Myrvold

12.1 Introduction

There is a long tradition, very much alive in the present day, of irrealism about quantum states—that is, of denying that quantum states represent anything in physical reality.[1]

In this chapter, I will argue that the grounds we have for taking quantum states to represent physical properties of the systems to which they are ascribed are as strong as the grounds we have for taking atoms or electromagnetic waves to be real and to have something like the properties we ascribe to them. I will take it for granted that we do, indeed, have sufficient grounds for belief in the reality of atoms and electromagnetic waves. It is not my intention to try to convince a committed scientific anti-realist to make an exception for quantum states. The issue at hand is orthogonal to the age-old struggle between scientific realists and anti-realists. My targets here are those who deny that quantum states represent anything in physical reality from a standpoint that holds that one can, indeed, under certain circumstances ascribe reality to entities that are not directly observable, but take it that there are reasons specific to the quantum context for denying ontological import to quantum states.

[1] This tradition goes back at least as far as Bohr and Einstein, who agreed that quantum mechanics should not be taken as descriptive of physical reality, though they disagreed on the propriety of seeking a theory that would be. For Bohr, all description of physical reality had to be couched in classical terms, and the limits of classical physics were the limits of physical description. Einstein argued, in several places (see, *e.g.*, Einstein, 1936), that quantum states should be regarded as akin to the probability distributions of classical statistical mechanics, that is, as representing incomplete knowledge of some deeper underlying physical state. Contemporary representatives of this tradition include those who call themselves *QBists* (formerly 'Quantum Bayesians') (Caves, Fuchs, and Schack 2002; Fuchs, Mermin, and Schack 2014; Fuchs and Schack 2015). The central tenet of QBism is that a quantum state assignment is nothing more than a way of encoding an agent's subjective degrees of belief about that agent's own future experiences. Views that take quantum states to be representations of a state of knowledge, rather than physical reality, are often called ψ*-epistemic* views. A prominent exponent of an epistemic view of quantum states is Rob Spekkens (2007; 2012). Richard Healey (2012, 2017a, 2017c, this volume, Chapter 7) advocates a pragmatic view of quantum states, which denies that quantum states are representational. See Healey (2017b) for an overview of views of this sort.

Wayne C. Myrvold, *On the Status of Quantum State Realism* In: *Scientific Realism and the Quantum*. Edited by: Steven French and Juha Saatsi, Oxford University Press (2020). © Wayne C. Myrvold.
DOI: 10.1093/oso/9780198814979.003.0012

The question of the representational status of quantum states is a question that can be addressed even though we know that quantum mechanics is not a fundamental theory, but rather, a non-relativistic, low-energy approximation to a quantum field theory, and even though we have good reason to believe that even our best quantum field theories are effective theories, low-energy approximations to some deeper theory that would incorporate gravitation. Electromagnetic fields are real, even though classical electromagnetic theory is an approximation, valid within a limited regime, to a more fundamental theory. Classical physicists had good reason to believe that any deeper theory would include electromagnetic fields in its ontology, even if these fields are not precisely as classical electromagnetic theory conceives them to be. A successful argument that quantum states are real would not be one that depended crucially on a fiction that quantum mechanics is exactly right. What is required is an argument that we can expect any theory that recovers the predictions of quantum mechanics, or at least a close approximation to them, within the known domain of applicability of quantum mechanics, to have something corresponding to quantum states in its ontology, either as fundamental ontology or emergent from something more fundamental. As we shall see, this imposes a non-trivial constraint, as it would not do to take as a premise of the argument some condition that is violated by quantum field theory.

In Section 12.3 we will examine some theorems that circumscribe the realm of possible physical theories that can account for quantum phenomena. The first to be considered is the result of Barrett, Cavalcanti, Lal, and Maroney (2014), which shows that quantum states cannot be construed as some have hoped they could be, as probability distributions over an underlying state space, in such a way that operational indistinguishability of quantum states can be accounted for in terms of overlap of the corresponding probability distributions. We will then consider the theorem of Pusey, Barrett, and Rudolph (2012), which demonstrates, on the basis of an assumption about independent preparations performed on distinct systems, known as the Preparation Independence Postulate (PIP), that distinct pure quantum states are ontologically distinct.[2]

In accordance with the requirement that the ontological lessons we draw from physical theory rely only on premises that can reasonably be expected to be preserved under the transition to a successor theory, we should ask whether the PIP passes muster in that respect. And, indeed, there is an aspect of it that is problematic, in light of quantum field theory. The Postulate assumes that, for a system consisting of two or more spatially separated subsystems, for appropriate preparations the resulting state of the whole can be regarded as consisting merely of a list of states of the component subsystems. This assumption is called the Cartesian Product Assumption (CPA). This holds in classical physics, but is violated in any theory that is realist about quantum states. It holds within a fragment of quantum mechanics in which the states prepared are product states (that is, in which there is no entanglement between the spatially

[2] In labelling the Preparation Independence Postulate and its relatives, we follow the terminology of Leifer (2014).

separated parts). However, quantum field theory gives us incentive to doubt whether product states can reliably be prepared (see Clifton and Halvorson, 2001).

This gives us motivation to formulate a substitute for the PIP that does not invoke or presuppose the CPA. In Section 12.4, I present a condition that holds whenever the PIP holds but which is strictly weaker than it, which I call the *Preparation Uninformativeness Condition* (PUC). This condition requires no assumption about the structure of the state space of composite systems or its relation to subsystem state spaces. On the basis of the assumption of this condition, it can be shown that distinct quantum states—as long as they aren't too close together—must be ontologically distinct.

All of these theorems are couched within the ontological models framework. This framework, explicitly formulated by Harrigan and Spekkens (2010), codifies reasoning implicit in the practice of information theory, quantum or otherwise, and, indeed, in much of science and everyday life. Aspects of the framework could be rejected, but, as the sort of reasoning invoked is implicit in so much of science, strong grounds would be required for doing so. Now, it is, of course, possible that the methods of inference that we routinely employ in other domains of science lead us astray when it comes to investigating the quantum domain. One thing that is, after all, uncontroversially true is that any realist construal of quantum mechanics entails rejection of some one or other tenet of classical physics that one might have otherwise thought could be taken for granted. I acknowledge this, and, indeed, I accept that, if we had strong evidence that these methods of inference lead us astray when applied to the quantum domain, it would be reasonable to reject them. What is *not* reasonable, and not consistent with an earnest investigation of the world around us, is to reject methods of inference simply because their application would lead to conclusions that one finds unpalatable. The claim I am advancing in this chapter is that we do not have grounds for doubting the conclusion that quantum states represent something in physical reality that are sufficient to undermine the premises and modes of reasoning that lead to that conclusion.

12.2 Arguments for Anti-Realism about Quantum States

First, let us look at some of the reasons that have been given for denying that quantum states represent something physically real. There are two ways that one could take these. One could take them as motivating pursuit of a project of trying to develop a theory in whose ontology quantum states do not appear. Another way would be to take them as arguments for the conclusion that quantum states do not represent anything physically real. The difference matters, because the criteria for success of the arguments are different, depending on what the conclusion is taken to be. Upon undertaking a research project, say, to attempt to find a theory of a certain sort, one does not require assurance that the project can reach its goal. All that is needed is that it appears to be a promising line of research, whose goal, if reached, would constitute an advancement in understanding. Moreover, unsuccessful attempts themselves may lead to deeper understanding, if they help us to understand why they were unsuccessful—especially if we learn that the goal could not be reached.

If, on the other hand, we had strong evidence for the conclusion that quantum states are not representational, then, faced with arguments, such as those to be considered later, that they are, this evidence might afford us reason to be suspicious of, and perhaps even reject, the premises that lead to the conclusion. I do not think we are in such a position. The sorts of arguments that are given for anti-realism about quantum states serve their purpose well if they are taken to provide motivation for a certain line of theory pursuit. If, however, one were to take them as providing evidence for the conclusion that quantum states do not represent anything physically real, the evidence provided is weak at best, and certainly not sufficient to cast doubt on mundane assumptions that otherwise would be accepted without question.

In my opinion, the project of constructing an empirically adequate physical theory whose ontology would dispense with quantum states was, indeed, a worthwhile and well-motivated project, and, moreover, one that has been fruitful, precisely because it has led to deeper insight into the obstacles that such an endeavour faces. The situation bears some resemblance to the question of the viability of local hidden-variables theories. Inspired by the EPR argument for the incompleteness of quantum mechanics, J. S. Bell (1966) raised the question of whether there could be a hidden-variables theory that did not share the nonlocality of the de Broglie–Bohm theory, noting that, as far as he knew, there was "no *proof* that *any* hidden variable account of quantum mechanics *must* have this extraordinary character." What happened was that the quest for a local hidden-variables theory led to an impossibility proof. Bell's proof rests on assumptions, as any proof must. One of these is the so-called 'no-conspiracy' assumption, namely, that it is possible to create an experimental set-up in which the instrumental settings are effectively independent of the prepared state of the system to be experimented upon. One can, without logical contradiction, reject this assumption. But a rejection of this sort is a blunt instrument; it could be used to reject *any* experimental conclusion one doesn't like. The relevant question is: in the case of the Bell experiments, do we have evidence of conspiracies of this sort? If the answer is simply that they are being invoked to avoid an unwelcome conclusion, then it seems not unfair to say that those who invoke them have abandoned the sincere quest for knowledge about the world.

Leifer (2014) has helpfully compiled some of the chief arguments that have been advanced in favour of rejecting realism about quantum states. Leifer regards these as sufficiently strong that a reader who appreciates their force should find the ψ-ontology theorems surprising.[3] As already mentioned, I think that these considerations are better thought of as providing motivation for a project of constructing a theory that does not include quantum states in its ontology, rather than as positive evidence for the unreality of quantum states.

[3] 'ψ-ontology', with pun intended, is a term that has gained currency among physicists who discuss these matters for views that hold that quantum states represent something in physical reality. It is attributed to Christopher Granade, who was a student in Rob Spekkens' quantum foundations course at Perimeter Institute for Theoretical Physics in 2010 (see Leifer, 2014, p. 71).

The first argument Leifer considers stems from the Rob Spekkens' toy theory (Spekkens, 2007). With a remarkably simple construction, Spekkens demonstrates that a number of phenomena that we might have thought of as distinctively quantum can be captured by a model that is essentially classical, with restrictions on state preparation and on access to information about the state of the system. An elementary system in this toy theory is a set of four boxes with a ball that can be in one of them. The preparable states of individual systems are restricted in such a way that the most one could know is that the ball is in one of two of the boxes, with equal probability for each. For a pair of elementary systems, in addition to the product states, there are also entangled states, in which, for each of the subsystems, the ball is equally likely to be in each of the boxes, but there is perfect correlation between the two systems.

Features of quantum theory that are reproduced in the toy theory include existence of pure states that cannot reliably be distinguished, no-cloning, and (an analogue of) interference. Quantum phenomena that are provably not reproduced in the toy theory include violations of Bell inequalities, and the possibility of obtaining a Kochen–Specker obstruction.[4] It is suggested that the quantum-like phenomena that the toy theory reproduces are evidence that quantum theory itself is a theory of the type instantiated by the toy theory, that is, a theory with an essentially classical state space and a restriction on possible state preparations. An alternative moral that could be drawn is that it was a mistake to think of those phenomena as distinctively quantum (Myrvold, 2010, p. 182); such phenomena are, at most, *weakly* nonclassical (Spekkens, 2016, p. 92).

Support for this latter moral is found in work on generalized probabilistic theories. The framework of these theories encompasses a wide variety of probabilistic theories.[5] The scope of this framework is wide enough to include classical probabilistic theories, with or without restrictions on state preparations, quantum theories, and, in addition, a whole host of theories neither classical nor quantum. The theory whose state space consists of the states allowed in Spekkens' toy theory and probabilistic mixtures of these states falls within the scope of this framework.

Within the class of generalized probabilistic theories, there is a distinguished class, those with a state space that is a simplex, meaning: any mixture has a unique decomposition into pure states. These theories are the classical theories with no restrictions on state preparation. Call these *fully classical* theories, to distinguish them from classical theories with preparation restrictions, which exhibit some features usually thought of as non-classical. Theories that are not fully classical have in common the feature of having pure states that are not distinguishable.[6] *Ipso facto* they have all of the consequences of that condition, such as no-cloning. It is no surprise that quantum theory, being one of the theories that are not fully classical, shares with the Spekkens toy theory the features that are shared by all theories that are not fully classical. This is no reason to think that there should be some commonality in the physical interpretation of all such theories.

[4] If this term is unfamilier to you, see Held (2018) for background.

[5] See Janotta and Hinrichsen (2014) for an introduction and overview.

[6] Provided that the theory's state space is the convex hull of its set of pure states.

Leifer suggests that the fact that quantum theory falls within this broad framework, which also includes classical probability models, is evidence for an epistemic view of quantum states. "In this theory, quantum states are playing the same role in the quantum case that probability measures play in the classical case, and so it is natural to interpret quantum states and classical probabilities as the same kind of entity" (Leifer, 2014, p. 76). Thinking along these lines, it seems to me, fails to do justice to the generalized probabilistic theories framework. The framework was constructed to embrace a wide range of theories, and arguably what it presumes is just the minimum one would expect of any physical theory. It is *completely* neutral as to the physical interpretation of the states of the theory, and in this neutrality is its strength, as it is the source of generality.

The second argument considered by Leifer concerns fragments of quantum theory that can be recovered in a classical model. For instance, under suitable restrictions on quantum state preparations and evolutions, the Wigner function, a function on classical state space definable in terms of the quantum state, is positive and acts like a density function for a probability distribution over a classical phase space.

Is this surprising, on the assumption that quantum states are real? I think not. Suppose that you had never heard of quantum theory, but had become convinced that classical physics is inadequate in certain ways, because its predictions in certain domains are incorrect. You would, quite reasonably, expect any successor theory to recover the successes of classical physics. This means that you would expect to obtain something like classical behaviour in the relevant domains, or, to put it another way, that there be fragments of the theory that exhibit classical or quasi-classical behaviour. Studies of the classical limit of quantum mechanics take positivity of the Wigner function as an indication of classicality. That quantum mechanics exhibits classical-like behaviour in certain domains—that is, that quantum theory has a quasi-classical limit—is not evidence that quantum states are unreal, but, rather, a precondition for taking quantum mechanics as a serious candidate for a comprehensive theory.

Leifer also takes, as a strength of the epistemic view of quantum states, the fact that it bypasses the notorious quantum state collapse. Certainly, it is an attractive idea, one that has no doubt occurred to many, that collapse of the quantum state be thought of as nothing more than updating of information upon learning the result of a measurement. The question is whether this can be made to work. Any approach to the so-called 'measurement problem', including one that denies that quantum states represent physical reality, owes us an account of what happens in an experiment. The mainstream approaches—hidden-variables theories, dynamical collapse theories, and Everettian interpretations—all provide such accounts. Each of these deals with collapse in different ways. On the de Broglie–Bohm pilot wave theory, collapse of the effective wave function is a demonstrable consequence of the theory. On dynamical collapse theories, collapse is a real physical process. And Everettian theories can explain why, under appropriate circumstances, agents may be justified in disregarding other branches of the wave function other than their own, just as if there had been collapse. There is yet no worked-out proposal for a theory that embraces quantum phenomena on which quantum states are epistemic. At best what we have is a hope that an account of what happens during an experiment could be given on which quantum states play no part in the ontology. The situation, then, is that all of the main avenues of approach have rejected the problematic textbook version of collapse, with its reliance on a distinction between 'measurements' and other physical processes, and have provided a unified account of the goings-on of physical systems that makes no use of this distinction at the fundamental level and which nevertheless gives an account of why the textbook formulation works as a heuristic. In each case, this is accomplished by taking the quantum state as part of the ontology. Against this, the ψ-epistemicist offers only a hope that it could be accomplished without ontic quantum states. As long as this hope remains unfulfilled, consideration of the issues surrounding the measurement problem remain a problem to be solved by, rather than evidence in favour of, a ψ-epistemic view.

To sum up: if there were powerful evidence against the conclusion that quantum states correspond to something in physical reality, it might be reasonable to question the assumptions behind the arguments, to be considered in the next section, for the reality of quantum states. But the situation we find ourselves in seems to be the opposite; there is *no evidence at all* supporting irrealism about quantum states. At best we have considerations that suggest the pursuit-worthiness of the project of attempting to construct a plausible theory that accounts for quantum phenomena without ontic quantum states.

12.3 Arguments for an Ontic Construal of Quantum States

In this section, the arguments for ψ-ontology will be framed against the background of the ontological models framework. We will introduce this framework, then consider some arguments for ψ-ontology. The conclusions will differ in strength, depending on the strength of the auxiliary assumptions involved. We will first consider the theorem of Barrett, Cavalcanti, Lal, and Maroney (2014), which shows that indistinguishability of quantum states cannot be fully accounted for by overlap of probability distributions over an ontic state space. We will then consider the theorem of Pusey, Barrett, and Rudolph (2012), and then a variant of it that replaces the key assumption of this theorem, the Preparation Independence Postulate, with a weaker assumption, which we will call the Preparation Uninformativeness Condition (PUC).

12.3.1 *The ontological models framework*

Consider the following set-up, the sort of scenario with which information theory, be it quantum or classical, routinely deals. Alice has a message that she wants to convey to Bob. She has a physical system that she can send to him, after subjecting it to one of some set of available preparation procedures. Alice and Bob have an agreed-upon coding that associates possible messages with the available preparation procedures. Alice chooses her preparation procedure, subjects the system to it, and sends it to Bob, who performs an experiment, and takes the outcome as informative about what Alice has done.

In the simplest case, suppose that there are two procedures, which we will call P_1 and P_0, and that Bob has available to him an experiment with two possible results, R_1 and R_0, such that with probability one he obtains R_1 if Alice has performed P_1 and R_0

if she has performed P_0. In such a circumstance, it is hard to escape the conclusion that Alice's preparation has an effect on the state of the mediating system, and that the sets of states that could result from P_1 and P_0, respectively, are disjoint. Insisting that all probabilities be regarded as subjective judgements, as QBists do,[7] does not change the situation. If Bob's credences, or subjective degrees of belief, are such that he assigns probability one to the outcome R_1 conditional on the supposition of P_1, and probability one to the outcome R_0 conditional on the supposition of P_0, the natural explanation for this is that Bob believes that Alice's preparation has an effect on the state of the system that influences the outcome, and that he believes that the sets of states that could result from P_1 and P_0, respectively, are disjoint.

Could this conclusion be avoided? One could postulate that Alice's preparation has a direct influence on the result of Bob's future experiment, an influence unmediated by any influence on the state of the world between the preparation and the experiment. One could also stipulate that we are forbidden to theorize about the states of the mediating system. Moves of this sort would undermine the usual patterns of reasoning that underlie information theory, which presume that Bob, by doing an experiment on the system transmitted to him by Alice, gains information about what Alice did, mediated by the system that passed between them. One, could, perhaps imagine situations in which we had strong evidence for the unreliability of such patterns of reasoning, evidence strong enough to warrant rejecting them. *Perhaps*! But it should be noncontroversial that the mere fact that application of such inference schemes leads to the conclusion that the world is fundamentally nonclassical, or that it has features that some find unpalatable, does not constitute evidence for their unreliability.

It doesn't change matters much if we stipulate, as QBists do (Fuchs, Mermin, and Schack, 2014), that Alice is forbidden to even consider the effects of her choices on the probabilities of outcomes of an experiment performed by another agent. To make a stipulation of this sort is to abandon the very framework of information theory, but it doesn't block the inference, as Alice can send messages to her future self, as an aid to memory. Unless Alice believes that, when she looks tomorrow at the laptop she typed on today, she will gain information about what she wrote earlier, mediated by the effect on the internal state of the laptop of her choices made today, then it is hard to understand what she is doing, or why.

In cases in which the probabilities are different from zero and one, the reasoning is similar. Suppose that Alice has a choice between two coin-flipping procedures: P_0, which yields *heads* and *tails* with equal probability, and P_1, which yields *heads* with probability 2/3 and *tails* with probability 1/3. Alice chooses a preparation, flips the coin, and then passes it to Bob, who looks at it and sees *heads* or *tails*. He thereby gains information about the preparation procedure. If his prior credences in P_0 and P_1 are $Cr(P_0)$ and $Cr(P_1)$, respectively, and if his conditional credences $Cr(H|P_0)$, $Cr(H|P_1)$, are those just mentioned, then, in the event of seeing *heads*, an application of Bayes' theorem yields the result that his posterior credences in the two preparations should satisfy,

[7] See, in particular, Caves, Fuchs, and Schack (2007).

$$\frac{Cr(P_1|H)}{Cr(P_0|H)} = \frac{Cr(H|P_1)\,Cr(P_1)}{Cr(H|P_0)\,Cr(P_0)} = \frac{4}{3}\left(\frac{Cr(P_1)}{Cr(P_0)}\right). \tag{12.1}$$

That is, his credence in P_1 is increased, and his credence in P_0 diminished, in such a way that their ratio is increased by a factor of 4/3. In the event of seeing *tails*, his posterior credences in the two preparations should satisfy,

$$\frac{Cr(P_1|T)}{Cr(P_0|T)} = \frac{Cr(T|P_1)\,Cr(P_1)}{Cr(T|P_0)\,Cr(P_0)} = \frac{2}{3}\left(\frac{Cr(P_1)}{Cr(P_0)}\right). \tag{12.2}$$

Thus, the result of looking at the coin is, for Bob, informative about the preparation Alice chose; seeing *heads* boosts his credence in P_1, and lowers his credence in P_0, whereas seeing *tails* boosts his credence in P_0, and lowers his credence in P_1.

In this case, there are two disjoint classes of physical states that the coin can be in, corresponding to *heads* and *tails*. Corresponding to each preparation are probabilities for the state of the coin ending up in each of these classes, and it is assumed that Bob can ascertain which of these classes the state of the coin is in without error, simply by looking. This latter assumption is inessential. We might assume that Bob has some blurriness of vision, which introduces error at the readout stage, and that, in any state of the coin, there is some probability that he will see it as *heads*, and some probability that he will see it as *tails*. This changes little; as long as the net probability that he will see the coin as *heads* is higher given preparation P_1 than it is given preparation P_0, seeing heads should boost his credence that the preparation was P_1.

In reasoning of this kind there are two sorts of probabilities to be taken into account. We have *preparation probabilities*: we associate with each preparation procedure a probability distribution over the possible physical states that could result from the preparation. We also have *outcome probabilities*: for any experiment that can be performed on the system, for any physical state of the system we have, for each outcome, a probability of obtaining that outcome given that physical state.

It is worth noting that *nothing at all* in this sort of reasoning depends on whether these preparation and outcome probabilities are taken to be epistemic, or a matter of physics, or some mixture of epistemic and physical considerations. Bob might take it that the underlying physics is deterministic and that any uncertainty he might have about the result of a preparation stems from uncertainty about the details of what goes on in the preparation; all that matters is that the choice of preparation matters to his credences about the resulting state in a way that matters to his credences about the results of his experiment. If, on the other hand, the preparation and outcome probabilities are regarded as objective chances, then, provided that Bob knows what objective chance distribution to associate with a given preparation, and what distribution to associate with outcomes of a given experiment, and provided that his credences satisfy the Principle Principal,[8] his conditional credences will be such that he takes the outcomes of the experiment to be informative about the preparation.

[8] This is a principal often tacitly assumed in probabilistic reasoning, which was explicitly identified and named by David Lewis (1980). It requires a meshing between an agent's degrees of belief in a proposition A and her degrees of belief in propositions about possible chances of A. The Principle requires that an agent's degree of belief in A, conditional on the supposition that the chance of A is equal to x, be itself equal to x.

Those who wish to construe the probabilities that appear in the theorems to be considered below as purely epistemic are welcome to do so; the conclusion they will arrive at is that an agent whose credences about experimental outcomes conform to quantum mechanics ought to regard distinct quantum states as ontologically distinct.

These sorts of considerations, which have been more or less implicit in much of the discussions concerning the reality of quantum states, have been explicitly formulated by Harrigan and Spekkens (2010). We associate with any physical system a physical state space, or *ontic state space* Λ, and a set $\mathcal{L}$ of subsets of Λ that will be taken to be the measurable sets, that is, the ones that are candidates for ascribing a probability to.[9] With any preparation procedure ψ is associated a probability distribution P_ψ on the measurable space $\langle \Lambda, \mathcal{L} \rangle$. For any experiment E, with a set of outcomes o_k, $k = 1, \ldots, n$, there is a corresponding set of response functions f_k, such that $f_k(\lambda)$ is the probability of obtaining outcome o_k in ontic state λ. As these are probabilities for a set of mutually exclusive and jointly exhaustive alternatives, they must add up to one. Thus, for all $\lambda \in \Lambda$,

$$\sum_{k=1}^{n} f_k(\lambda) = 1. \tag{12.3}$$

With these in place, the probability that a system subjected to preparation ψ yields outcome o_k when experiment E is performed on it is the expectation value of f_k with respect to P_ψ.

$$P_\psi(o_k) = \int_\Lambda f_k(\lambda)\, dP_\psi(\lambda). \tag{12.4}$$

A few words of comment about the notion of preparation being invoked here. Note that we associate with each preparation procedure a corresponding probability distribution on the ontic state space. These probability distributions differ for different preparations, but it is assumed that, once the preparation performed has been specified, everything relevant to probabilities of outcomes of subsequent experiments has been specified. In particular, a preparation screens off such things as details of the past of the system that are not relevant to specification of the preparation. One could take this to be part of the meaning of 'preparation'—if you think that the past of a system continues to be relevant to future events, after a certain procedure has been performed, then you should take differing pasts to correspond to different preparations. Local preparations—that is, preparations taking place in a bounded region of space and time—are taken to screen off correlations between the prepared systems and the world outside.

The assumption that preparations of this sort are possible, and, indeed, are routinely performed in laboratories, is a substantive assumption, an assumption that does not follow from anything like a condition of causal locality. It is neither necessarily true nor knowable a priori. However, it *is* an assumption that lies deep at the heart of virtually all experimental science. If we ever come to a point at which we have reasons to doubt this sort of assumption, it will not come about as a result of experiments that

[9] We make the assumption, which is usual in probability theory, that $\mathcal{L}$ contains Λ and is closed under complementation and countable union; that is, we will take $\mathcal{L}$ to be a σ-algebra. See standard texts, such as Billingsley (2012).

presuppose it. And if we were presented with a reason to doubt this sort of assumption, it is hard to see how this doubt could be sufficiently contained so as not to undermine all of experimental science. Fortunately, we are not in such a position, as no one has offered grounds for doubting it.

Two preparations ψ, ϕ are said to be *distinguishable* if and only if there is an experiment E such that, for each outcome o_k of the experiment, either $P_\psi(o_k)$ is zero or $P_\phi(o_k)$ is zero. This means that, if a system is subjected to one of two preparations, but you don't know which, you can become certain of which it was by performing the experiment, as every outcome precludes one or the other of the possible preparations. This generalizes to larger sets of preparations: a finite set of preparations $\{\psi_i\}$ is said to be distinguishable if and only if there is an experiment E such that, for each outcome o_k, at most one of $\{P_{\psi_i}(o_k)\}$ is nonzero. Following Leifer (2014), we will say that a finite set of preparations $\{\psi_i\}$ is *antidistinguishable* if and only if there is an experiment E such that each outcome of E has zero probability on some preparation in the set. That is, no matter what the outcome of the experiment E is, it rules out at least one of the preparations.

Two preparations are said to be *ontologically distinct* if there is a measurable subset S of the ontic state space such that $P_\psi(S) = 1$ and $P_\phi(S) = 0$. It is a straightforward theorem that any distinguishable set of preparations is pairwise ontologically distinct. The converse might not hold; a pair of ontologically distinct preparations might not be distinguishable. This will be the case whenever there are limitations on what one can learn about the ontic state of a system in a single experiment.

If a set of preparations $\{\psi_i\}$ is antidistinguishable, this entails that the corresponding probability distributions $\{P_{\psi_i}\}$ have null joint overlap. That is, there is no subset S of the ontic state space such that $P_{\psi_i}(S) > 0$ for all ψ_i in the set.

In the coin-flip example, the two preparations, involving differing nonextremal probabilities for the outcomes *heads* and *tails*, are neither distinguishable nor ontologically distinct. In the absence of limitations on available experiments, on a classical theory, any pair of ontologically distinct states will be distinguishable. What I mean by this is: if, for every measurable subset S of the state space, there is an experiment that determines whether or not the state is in S, then every pair of ontologically distinct states is distinguishable.

In quantum mechanics, as is well known, nonorthogonal states are not distinguishable. If a pure quantum state is part of the ontology, then preparations of distinct pure states will be ontologically distinct, and so there will be ontologically distinct preparations that are not distinguishable.

The question arises whether nonorthogonal quantum states are analogous to classical states, in which indistinguishability of preparations corresponds to overlap in the associated probability distribution on the ontic state space. If this is the case, then one and the same ontic state is compatible with distinct quantum states, which is to say: the ontic state does not uniquely determine the quantum state. If, on the other hand, the quantum state supervenes on the ontic state, then preparations corresponding to distinct pure quantum states will be ontologically distinct. If an ontological model of quantum state preparations and experiments is such that, for any two distinct quantum states $|\psi\rangle$, $|\phi\rangle$, any pair of preparations ψ, ϕ that prepare those states are ontologically distinct, the model is said to be *ψ-ontic*. Harrigan and

Spekkens (2010) define *ψ-epistemic* as the negation of ψ-ontic. In their terminology, therefore, a model is ψ-epistemic if there is even one pair of distinct quantum states that are not ontologically distinct. A stronger notion is that of a *pairwise ψ-epistemic* model, in which no pair of nonorthogonal pure states is ontologically distinct.

The terminology 'ψ-ontic' is apt. If preparations corresponding to two quantum states $|\psi\rangle$, $|\phi\rangle$ are always ontologically distinct, this means that the ontic state always reflects which of these states was prepared. To be physically real, it is not required that quantum states be part of the fundamental ontology of the theory; states that supervene on the fundamental ontology are no less real for not being fundamental. An analogy: suppose that I specify some lighting and viewing conditions, and consider the set of things that, under those conditions, look yellow to me, and the set of things that, under those conditions, look blue to me. These sets are, presumably, ontologically distinct. The two sets would not be simply describable in physical terms, and it would be difficult to explain to anyone why physical things are being lumped together in these ways without reference to the visual system of creatures like me. But they are ontologically distinct nonetheless, and the distinction reflects a distinction in reality.

On the other hand, taking 'ψ-epistemic' to be simply the negation of 'ψ-ontic' seems to me to be potentially misleading. Consider, for example, a classical system, whose ontic state is represented by a point in its phase space. Suppose that one could learn either its position, or its momentum, but not both, though it always has determinate position and momentum. Any position is compatible with any momentum, and hence, for any position x and momentum p, the set of ontic states corresponding to position x overlaps with the set of states corresponding to momentum p. That doesn't mean that there is anything epistemic about position or momentum.

In addition, to call a model 'ψ-epistemic' if there are distinct quantum states whose associated probability distributions have *some* overlap, no matter how small, is potentially misleading, as it might suggest that the goal of constructing an interpretation on which quantum states are like classical probability distributions has been achieved. This, however, would require that the model be what has been called a *maximally ψ-epistemic* model (Barrett et al., 2014). On such a model, the indistinguishability of quantum states is fully explained by overlap of the corresponding probability distributions on ontic state space.

In addition to preparations that are perfectly distinguishable, there are also preparations that come close. A coin-flipping procedure that yields *heads* with probability very close to unity is distinguishable, not with complete certainty, but with high probability, from a coin-flipping procedure that yields *tails* with probability close to unity. For one way to quantify this, imagine the following game. A system is subjected to one of a pair of preparations, ψ, ϕ, with equal probability. You are presented with the prepared system, and are allowed to perform any experiment that you like. On the basis of the outcome of the experiment, you make a guess as to which preparation was performed. We ask: if you choose your experiment wisely, how high can the probability of your making a correct guess be? In the best case, there is an experiment that is certain to yield differing results depending on which preparation was applied, and the probability of correctly identifying the preparation is unity. In the worst case, any outcome of any experiment you can do has the same probability on both preparations, and the probability of correctly identifying the preparation is no greater than one half. In general, the probability of correctly identifying the preparation, on an optimal strategy, is

$$P(\text{correct guess}) = \frac{1}{2}\left(1 + \sup\left|P_\psi(o) - P_\phi(o)\right|\right). \tag{12.5}$$

Here 'sup' means *supremum*, that, is, the maximum value, taken over all outcomes o of experiments that can be performed on the system in question, or if there is no maximum value but only an increasing sequence of values that approaches some limiting value, this limiting value. We define the *distinguishability* of the preparations as

$$d(\psi,\phi) = \sup\left|P_\psi(o) - P_\phi(o)\right|. \tag{12.6}$$

The distinguishability $d(\psi,\phi)$ ranges between 0, for the case in which ψ and ϕ are indistinguishable, and 1, for perfectly distinguishable preparations. If the preparations correspond to quantum states $|\psi\rangle$, $|\phi\rangle$, then, if there are no restrictions on the permitted experiments (that is, if every experiment that, according to quantum mechanics, is possible, is permitted), we have

$$d(\psi,\phi) = \sqrt{1 - |\langle\phi|\psi\rangle|^2}. \tag{12.7}$$

We will want also a notion of approximate ontological distinctness. Given two probability distributions P, Q on a measurable space $\langle\Lambda, \mathcal{L}\rangle$, we define the *statistical distance*, also known as the *total variation distance*, between P and Q as

$$\delta(P,Q) = \sup_{A\in\mathcal{L}} |P(A) - Q(A)|. \tag{12.8}$$

Its value ranges between 0, when $P = Q$, and 1, when P and Q have disjoint supports. We define the *classical overlap* of two probability distributions by

$$\omega(P,Q) = 1 - \delta(P,Q). \tag{12.9}$$

Clearly, for any preparations ψ, ϕ, we will always have

$$d(\psi,\phi) \le \delta(P_\psi, P_\phi). \tag{12.10}$$

That is, distinguishability of two preparations can never be greater than their ontological distinctness. In the classical case, if there are no restrictions on experiments—that is, if, for any measurable subset S of the state space, there is an experiment that determines whether or not the state is in S—then we have equality in (12.10). In this case, all indistinguishability of two preparations is accounted for by overlap between the corresponding state-space probability distributions. Following Barrett et al. (2014), we will say that an ontological model of some fragment of quantum mechanics is *maximally ψ-epistemic* if and only if, for every pair of states $|\psi\rangle, |\phi\rangle$,

$$d(\psi,\phi) = \sqrt{1 - |\langle\phi|\psi\rangle|^2} = \delta(P_\psi, P_\phi), \tag{12.11}$$

or, equivalently,

$$\omega(P_\psi, P_\phi) = 1 - \sqrt{1 - |\langle\phi|\psi\rangle|^2}. \tag{12.12}$$

12.3.2 The BCLM theorem

Following Barrett, Calvalcanti, Lal, and Maroney (2014) (BCLM), we define the *quantum overlap* of two pure quantum states $|\psi\rangle, |\phi\rangle$ as

$$\omega_Q(|\psi\rangle, |\phi\rangle) = 1 - d(\psi, \phi) = 1 - \sqrt{1 - |\langle\phi|\psi\rangle|^2}. \tag{12.13}$$

This is zero for orthogonal states ($\langle\phi|\psi\rangle = 0$), and unity when $|\psi\rangle = |\phi\rangle$. If one had a theory on which quantum states were like classical probability distributions, and indistinguishability of quantum states could be fully accounted for by overlap of the corresponding distributions on an ontic state space—that is, a maximally ψ-epistemic theory—one would have, for all preparations ψ, ϕ that prepare pure quantum states $|\psi\rangle, |\phi\rangle$,

$$\omega(P_\phi, P_\psi) = \omega_Q(|\psi\rangle, |\phi\rangle). \tag{12.14}$$

This can be achieved for some fragments of quantum theory, which is what gives impetus to the project of attempting to construct a comprehensive theory of this sort that accounts for all quantum phenomena. However, as Barrett et al. (2014) demonstrate, it cannot be achieved for a model that fully reproduces quantum mechanics on a Hilbert space of dimension greater than three.

Here, in a nutshell, is the argument. Consider a Hilbert space $\mathcal{H}_d$ of a dimension $d = p^n$, greater than 3, that is a power of some prime number p. BCLM show that, for any $|\phi\rangle \in \mathcal{H}_d$, one can construct a set of state vectors, $\Psi = \{|\psi_i\rangle, i = 1, \dots, d^2\}$ with the following properties.

(a) For all $|\psi_i\rangle \in \Psi$, $|\langle\phi|\psi_i\rangle| = 1/\sqrt{d}$.
(b) For any pair $|\psi_i\rangle, |\psi_j\rangle$ of distinct elements of Ψ, either
   (i) $|\psi_i\rangle$ and $|\psi_j\rangle$ are orthogonal to each other, and hence the corresponding preparation distributions have null overlap, or
   (ii) the triple $\{|\phi\rangle, |\psi_i\rangle, |\psi_j\rangle\}$ is an antidistinguishable set, and hence the corresponding preparation distributions have null joint overlap.

On either of the alternatives, there is no joint overlap of $\{P_\phi, P_{\psi_i}, P_{\psi_j}\}$.

Now consider the average value of the overlap $\omega(P_\phi, P_{\psi_i})$, averaged over all elements of the set Ψ. Call this $\bar{\omega}(P_\phi, \Psi)$.

$$\bar{\omega}(P_\phi, \Psi) = \frac{1}{d^2} \sum_{i=1}^{d^2} \omega(P_\phi, P_{\psi_i}). \tag{12.15}$$

From the fact that no pair of distinct elements P_{ψ_i}, P_{ψ_j} of Ψ have non-null joint overlap with P_ϕ, it follows that

$$\sum_{i=1}^{d^2} \omega(P_\phi, P_{\psi_i}) \leq 1, \tag{12.16}$$

and hence that

$$\bar{\omega}(P_\phi, \Psi) \leq \frac{1}{d^2}. \tag{12.17}$$

Since, for each $|\psi_i\rangle$ in Ψ, $|\langle\phi|\psi_i\rangle| = 1/\sqrt{d}$, the value of the quantum overlap with ϕ is the same for each.

$$\omega_Q(|\phi\rangle, |\psi_i\rangle) = 1 - \sqrt{1 - 1/d}. \tag{12.18}$$

Call this value of the quantum overlap with $|\phi\rangle$, which is the same for all members of Ψ, $\omega_Q(|\phi\rangle, \Psi)$. From (12.17), with a little bit of arithmetic, we get

$$\bar{\omega}(P_\phi, \Psi) \leq \frac{1}{d}\left(1 + \sqrt{1 - 1/d}\right)\, \omega_Q(|\phi\rangle, \Psi) < \frac{2}{d}\, \omega_Q(|\phi\rangle, \Psi). \tag{12.19}$$

So, for example, for the case $d = 4$, the lowest-dimensional Hilbert space to which this theorem applies, for any vector $|\phi\rangle$ there is a set Ψ of 16 vectors such that the average overlap of P_ϕ with distributions corresponding to elements of Ψ satisfies

$$\bar{\omega}(P_\phi, \Psi) \leq \frac{1}{4}(1 - \sqrt{3}/2)\, \omega_Q(|\phi\rangle, \Psi) \approx 0.47\, \omega_Q(|\phi\rangle, \Psi). \tag{12.20}$$

Now, the average of the overlap $\omega(P_\phi, P_{\psi_i})$ taken over the set Ψ cannot be less than the smallest value of this overlap for $|\psi_i\rangle$ in that set. Therefore, there must be at least one $|\psi_i\rangle$ in Ψ such that

$$\omega(P_\phi, P_{\psi_i}) < \frac{2}{d}\, \omega_Q(|\phi\rangle, |\psi_i\rangle). \tag{12.21}$$

No ontological model for quantum mechanics can come close to the dream of having quantum states be like classical probability distributions. Even in a 4-dimensional Hilbert space, any ontological model must have, for some $|\phi\rangle$, $|\psi\rangle$, an overlap between the corresponding probability distributions that is less than half of the quantum overlap between these states. For larger Hilbert spaces, the minimum value of the ratio of classical overlap ω to quantum overlap ω_Q must be even smaller, and, for an ontological model of quantum mechanics on an infinite-dimensional Hilbert space, this ratio can have no minimum value greater than zero. This completely dashes the hope that provides much of the impetus for the project of constructing a theory on which quantum states are not ontic.

12.3.3 *The PBR theorem*

The BCLM theorem applies to any ontological model of quantum mechanics, and shows that no such model can be fully ψ-epistemic. There can be no theorem of this sort, which places no conditions on the ontological model, that has the conclusion that distinct quantum states are always ontologically distinct, as it is possible to construct ontological models in which any pair of distinct quantum states have some overlap in their corresponding probability distributions (Aaronson et al., 2013). A ψ-ontology theorem, therefore, must make some assumptions about the ontological model. These assumptions should not be arbitrary, but should be physically well motivated. In this section we consider the theorem of Pusey, Barrett, and Rudolph (2012) (PBR), which imposes an independence condition on probability distributions corresponding to product states.

Consider a pair of systems, A, B, each of which is to subjected to one of two distinct quantum state preparations, $|\psi\rangle$, $|\phi\rangle$, with $|\langle\psi|\phi\rangle| \leq 1/\sqrt{2}$. Consider now the set of states

$$\{|\psi\rangle_A\,|\psi\rangle_B, |\psi\rangle_A\,|\phi\rangle_B, |\phi\rangle_A\,|\psi\rangle_B, |\phi\rangle_A\,|\phi\rangle_B\}.$$

It can be shown that this set of states is antidistinguishable (Moseley, 2013). That is, there is an experiment E such that each outcome of the experiment is precluded by one of these preparations. This entails that there is no four-way joint overlap between the probability distributions on ontic state space corresponding to these four states.

Now suppose we add a further postulate, the Preparation Independence Postulate (PIP). This is actually the conjunction of two postulates. The first postulate, which, following Leifer, we call the *Cartesian Product Assumption* (CPA), is the condition that, when a pair of systems are independently subjected to pure-state preparations, the set of ontic states that can result from the preparation can be represented as a subset of the Cartesian product of state spaces of the individual systems. That is, the ontic state λ can be represented as an ordered pair $\langle\lambda_A, \lambda_B\rangle$, where λ_A represents the ontic state of A and λ_B represents the ontic state of B. The second postulate, the *No Correlations Assumption* (NCA), is the condition that, for appropriate preparations, the probability distributions corresponding to the four joint preparations are simply products of local distributions. That is, there are probability distributions P^A_ψ, P^A_ϕ on the state space of A, and probability distributions P^B_ψ, P^B_ϕ on the state space of B, such that, for any measurable subsets Δ_A of A's state space and Δ_B of B's state space, the probability, on the joint distribution $P_{\psi,\psi}$ corresponding to $|\psi\rangle_A\,|\psi\rangle_B$, that λ_A is in Δ_A and λ_B is in Δ_B, is simply the product of $P^A_\psi(\Delta_A)$ and $P^B_\psi(\Delta_B)$, and the probability, on the joint distribution $P_{\psi,\phi}$ corresponding to $|\psi\rangle_A\,|\phi\rangle_B$, that λ_A is in Δ_A and λ_B is in Δ_B, is the product of $P^A_\psi(\Delta_A)$ and $P^B_\phi(\Delta_B)$, and so on, for the other possible preparations.

The NCA can itself be regarded as a conjunction of two assumptions. The first, which we will call *Ontic Parameter Independence*, is that, for a given choice of preparation on A, the marginal distribution of λ_A—that is, the distribution obtained from the joint distribution over $\langle\lambda_A, \lambda_B\rangle$ obtained by averaging over λ_B—is the same for each choice of preparation on B. The second is the condition that, for any choice of preparations on the two systems, the corresponding probability distribution is one on which λ_A and λ_B are independently distributed. This assumption may well be called the No Correlations Assumption, but, since that label is already in use, we will call it the *Independence Assumption*.

The Ontic Parameter Independence assumption is a causal locality assumption, and is required for compatibility with relativistic causality. If it is violated, a choice of preparation on one system influences the probability of the result of the other preparation, even if we do not have the epistemic access to the ontic state of the system required to exploit this for signalling. The Independence Assumption is not required by causal locality, as there may be correlations between the states of the two systems that are due to influences in their common past. The assumption really amounts to the assumption that there is some way to effect the preparations so that such correlations are effectively screened off. It is not required that *every* procedure that we would regard as preparing the requisite quantum product state effect this screening off, only that there be *some* way to do this. Though this is not required by any sort of condition of causal locality, it is the sort of assumption that is pervasive in experimental science.

With the PIP in place, the condition that there be no four-way overlap between the four distributions considered entails ontological distinctness of P_ψ and P_ϕ on the state spaces of the subsystems A and B. To see this, assume the contrary: suppose there is a subset $\Delta_A \subseteq \Lambda_A$ that is assigned nonzero probability by both P^A_ψ and P^A_ϕ, and that there is a subset $\Delta_B \subseteq \Lambda_B$ that is assigned nonzero probability by both P^B_ψ and P^B_ϕ. Then, if the joint probability distributions satisfy the PIP, the set $\Delta_A \times \Delta_B$, which consists of pairs $\langle \lambda_A, \lambda_B \rangle$, with $\lambda_A \in \Delta_A$ and $\lambda_B \in \Delta_B$, is assigned nonzero probability by all four preparations, which is incompatible with the antidistinguishability of the set of preparations. We therefore, conclude that either P^A_ψ and P^A_ϕ are ontologically distinct, or P^B_ψ and P^B_ϕ are. However, if these are systems of the same type, subjected to the same choices of preparations, it seems reasonable to assume that the probability distributions are unchanged under an exchange of A and B, from which it follows that P^A_ψ is ontologically distinct from P^A_ϕ, and P^B_ψ from P^A_ϕ.

This argument applies to any pair of states with $|\psi\rangle$, $|\phi\rangle$, with $|\langle\psi|\phi\rangle| \leq 1/\sqrt{2}$. For a pair of distinct states with a larger quantum overlap (that is, with $|\langle\psi|\phi\rangle| > 1/\sqrt{2}$), we consider a larger set of systems of the same type. Consider a system consisting of $2n$ subsystems. Divide them into two equal subsets, which we will call A and B. Our choice of preparations consists of a choice between subjecting all of the systems in A to the $|\psi\rangle$-preparation, or subjecting all of them to the $|\phi\rangle$-preparation. We make the same choice for B. Since $|\psi\rangle$ and ϕ are distinct, $|\langle\psi|\phi\rangle| < 1$, and, for sufficiently large n, the n-fold product state $|\psi\rangle_1 \ldots |\psi\rangle_n$ has sufficiently small overlap with the n-fold product $|\phi\rangle_1 \ldots |\phi\rangle_n$ for the theorem to apply. With the PIP in place, we conclude ontological distinctness of these n-fold product states, and, using the PIP again, of the states $|\psi\rangle_i$, $|\phi\rangle_i$ of the individual subsystems.

The result is robust under elimination of the idealization of perfect preclusion, as it must be, to be taken seriously as telling us something about the actual world. Suppose that there exists an experiment such that, for some small ε, for each outcome, there is a preparation that ascribes a probability less than ε to that outcome. On the assumption of the PIP, it follows that the overlap $\omega(P_\phi, P_\psi)$ between probability distributions corresponding to the two preparations is less than $2\sqrt{\varepsilon}$. See the Supplementary Information section of Pusey, Barrett, and Rudolph (2012) for details of the proof.

12.4 Doing without the Cartesian Product Assumption

12.4.1 *The preparation uninformativeness condition*

We have so far not discussed the status of the CPA. It is, in fact, violated in relativistic quantum field theories as we now have them. Suppose that Alice and Bob perform operations on two systems A and B, these operations taking place within bounded spacetime regions, at spacelike separation from each other. We assume that the effects of Alice and Bob's operations on the quantum state can be represented by operators operating on the quantum state, and adopt the usual assumptions, required to ensure compatibility with relativistic causality, that the operators representing Alice's operations commute with operators representing observables at spacelike separation from her operations, and with those representing Bob's. On these assumptions, in the context of quantum field theory, we cannot assume that there is an operation that can be counted on to *completely* remove all entanglement between the systems A and B, and prepare them in a state that is exactly a product state.[10] A product state can be approximated as closely as we like, but cannot be reliably achieved exactly. In light of this, we need an independence postulate that does not presume that it is possible to prepare product states. It is best not to make any assumption about the structure of the state space at all, as we cannot expect to anticipate what sorts of state descriptions future theories might bring.

The assumption we will adopt in place of the PIP is one that we will call the Preparation Uninformative Condition (PUC). The PUC is meant to capture as much as we can of the content of the assumption that local state-preparations are possible without presupposing anything at all about the structure of the state spaces of composite systems. To state the assumption, we consider the following set-up. Suppose that, for systems A, B, we have some set of possible preparations of the individual systems. Suppose that the choice of preparation for each of the subsystems is made independently. Following the preparation of the joint system, which consists of individual preparations on the subsystems, you are not told which preparations have been performed, but you are given a specification of the ontic state of the joint system. On the basis of this information, you form credences about which preparations were performed. In the case of ontically distinct preparations, you will be certain about what preparation has been performed; if the preparations are not ontically distinct, you may have less than total information about which preparations were performed.

We ask: under these conditions, if you are now given information about which preparation was performed on one system, is this informative about which preparation was performed on the other? The Preparation Uninformative Condition is the assumption that it is not. This condition is satisfied in any model that satisfies the PIP. It is also satisfied whenever the preparations are ontically distinct. In such a case, given the ontic state of the joint system, you know precisely which preparations have been performed, and being told about the preparation on one system does not add to your stock of knowledge.

One way in which the PUC can be violated is to have the ontic state space of the joint system to be the Cartesian product of the subsystem ontic spaces, and for the joint probability distributions to be ones in which the states of the subsystems are correlated. It is also violated, as we shall see, by models, such as those constructed by Aaronson et al. (2013), on which nonorthogonal quantum states are never ontically distinct.

The PUC is implied by the PIP, but it is strictly weaker. Even if the CPA is assumed, it is possible to construct models for the PBR set-up, outlined in Section 12.3, in which the PUC is satisfied but $P_{\psi,\psi}$ and $P_{\psi,\phi}$ have nonzero overlap, which by the PBR theorem, is ruled out for models that satisfy the NCA. See Myrvold (2018) for one such construction.

[10] See Clifton and Halvorson (2001) for further discussion.

The PUC, it seems to me, is a necessary condition for the operations considered to count as local state-preparations. The substantive physical assumption made is that it is that such preparations are achievable, with sufficient effort, or, failing that, that it is possible to achieve approximate satisfaction of the condition, to as high a degree of approximation as is desired. This is all that is needed for the conclusions we will be drawing. It is not assumed that arbitrary operations satisfy the condition; only those that are to be counted as local state-preparations.

It should be emphasized that the PUC is *not* a causal locality condition; there are operations that violate it without violating causal locality. To see this, consider the following example, from quantum mechanics. Consider two systems, A, B, with associated Hilbert spaces $\mathcal{H}_A$ and $\mathcal{H}_B$. Take a pair of orthogonal state vectors $|0\rangle, |1\rangle$ from each Hilbert space, and form an entangled state vector,

$$\left|\Phi^+\right\rangle = \frac{1}{\sqrt{2}}\left(|0\rangle_A\,|0\rangle_B + |1\rangle_A\,|1\rangle_B\right). \tag{12.22}$$

Alice and Bob will each choose between two operations, and perform them, after which you will be told the resulting quantum state (assumed ontic for the purpose of this example). Suppose that Alice and Bob each have the choice between doing nothing, or performing a bit-flip operation that interchanges $|0\rangle$ and $|1\rangle$. You are told that the resulting state is just $\left|\Phi^+\right\rangle$, the same state that they started with. You are undecided between two alternatives: either Alice and Bob both did nothing, or they both did a bit flip. Clearly, in this situation, given the ontic state, information about Alice's choice of operation tells you something about Bob's. But this is a symptom of the fact that this is not a situation that counts as a pair of local state-preparations. The systems start out in an entangled state, and remain entangled. If, on the other hand, Alice and Bob's choices are between operations guaranteed to disentangle the systems, then, given the resulting quantum state, which would be a product state, information about Alice's choice of operation would tell you nothing about Bob's. Operations of that sort *are* candidates for being regarded as local state-preparations.

12.4.2 *A ψ-ontology result without the CPA*

We consider a set-up consisting of two subsystems A, B, with a choice of preparations ψ, ϕ to be made on each. Suppose that the set of four states arising from the two choices of preparation on the two subsystems is antidistinguishable. It follows from this that there is no joint overlap between the probability distributions corresponding to the four preparations. If, now, we impose the Preparation Uninformativeness Condition, it follows that, given the ontic state of the joint system, you will be undecided about at most one of the preparations performed on the subsystems. That is, either the ontic state allows you to uniquely determine the preparation of A, or it allows you to uniquely determine the preparation of B. This means that $P_{\psi,\psi}$ and $P_{\phi,\phi}$ have null overlap, as do $P_{\psi,\phi}$ and $P_{\phi,\psi}$.

To see this, suppose the contrary. Suppose that, given the ontic state λ, you are undecided about the preparations of both subsystems. This indecision can obtain only when λ is either in a joint overlap of $P_{\psi,\psi}$ and $P_{\phi,\phi}$, or in a joint overlap of $P_{\psi,\phi}$ and $P_{\phi,\psi}$. Suppose it is the former. Then, since there is no four-way overlap between the four preparation distributions, the ontic state must be incompatible with at least one of the other two preparations. Suppose that it is incompatible with $\langle\psi,\phi\rangle$. Then you are undecided between the preparations $\langle\psi,\psi\rangle$ and $\langle\phi,\phi\rangle$, but have zero credence in $\langle\psi,\phi\rangle$. Suppose, now, you are told that the A-preparation was ψ, and you update your credences on that information. You have now become certain that the B-preparation was ψ, in violation of the PUC. Similarly, if the ontic state λ is incompatible with $\langle\phi,\psi\rangle$, being told that the A-preparation was ϕ is informative about the B-preparation, in violation of the PUC. The same reasoning holds, of course, for an overlap of $P_{\psi,\phi}$ and $P_{\phi,\psi}$. Therefore, from antidistinguishability of the four preparations and the PUC it follows that the ontic state λ must uniquely determine either the quantum state of A or the quantum state of B.

This result is robust under de-idealization. If, for some small ε, there is a 4-outcome experiment E such that each preparation accords probability less than ε to some outcome of E, then, on the assumption of the PUC, $P_{\psi,\psi}$ and $P_{\phi,\phi}$ have small overlap, as do $P_{\psi,\phi}$ and $P_{\phi,\psi}$ (See Myrvold, 2018 for details):

$$\begin{aligned} \omega(P_{\psi,\psi}, P_{\phi,\phi}) &\leq 4\sqrt{\varepsilon}; \\ \omega(P_{\psi,\phi}, P_{\phi,\psi}) &\leq 4\sqrt{\varepsilon}. \end{aligned} \tag{12.23}$$

Now consider a large number N of systems, each subject to a choice of ψ or ϕ preparations. Call this large system, consisting of N subsystems, Σ_N. Among the experiments that can be performed on Σ are, for each pair of subsystems $\langle i,j\rangle$, an experiment that antidistinguishes the four alternatives $\{\langle\psi_i,\psi_j\rangle, \langle\phi_i,\psi_j\rangle, \langle\psi_i,\phi_j\rangle, \langle\phi_i,\phi_j\rangle\}$ for those subsystems. We require of our theory that it reproduce the quantum probabilities for any experiment that might be performed on the system. Then, on the assumption of the PUC, this entails that the ontic state of Σ_N must be such that, for each pair of subsystems, this ontic state uniquely determines the quantum state of at least one of them. It follows from this that the ontic state of Σ_N must uniquely determine the quantum state of at least $N-1$ of the subsystems. Therefore, in a large array of systems of this sort, the ontic state of the whole must uniquely determine the quantum state of the vast majority of them, with at most one exception. By taking N large enough, we can make the probability that a randomly chosen subsystem has its quantum state uniquely determined by the ontic state of Σ_N as close to unity as we like.

Let us now impose a *Principle of Extendibility*. This is the requirement that the ontic state of the system Σ_N be compatible with regarding the system as a subsystem of a larger system $\Sigma_{N'}$ consisting of N' subsystems subjected to ψ or ϕ preparations, for arbitrarily large N'. With this assumed, the probability that *all* of Σ_N's subsystems have their preparations uniquely determined by the ontic state of Σ_N must be greater than p for all $p < 1$. That is, with probability one, all of the subsystems of Σ_N must be such that their quantum states are uniquely determined by the ontic state of Σ_N.

This result is, again, robust under relaxation of the assumption of perfect antidistinguishability. It can be shown that, if, for each pair of subsystems, there is an experiment such that each of the outcomes has probability less than ε, then the ontic state of Σ_N must be such that it permits almost certain identification of the quantum state of a randomly selected subsystem. Once again, see Myrvold (2018) for details.

This result holds for pure-state preparations $|\psi\rangle$, $|\phi\rangle$, with $|\langle\psi|\phi\rangle| \leq 1/\sqrt{2}$, and, because it is robust under approximations, for any pair of preparations that approximate such states closely enough to permit approximate antidistinguishability. For quantum states with a greater quantum overlap, we can use the result to show that for sufficiently large n, the n-fold product $|\psi\rangle_1 \ldots |\psi\rangle_n$ is ontologically distinct from $|\phi\rangle_1 \ldots |\phi\rangle_n$. Unlike the PIP, the PUC does not permit us to conclude, straightaway, that for individual subsystems $|\psi\rangle$ is ontologically distinct from $|\phi\rangle$. But it is hard to believe that there is a theory worth taking seriously which is such that $|\psi\rangle$ and $|\phi\rangle$ are ontologically distinct whenever $|\langle\psi|\phi\rangle| \leq 1/\sqrt{2}$ and, though $|\psi\rangle$ and $|\phi\rangle$ are not ontologically distinct for individual systems, for a sufficiently large collection of systems the set of states that can arise from subjecting all of them to the $|\psi\rangle$-preparation is ontologically distinct from the set of states that care arise from subjecting all of them to the $|\phi\rangle$-preparation. To echo Bell (1977), if someone presents me with a candidate for such a theory, I will not refuse to listen, but I will not myself try to make such a theory.

12.5 Conclusion

The PUC is a fairly weak condition, consistent with pervasive nonseparability of state descriptions, and satisfied even in manifestly nonlocal theories of quantum phenomena, such as the de Broglie–Bohm theory. Though, of course, it is possible to consider theories on which it does not hold, it must be admitted that we have *no evidence whatsoever* that this condition is not satisfied in the actual world. Any theory that satisfies this condition must have it that distinct states with an inner product not greater than $1/\sqrt{2}$ are ontologically distinct. Furthermore, no theory whatsoever, whether it satisfies the PUC or not, can both reproduce the quantum probabilities for results of experiments and satisfy the *desideratum* that indistinguishability of states be fully accounted for by overlap of the corresponding probability distributions. For these reasons, though the project of constructing a theory, of the sort envisaged by Einstein, in which quantum states are analogous to probability distributions in classical statistical mechanics, was well-motivated, we have to admit that the fruit it has borne consists of insight into why the goal cannot be achieved.

We have reached the point, it seems to me, at which anyone concerned with understanding what the empirical success of quantum theory is telling us about the world should acknowledge that it is telling us that the furniture of the world includes something corresponding to quantum states. This, of course, does not come close to settling the question of what a complete account of the world might be like, and the old questions remain about how best to understand what it is that quantum theory is telling us about the world we live in.

Acknowledgements

Support for this research was provided by Graham and Gale Wright, who sponsor the Graham and Gale Wright Distinguished Scholar Award at the University of Western Ontario.

This research was also supported by Perimeter Institute for Theoretical Physics. Research at Perimeter Institute is supported by the Government of Canada through the Department of Innovation, Science and Economic Development and by the Province of Ontario through the Ministry of Research and Innovation.

I am grateful to the members of the Philosophy of Physics Reading Group at the University of Western Ontario for helpful feedback and discussions.

References

Aaronson, S., A. Bouland, L. Chua, and G. Lowther (2013). ψ-epistemic theories: the role of symmetry. *Physical Review A* 88, 032111.

Barrett, J., E. G. Cavalcanti, R. Lal, and O. J. E. Maroney (2014). No ψ-epistemic model can fully explain the indistinguishability of quantum states. *Physical Review Letters* 112, 250403.

Bell, J. S. (1966). On the problem of hidden variables in quantum mechanics. *Reviews of Modern Physics* 38, 447–52. Reprinted in Bell (1987), pp. 1–13.

Bell, J. S. (1977). Free variables and local causality. *Epistemological Letters* 15, 79–84. Reprinted in *Dialectica* 39 (1985), pp. 103–16, and in Bell (1987), pp. 100–4.

Bell, J. S. (1987). *Speakable and Unspeakable in Quantum Mechanics*. Cambridge: Cambridge University Press.

Billingsley, P. (2012). *Probability and Measure, Anniversary Edition*. Hoboken, NJ: Wiley.

Caves, C., C. Fuchs, and R. Schack (2007). Subjective probability and quantum certainty. *Studies in History and Philosophy of Modern Physics* 38, 255–74.

Caves, C. M., C. A. Fuchs, and R. Schack (2002). Quantum probabilities as Bayesian probabilities. *Physical Review A* 65, 022305.

Clifton, R. and H. Halvorson (2001). Entanglement and open systems in algebraic quantum field theory. *Studies in History and Philosophy of Modern Physics* 32, 1–13.

Einstein, A. (1936). Physik und Realität. *Journal of the Franklin Institute* 221, 349–382. English translation in Einstein (1954).

Einstein, A. (1954). Physics and reality. In *Ideas and Opinions*, pp. 290–323. New York: Crown Publishers, Inc. Translation of Einstein (1936).

Fuchs, C. A., N. D. Mermin, and R. Schack (2014). An introduction to QBism with an application to the locality of quantum mechanics. *American Journal of Physics* 82, 749–54.

Fuchs, C. A. and R. Schack (2015). QBism and the Greeks: why a quantum state does not represent an element of physical reality. *Physica Scripta* 90, 015104.

Harrigan, N. and R. W. Spekkens (2010). Einstein, incompleteness, and the epistemic view of quantum states. *Foundations of Physics* 40, 125–57.

Healey, R. (2012). Quantum theory: a pragmatist approach. *The British Journal for the Philosophy of Science* 63, 729–71.

Healey, R. (2017a). Quantum states as objective informational bridges. *Foundations of Physics* 47, 161–73.

Healey, R. (2017b). Quantum-Bayesian and pragmatist views of quantum theory. In E. N. Zalta (Ed.), *The Stanford Encyclopedia of Philosophy* (Spring 2017 edn). Metaphysics Research Lab, Stanford University. https://plato.stanford.edu/entries/quantum-bayesian/

Healey, R. (2017c). *The Quantum Revolution in Philosophy*. Oxford: Oxford University Press.

Held, C. (2018). The Kochen-Specker theorem. In E. N. Zalta (Ed.), *The Stanford Encyclopedia of Philosophy* (Spring 2018 edn). Metaphysics Research Lab, Stanford University. https://plato.stanford.edu/archives/spr2018/entries/kochen-specker/.

Janotta, P. and H. Hinrichsen (2014). Generalized probability theories: what determines the structure of quantum theory? *Journal of Physics A: Mathematical and Theoretical* 47, 323001.

Leifer, M. S. (2014). Is the quantum state real? An extended review of ψ-ontology theorems. *Quanta* 3, 67–155.

Lewis, D. (1980). A subjectivist's guide to objective chance. In R. C. Jeffrey (Ed.), *Studies in Inductive Logic and Probability*, Vol. 2, pp. 263–93. Berkeley: University of Caifornia Press.

Moseley, C. (2013). Simpler proof of the theorem by Pusey, Barrett, and Rudolph on the reality of the quantum state. https://arxiv.org/abs/1401.0026.

Myrvold, W. C. (2010). From physics to information theory and back. In A. Bokulich and G. Jaeger (Eds), *Philosophy of Quantum Information and Entanglement*, pp. 181–207. Cambridge: Cambridge University Press.

Myrvold, W. C. (2018). ψ-ontology result without the Cartesian Product Assumption. *Physical Review A* 97, 052109.

Pusey, M., J. Barrett, and T. Rudolph (2012). On the reality of the quantum state. *Nature Physics* 8, 475–8.

Spekkens, R. W. (2007). Evidence for the epistemic view of quantum states: a toy theory. *Physical Review A* 75, 032110.

Spekkens, R. W. (2012). Why I am not a psi-ontologist. Lecture, video available at http://pirsa.org/12050021/.

Spekkens, R. W. (2016). Quasi-quantization: classical statistical theories with an epistemic restriction. In G. Chiribella and R. W. Spekkens (Eds), *Quantum Theory: Informational Foundations and Foils*, pp. 83–135. Dordrecht: Springer.

PART V

Scientific Realism and Quantum Field Theory

13

The Non-Miraculous Success of Formal Analogies in Quantum Theories

Doreen Fraser

13.1 Introduction

Historical case studies fuel the scientific realism–anti-realism debate. Core examples are the luminiferous ether models in the nineteenth century, which have been variously interpreted as (literally) false because the term 'ether' fails to refer (Laudan, 1981), as giving an approximately correct representation of the structure of "the oscillations of something or other at right angles to the light" on which the modeled optical phenomena depend (Worrall, 1989), as approximately true because the term 'ether' actually refers to the electromagnetic field (Psillos, 1999), or as approximately true because light has the higher-order (i.e., multiply realizable), causal-nomologically specified property of satisfying a principle of superposition with respect to spatial components (Saatsi, 2005). Mechanical ether models were one facet of a heuristic strategy of using analogies to develop a theory for electromagnetism. William Thomson initially drew analogies to Fourier's theory of heat to model electrostatics, and then James Clerk Maxwell picked up on Thomson's method of analogical reasoning and constructed analogue incompressible fluid and other mechanical models for electromagnetic phenomena. Later mechanical ether models were used as analogue models for electromagnetism. Analogical reasoning has also been widely used in quantum theory. However, the analogies that are employed in at least some cases in quantum theory are different in kind from those that figured in the development of electromagnetism. As a result, these case studies of analogical reasoning in quantum theory carry different morals for the scientific (anti)-realism debate than the familiar luminiferous ether models.

My primary case study of the use of analogies in quantum theory is the development of the Higgs model by analogy to models of superconductivity. Quantum field theory (QFT) is the framework theory for the Higgs model and quantum statistical mechanics (QSM) supplies the framework for the pertinent models of superconductivity. Adam Koberinski and I have argued that the analogies drawn in this case are formal analogies that are not accompanied by physical analogies

Doreen Fraser, *The Non-Miraculous Success of Formal Analogies in Quantum Theories* In: *Scientific Realism and the Quantum*. Edited by: Steven French and Juha Saatsi, Oxford University Press (2020).
 DOI: 10.1093/oso/9780198814979.003.0013

(Fraser and Koberinski, 2016). That is, the analogical mappings from the source domain (superconductivity) to the target domain (the electroweak interaction) are *formal analogies* because they map elements of the models that play the same formal roles in the formal (mathematical) structures of the models. This is the principle on which the analogical mappings are determined, but of course this is compatible with the mapped elements also being physically similar in some respect, which would constitute accompanying *physical analogies*. However, in the Higgs case, the analogies are *purely formal analogies* (i.e., not also physical). This is a departure from the earlier use of analogical reasoning in the development of electromagnetism, in which there are both formal and physical analogies.

Cases of the successful use of formal analogies necessitate some modifications to our high-level picture of how science works. The philosophical literature on analogies in science has been largely skeptical about the usefulness of purely formal analogies in science (e.g., Hesse, 1966; Holyoak and Thagard, 1995; Bartha, 2010). The reasons for skepticism about the efficacy of formal analogies are complex and varied, but one source is broadly realist intuitions about the relationship between success and truth in science that are captured by the 'No Miracles' Argument (NMA): the best explanation of the success of science is getting something approximately right about the world. The main line of argument developed in this chapter is that the successful use of purely formal analogies undermines this realist intuition by breaking the connection between success and approximate truth. Physical analogies are compatible with this realist intuition because they pick out physical similarities. Learning that a physical analogy holds between two systems confers knowledge of a physical fact about the target system. In contrast, learning that a purely formal analogy obtains does not deliver any substantial information about the physical world because a purely formal analogy does not rely on any physical similarities between the source and target domains.[1] Consequently, contrary to the NMA, the successful employment of purely formal analogies is not explained by the fact that we got something right about the world. The explanation held up by the NMA as the best explanation is not a candidate explanation in this case. However, this should not be taken to entail that the successful use of formal analogies is miraculous. I shall argue that there is a perfectly satisfactory explanation for the success of formal analogical reasoning, but that it does not involve getting something right about the world.

The central argument in this chapter is that the use of purely formal analogies in the development of the Higgs model (Section 13.3) is a counterexample to two versions of the NMA. The first version is Stathis Psillos' Refined Explanationist Defense of Realism, which concerns scientific methodology (Sections 13.2 and 13.4). The second version is the Argument from the History of Science for structural realism (Frigg and Votsis, 2011), which incorporates an inference to the approximate truth of the mathematical structure of theories (Section 13.5). When invoking either version of the NMA, it is tempting to cast the approximate truth of pertinent theories as the only correct explanation for the instrumental success of science. This position is ruled out by the successful use of purely formal analogies. Sections 13.4.1, 13.5.1, and 13.6 consider

[1] This inference does not apply to Pythagorean views such as Tegmark's according to which the physical world is constituted by mathematics (Tegmark, 2008). Such views will be set aside in this chapter.

possible responses to these arguments open to scientific realists who want to retain the NMA. It is the use of purely formal analogies that differentiates the Higgs model case study (and plausibly other cases in quantum theory (Section 13.3)) from the historical cases of the use of analogical reasoning such as electromagnetism, which informed these versions of the NMA. Thus, attention to quantum theories can contribute to the realism–anti-realism debate by bringing to light novel scientific methods not used in familiar historical cases.

13.2 The Refined Explanationist Defense of Realism and Physical Analogies

To motivate the analysis of the Higgs case study in the next section, consider the version of the 'No Miracles' Argument (NMA) advanced in Psillos (1999) and how it is supported by the historical development of electromagnetism. Psillos' Refined Explanationist Defense of Realism takes Boyd's explanationist defense of realism as its starting point (78). In accordance with a commitment to naturalism, the argument takes the form of an inference to the best explanation, a form of argument that is used in science, and the explanandum is the instrumental reliability of scientific methodology in general. Boyd asserts that all aspects of scientific methodology fall within the scope of the argument and offers as examples—in addition to the paradigm example of measurement procedures for theoretical magnitudes—"principles of experimental design, choices of research problems, standards for the assessment of experimental evidence and for assessing the quality and methodological import of explanations, principles governing theory choice, and rules for the use of theoretical language" (Boyd, 1990, 361). Psillos concurs that "all aspects of scientific methodology are deeply theory-informed and theory-laden," and thus fall within the scope of the argument (78). Psillos refines Boyd's argument by explicitly recognizing that scientific methods sometimes fail and by localizing the inference to the approximate truth of only those aspects of theories that explain instrumental success (80). Here is Psillos' summary of his refined version of the argument:[2]

(REDR) The best explanation of the instrumental reliability of scientific methodology is that: the theoretical statements which assert the specific causal connections or mechanisms by virtue of which scientific methods yield successful predictions are true. (78)

The core intuition is that the approximate truth of the theories that underwrite the methodologies is responsible for the instrumental success of the methodologies because the methods are theory-dependent.

Psillos argues that this core intuition and the REDR are borne out in the historical example of the development of electromagnetism. His interpretation of this case differs from the structural realist interpretation, but (as Psillos acknowledges) there

[2] More recently, Psillos has placed greater emphasis on the REDR being a two-stage argument in order to clarify the role of inference to the best explanation reasoning (Psillos, 2011). The arguments concerning formal analogies made here also apply to this articulation of the argument.

is common ground between the two interpretations.[3] This common interpretation of the electromagnetism case study includes the noteworthy features that differentiate it from the Higgs model case study. Following Stein and Poincaré, Psillos emphasizes the role that analogical reasoning played in Maxwell's development of electromagnetism (137–45). Maxwell relied on the construction of analogue models for electromagnetic phenomena. For example, Maxwell (1861 [1890]) develops a mechanical model for a range of electromagnetic phenomena constructed out of molecular vortices connected by idle wheels. At the time, Maxwell regarded this model as a possible causal model for electromagnetic phenomena (i.e., that it is possible that electromagnetic phenomena are caused by the motions of the molecular vortices and idle wheel particles). If this possibility were actualized, then the model would not be merely an analogue of the model for electromagnetic phenomena, but identical with it. However, Maxwell believed that this was unlikely. Nevertheless, while this particular mechanical model was unlikely to be the correct model, he maintained that there was "good evidence" that electromagnetic phenomena have a mechanical cause (Maxwell 1873 [1954], 416–17).[4]

From our current perspective, Maxwell's vortex–idle wheel model incorrectly represents electromagnetic phenomena as having a mechanical cause. Nevertheless Maxwell was able to arrive at his equations for electromagnetism using this model. How? There is a *physical analogy* between his vortex–idle wheel model and the true model of classical electromagnetism: the causal relations between the centrifugal force of the molecular vortices and the rotation of light map to the causal relations between the magnetic force field and the rotation of the light. To more precisely characterize this analogy, the distinctions in Mary Hesse's 1966 account of analogies are useful. A *horizontal relation* is the analogical mapping of an element of the source model to an element of the target model. A *vertical relation* is a relation between elements within a single model (i.e., for either the source domain or the target domain). Hesse's paradigm example of vertical relations is causal relations.[5] In these terms, Maxwell's analogy maps vertical relations in the source domain (viz. the centrifugal force of the molecular vortices and the rotation of light) to vertical relations in the target domain (viz. the causal relations between the magnetic force field and the rotation of the light). This is a *physical analogy* because the mapped vertical relations are physically similar (i.e., both causal relations). In contrast, a *formal analogy* maps vertical relations that play similar formal (e.g., mathematical) roles. It is possible for analogies to be both physical and formal. *Purely formal analogies* are exclusively formal analogies. The completion of the treatment of electromagnetism within Lagrangian mechanics

[3] And there are further differences between interpretations favored by adherents of different variants of structural realism. Psillos addresses Worrall's variant of structural realism.

[4] According to this view of Maxwell's, there is a physical analogy between vertical relations: the causal relations in the vortex–idle wheel model are mapped to causal relations in the true mechanical ether model. For example, in the vortex–idle wheel model, the centrifugal force of the vortices causes the magnetic rotation of light in the magneto-optical (aka Faraday) effect; there is a mechanical analogue of the centrifugal force in the true mechanical model (Harman, 1998, 98–106).

[5] Hesse also uses the development of electromagnetism as a case study of analogical reasoning, but, for various reasons, her treatment differs from those discussed here.

at the end of the nineteenth century supports this characterization of the physical analogy. It was recognized that an abstract Lagrangian can admit concrete models of different types, including both mechanical and non-mechanical models. The shared Lagrangian representation establishes that the models of the two types share the same causal structure. The abstract Lagrangian formulation of mechanics allows the posited energy of the ether to be expressed as a function of the classical fields (Stein, 1989, 62).

It is common ground between Psillos and structural realists that Maxwell's reasoning invoked a physical analogy between vertical physical relations in the vortex–idle wheel model and vertical physical relations in the true model of classical electromagnetism. Psillos and structural realists part ways in their further specifications of what Maxwell got right. Psillos agrees that the *divide et impera* strategy of differentiated epistemic commitment to elements of theories is the right strategy for defending realism, but disagrees with an across-the-board epistemic commitment to structure (as opposed to content). He argues that the correct way to understand analogue ether models is to interpret the term 'ether' as having the same referent as the term 'electric field' in the fully developed theory of electromagnetism (296–7). The rationale for this interpretation is supplied by Psillos' casual descriptive theory of reference. The core natural kind-constitutive properties are shared by ether and the electric field, which allows them to play the same causal role in the production of electromagnetic phenomena. This diagnosis of the continuity between ether models and the full-blown classical field theory of electromagnetism also underwrites the application of the REDR. The method is the use of analogue models and the explanation for its success is that—when correctly interpreted—the term 'ether' in the analogue models and the term 'electric field' in the theory refer to the same entity, and thus generate approximately true statements about the causal mechanism that produces electromagnetic phenomena. It is essential to Psillos' interpretation that the ether model and the full theory of electromagnetism plausibly represent the same casual connections. This essential condition is satisfied because the analogies that Maxwell draws track similarities between the structures of the causal relations that are captured by the abstract Lagrangian framework.

Psillos' REDR incorporates an explanation of the instrumental success of science that is well-suited to the method of using physical analogies that map (vertical) causal relations in the source model to (vertical) causal relations in the target model. However, analogies that do not map causal relations (or mechanisms) would not be explained by the REDR. Purely formal analogies fall into this category. Thus, from the perspective of REDR, the use of purely formal analogies should not be an instrumentally reliable scientific method.[6] As we shall see in the next section, the historical development of the Higgs model does not conform to this expectation.

[6] Psillos (following Hesse and Achinstein) includes formal analogies in his exposition of analogical reasoning, but purely formal analogies play no role in his defense of realism (141–2). This is understandable because the electromagnetism case study and the other historical cases that he considers use physical analogies.

13.3 Formal Analogies in the Development of the Higgs Model

13.3.1 *Overview*

The use of analogies to models of superconductivity in the development of the Higgs model is a very sophisticated example of analogical reasoning. The complexity of this case means that tracing the core sets of analogies involved is a non-trivial exercise. One complication is that analogies are drawn to two models of superconductivity, the phenomenological Ginzburg–Landau (GL) model and the dynamical Bardeen–Cooper–Schrieffer (BCS) model (Ginzburg and Landau, 2009; Bardeen, Cooper, and Schrieffer, 1957). Fraser and Koberinski (2016) offer a detailed analysis of this case study. To simplify matters for the present purpose of assessing the NMA, this overview will focus on the tighter of the two sets of analogies, those between the GL model and the Higgs model.

Initially, the analogy to superconductivity was motivated by the presence of effectively massive photons and effectively massive plasmons and the absence of massless bosons in the superconducting phase in the BCS model. In the superconducting phase, the global $U(1)$ gauge symmetry is broken. These features were noteworthy because they contradicted widely held beliefs in the particle physics community. At the time, particle physicists believed that the gauge bosons in Yang–Mills-type theories—such as the photon in electromagnetism—are necessarily massless (Jona-Lasinio, 2002). Furthermore, on the strength of Goldstone's theorem, particle physicists believed that broken symmetries are necessarily accompanied by massless Goldstone bosons, and experiments on electroweak interactions had not turned up unaccounted for massless particles. The BCS model of superconductivity was suggestive, but was not regarded as decisive evidence that models of particle physics with massive gauge bosons and without massless Goldstone bosons are possible because the models of superconductivity are non-relativistic and models in particle physics are relativistic (Higgs, 2014, 851). The analogies between effectively massive particles in superconductivity and genuinely massive particles in electroweak interactions supplied a starting point, but it was the set of formal analogies that followed that led to the construction of the Higgs model. It was counterexamples such as this, constructed by formal analogy to superconductors, which supplied the convincing evidence that spontaneous symmetry breaking (SSB) in relativistic particle physics need not be accompanied by massless Goldstone bosons.[7]

At the heart of the set of formal analogies are the order parameters and the infamous Mexican hat diagram (Table 13.1). In the GL model, the order parameter that distinguishes the 'normal' state of the metal from the superconducting state is $\psi(\mathbf{x})$, which is the effective collective wave function for the superconducting electrons. ($\mathbf{x}$ represents space.) $|\psi(\mathbf{x})|^2$ is zero in the symmetric phase and non-zero in the broken symmetry phase. In the Higgs model, the analogue of $\psi(\mathbf{x})$ is the complex scalar quantum field $\phi(x)$ which (after manipulations) is associated with the Higgs

[7] Of course, this claim should not be read as Whiggish take on the history. See Wells (2018) for a discussion of the negative and skeptical attitudes with which the Higgs model was received.

boson. (x represents spacetime.) In the GL model, the central equation is the following expression for the free energy at the critical temperature:

$$\mathcal{F}_s = \mathcal{F}_n + a\,|\psi(\mathbf{x})|^2 + \frac{b}{2}\,|\psi(\mathbf{x})|^4 \tag{13.1}$$

where $\mathcal{F}_n$ represents the free energy density of the 'normal' metal. Parameters a and b are both functions of temperature T; SSB is only possible when $a < 0$. In the simpler Abelian Higgs model,[8] the formal analogy between the Lagrangian $\mathcal{L}$ and the free energy in the GL model is particularly transparent. The Higgs potential terms in $\mathcal{L}$ are

$$V_H = \mu^2|\phi(x)|^2 + \lambda|\phi(x)|^4 \tag{13.2}$$

V_H takes the same form as the final two terms in $\mathcal{F}_s$. Symmetry is broken when $\mu^2 < 0$. The Mexican hat diagram for the Higgs model represents $V_H(\phi)$ when symmetry is broken.

Table 13.1 summarizes the analogical mappings between the GL and Higgs models.[9] The rows of the table lay out the horizontally related elements. The formal analogies are apparent in the mathematical expressions for the symmetries (row 1), Mexican hat terms of the free energy (GL) or Lagrangian (Higgs) (row 5), and order parameters (row 6), which are mathematical expressions of the same type. The evidence that supports the conclusion that this set of analogies is purely formal comes from considering how the Mexican hat terms (i.e., Equations (13.1) and (13.2)) are derived and noting the substantial physical dissimilarities between the mapped elements. In both instances, the Mexican hat terms are produced by taking a symmetric Taylor expansion about some non-zero value of the order parameter. Coefficients multiplying odd powers of the order parameter are set to zero to ensure symmetry, and expansion to the

Table 13.1. Analogies between superconductor models and the Higgs model.

	Ginzburg–Landau (GL) model	Higgs model
1	$U(1)$ broken (global) gauge symmetry group	$SU(2) \times U(1)$ broken (local) gauge symmetry group
2	(limited-range) photon with effective mass (two transverse components)	massive W, Z bosons
3	plasmon with massive longitudinal component	massive Higgs boson
4	free energy density of superconducting state $\mathcal{F}_s$	Lagrangian $\mathcal{L}$
5	$a\|\psi(\mathbf{x})\|^2 + \frac{b}{2}\|\psi(\mathbf{x})\|^4$	$V_H = \mu^2\|\phi(x)\|^2 + \lambda\|\phi(x)\|^4$
6	collective wave function for superconducting electrons $\psi(\mathbf{x})$ as the order parameter	scalar particle field $\phi(x)$ as the order parameter
7	T	*no analogue*

[8] In the Abelian model, $U(1)$ symmetry is broken instead of $SU(2) \times U(1)$ symmetry. See Fraser and Koberinski (2016, 78) for details.

[9] Again, to minimize the technicalities, some entries in this table are drawn from the Abelian Higgs model.

fourth order is the lowest order expansion that leads to SSB. (In the case of the Higgs mechanism, this is also the largest renormalizable interaction term.) Thus, the formal similarities are largely due to the use of an approximation procedure that does not reflect any deep physical similarities between the two systems.

The mapped elements are formally similar, but there are substantial physical disanalogies because the mapped elements have different physical interpretations. For example, $\psi(\mathbf{x})$ is defined over space and is a collective (non-relativistic quantum mechanical) wave function that represents the collection of superconducting electrons; in contrast, $\phi(x)$ is defined over spacetime and is a (relativistic quantum field theoretic) scalar quantum field. More significant, however, are the disanalogies in the vertical relations. A noteworthy feature of the formal analogical mappings from the GL to the Higgs model is that they do not map vertical relations to vertical relations with the same physical interpretation. To see this, consider spontaneous symmetry breaking. In the GL model, SSB is a temporal process. For example, the model describes a temporal process in which a system at time t_1 is in a non-superconducting state and then transitions to a superconducting state at time t_2. This transition can be brought about by reducing the temperature of the system from the initial temperature T to a temperature below the critical temperature T_c, or vice versa. A popular tabletop demonstration of this effect is a maglev (i.e., magnetic levitating) train. The magnetic train track is initially cooled by placing it in dry ice, and then the magnetic train levitates above the track and moves almost frictionlessly along it. When the train is operated at room temperature, the track gradually warms until the train no longer levitates. This is the point at which a phase transition occurs. The phase in which the train levitates is the broken symmetry phase; the phase in which the train does not levitate is the symmetric phase. There is no analogue of this temporal process in the Higgs model. The analogical mappings laid out in Table 13.1 do not map the temporal process in which symmetry is spontaneously broken in the GL model to a temporal process in the Higgs model.[10] Furthermore, the causal process of SSB in the GL model is not mapped to a causal process of SSB in the Higgs model. This follows from the fact that (whatever one's philosophical views on the relationship between temporal and causal order) causal processes are temporal processes. That the set of analogical mappings does not assign an analogue of the temperature T is a symptom of the fact that the analogical mappings do not respect the causal structure of the GL model.

Yet a further physical disanalogy is that the modal relations in the GL model are not mapped to modal relations of the same type in the Higgs model. For a given condensed matter system represented using the GL model, symmetric ground states and states in which the symmetry is broken can both be physically possible. In contrast, for a given particle physics system represented using the Higgs model, ground (i.e., vacuum) states which are symmetric and states in which the symmetry is broken are not both physically possible.[11]

[10] The differences in the manners in which time is represented are particularly transparent in the algebraic frameworks for QSM and QFT. See Fraser and Koberinski (2016, 80–1).

[11] Note that this observation about the relationship between symmetric and broken symmetry vacuum states is separate from the much-discussed question in the philosophy of physics literature of whether, in the standard textbook presentation, the broken symmetry vacuum states related by the gauge transformation

13.3.2 *Explaining the successful development of the Higgs model*

Anticipating the discussion of the 'No Miracles' Argument, what is the explanation for the successful application of formal analogical reasoning in this case? The contribution that this set of analogies made to the development of the Higgs model was to correct misconceptions that were prevalent in particle physics circa 1960, namely that it is not possible to have massive Yang–Mills bosons and that SSB is necessarily accompanied by massless Goldstone bosons. In the case of the massless Goldstone bosons, Goldstone's theorem appeared to show that it was not possible to construct consistent mathematical models with the desired features. The formal analogies allowed particle physicists to recognize that the space of mathematically possible models was larger than they had realized. For probing mathematical or logical possibility, formal analogies are sufficient; a physical analogy is not necessary. Moreover, in this case the physical disanalogies actually contributed to the heuristic usefulness of the formal analogies. As a result of the physical disanalogies, key features of the superconductivity model were accessible to experiment. For example, the Meissner effect: beyond a small penetration depth, neither electric nor magnetic fields will be present within the superconductor, and the short range of the electromagnetic interaction is regarded as an indication of the effective mass of a photon. In contrast, it proved much more difficult to experimentally test the formally analogous features of the Higgs model. The genuinely massive W and Z bosons were not detected until 1983. Finally, this is admittedly speculative, but it seems plausible that the availability of an intuitive picture of SSB in superconductors aided the construction of the models. The collective behavior of the spins on the lattice as temperature varies is intuitively picturea ble, whereas there was no intuitive picture of the interaction of particles (or fields or ?) in QFT circa 1960. Recall that this was the heyday of the S-matrix program, which in its strongest form dispensed with field representations of the interacting system entirely.

I take this to be a satisfactory explanation of the instrumental success of formal analogies in this case. However, one might wonder, is there a post hoc explanation of the success of the analogical reasoning that reintroduces physical analogies? Table 13.1 lists no analogue for temperature T. Again Hesse's distinctions are useful: a *positive analogy* is an analogy that is supported by evidence (at a time t) and a *neutral analogy* is an analogy that it is neither supported nor undermined by empirical evidence (at a time t). Was there a neutral analogy between temperature T in the GL model and temperature T in the electroweak theory in the early 1960s that, with subsequent theorizing and experiment, has become a positive analogy? Contemporary wisdom is that SSB in the the electroweak sector occurred once in the early universe, when the temperature was approximately 200MeV (LeBellac, 1996, xi). Do theoretical developments since 1964 offer a way to introduce temperature into the Higgs model as the analogue of temperature in the GL model, and thereby introduce analogical maps from temporal and causal processes in the GL model to such processes in the Higgs model?

are physically equivalent. See Ruetsche (2011, ch. 14) for further discussion. Struyve (2011) points out that early treatments of the Abelian Higgs model were gauge invariant, including the derivation of the same effective Lagrangian as in standard textbook presentations.

The short answer is no. In brief, the problem with trying to consistently extend the set of analogies to include temperature (or causal or temporal processes more generally) is that time is already represented in the Higgs model. Within this model, it is not possible to add variables (e.g., T) that introduce new temporal processes because the time evolution of the system is already described by the Higgs model. Changing the description of the time evolution changes the model (i.e., describes a different system or the same system in different possible states). This would not help to explain why the original set of analogies between the Higgs model and the GL model is successful.[12]

The contrast with the historical case of electromagnetism is illuminating. In that case, there are physical analogies that map causal relations in Maxwell's vertex–idle wheel model to causal relations in the subsequent field model for classical electromagnetism. In contrast, in the GL–Higgs case, causal relations in the GL model are not mapped to causal relations in the Higgs model, and the analogies are purely formal. Furthermore, the abstract Lagrangian formulation of electromagnetism subsumes the concrete ether and classical field models. The fact that both are concrete models of one abstract Lagrangian entails that the kinetic and potential energies attributed to elements of Maxwell's vortex–idle wheel model are functionally related to the values

[12] At somewhat greater length: The account of the phase transition in the electroweak theory is given by relativistic field theory at finite temperature (aka thermal field theory), which was first proposed in 1965 and then rediscovered in the mid-1970s (LeBellac, 1996, xi). Relativistic field theory at finite temperature is statistical mechanics for relativistic field theories; QSM is statistical mechanics for non-relativistic quantum mechanics (NRQM). That is, relativistic field theory at finite temperature is to QFT as QSM is to NRQM. The relationship between relativistic field theory at finite temperature and QFT does not lend itself to an extension of the set of analogies to include temperature because relativistic field theory at finite temperature is an application of statistical mechanics, not an extension of QFT. The details of relativistic field theory at finite temperature are complex, but temperature and other statistical mechanical properties are introduced in ways familiar from other applications of statistical mechanics. For example, particle physics systems are modeled in contact with heat baths (LeBellac, 1996). This system is different from the system described by the Higgs model. An important difference in kind is that whereas the Higgs model describes a closed system, when a heat bath is added the Higgs model system is an open system (see Fraser and Koberinski, 2016, section 5.2 for further discussion). As a result, relativistic field theory at finite temperature does not help to add a temperature variable to the description of the system of interest, which is the system represented by the Higgs model. Put another way, relativistic field theory at finite temperature is applied to describe a statistical mechanical phase transition which breaks symmetry in a particle physics system, but this does not correspond to the representation of electroweak SSB in the Higgs model. To appeal to the statistical mechanical phase transition is to change the subject.

Another way of seeing that the set of analogical mappings cannot be consistently extended to include temperature is to trace the details of theoretical development of relativistic field theory at finite temperature from QSM. The strategy employed is to define a relativistic extension of an identity between the partition function in QSM and the trace in NRQM (LeBellac, 1996, chs 2 and 3). For simplicity consider one dimension (the time dimension):

$$QSM: \; Z(\beta) = Tr\left(e^{-\beta H}\right)$$

$$NRQM: \; Tr\left(e^{-iH(t_f - t_i)}\right)$$

where β is inverse temperature. The expressions are identified after the NRQM expression is transformed by (1) analytically continuing time ($t \to it$) and (2) setting $t_i = -\beta/2$ and $t_f = +\beta/2$. Time in relativistic field theory at finite temperature cannot both be identified with temperature in QSM *and* be taken (as a component of spacetime) as the analogue of space in QSM as it is in the analogies underlying SSB.

attributed to fields in the classical field model. In contrast, in the GL–Higgs case the models do not share a Lagrangian. The formal analogy is drawn between terms of the free energy of the superconducting state and the potential in the Lagrangian for the Higgs system (see row 5 of Table 13.1). This analogical mapping is intimately related to the fact that causal relations do not map to causal relations.[13]

Finally, a different approach to arguing that there is a physical analogy that lies behind the success of the reasoning in this case would be to contend that mass is the relevantly similar physical property of the two systems. After all, the motivation for drawing the analogy in the first place was that in the superconducting phase the superconductor has effectively massive photons and plasmons. However, while the physical property of effective mass motivated drawing analogies, the set of analogies that resulted from this starting point does not support the inference that there are relevant physical similarities between effective mass in the superconductor and genuine mass in the electroweak system. The physical interpretation of effective mass of the photon in the superconductor is tied to the physical process of mass gain (and shortening of range) when the system undergoes phase transitions, and there is no physical process in which mass is gained in the Higgs case. The physical interpretation of the mass of the plasmon comes from the collective behavior of the lattice of atoms. This is also a physical dissimilarity between the superconductor and particle physics models: there is not even a lattice of material particles in the particle physics case. Mass in the superconductor is not physically similar to mass in the Higgs model; however, alternatives to the Higgs model ('Beyond the Standard Model' models) have been proposed in which the Higgs is a composite particle analogous to the Cooper pair bosons in the BCS model of superconductivity (e.g., the minimal technicolor model). But, once again, these models are based on a different set of analogies; the analogies invoked are not a consistent extension of the set laid out in Table 13.1.

13.4 The Refined Explanationist Defense of Realism Revisited

Recall Psillos' summary of his Refined Explanationist Defense of Realism:

> *(REDR)* The best explanation of the instrumental reliability of scientific methodology is that: the theoretical statements which assert the specific causal connections or mechanisms by virtue of which scientific methods yield successful predictions are true. (78)

The Higgs case study defies an explanation of this type. The method used is purely formal analogical reasoning. The application of this method in this case has been instrumentally successful. The most direct instrumental success of the Higgs model has been the detection of a particle consistent with the Higgs boson at the Large

[13] The deeper justification for this analogical mapping comes from the effective action formalism for QFT (Jona-Lasinio, 1964, Peskin and Schroeder, 1995). This analysis compares the generating functional of correlation functions in statistical mechanics to the generating functional for propagators (or vacuum expectation values) in QFT. A further formal analogy is that, in both QFT and QSM, variational principles are used to determine the stable vacuum (QFT) or ground (QSM) states. The quantity in QFT that plays the formally analogous role to energy density in statistical mechanics is the Lagrangian density.

Hadron Collider in 2012, but the Higgs model has had many other more indirect instrumental successes. Even if the Higgs model were to eventually be supplanted by some 'Beyond the Standard Model' model, it presumably has already surpassed the minimal standards for instrumental success required by Boyd and Psillos. Thus, the instrumental success of the method of formal analogical reasoning in the Higgs case is an instance of scientific methodology, the instrumental reliability of which is what Boyd and Psillos seek to explain.

However, the best explanation proposed by the REDR is not a possible explanation for the success of the method of purely formal analogical reasoning in the Higgs model case study. If the analogical mappings respected the causal structure, then getting this shared causal structure approximately right in both the superconductor and the electroweak models would be a candidate explanation of the successful use of the analogies. In Psillos' terms, the explanation would be that the theoretical statements asserting the causal connections in both models are approximately true. However, the analogical mappings *do not* map causal processes in the GL model of superconductivity to causal processes in the Higgs model. There is no shared causal structure of SSB that is common to the GL and Higgs models; therefore, the truth or falsity of the theoretical statements asserting causal connections within each of the two models is not relevant to explaining the success of formal analogical reasoning in this case.

In order for this Refined Explanationist Defense of Realism to support Psillos' own brand of realism, which relies on a causal-descriptive theory of reference, the restriction to "specific causal connections or mechanisms" is essential. However, in the context of his discussion of the Explanationist Defense of Realism, Psillos allows that the explanans may have a broader scope that includes "truth-like descriptions of causal mechanisms, entities, and laws" (81). Expanding the scope of the explanans in this way seems unlikely to serve to make it applicable to the GL–Higgs case. Recall from Section 13.3 that the purely formal analogies invoked in the development of the Higgs model do not relate similar physical descriptions of entities. An in-depth discussion of what sorts of accounts of laws of nature would be robust enough to underwrite this explanation is beyond the scope of this chapter, but a significant obstacle to the appeal to laws is that the analogical mappings do not preserve the modal structure either. SSB in the GL model employs a notion of physical possibility that pertains to states dynamically accessible to a specified system. In contrast, SSB in the Higgs model employs a notion of physical possibility that does not pertain to states dynamically accessible to a specified system. Symmetric and broken symmetry vacuum states are possibilities for systems in different worlds which are not accessible from our world.

Ultimately, the problem with applying the explanationist defense of realism to the method of formal analogical reasoning does not lie with how causal connections, mechanisms, or laws of nature are spelled out. Consider this minimal version of the REDR:

> *(REDR′)* The best explanation of the instrumental reliability of scientific methodology is that: the theoretical statements by virtue of which scientific methods yield successful predictions are true.

Even this minimal version of the argument is inapplicable to the case study because the truth of statements in the GL model of superconductivity is not relevant to the success of the method for formulating the Higgs model for particle physics because there are no physical analogies between the models. Assume that we have gotten something approximately right about the superconductor system when we describe it with the GL model. We draw formal analogies to construct a new model for electroweak interactions in particle physics—the Higgs model. The Higgs model turns out to be instrumentally successful. The approximate truth of theoretical statements in the GL model *that describe superconductors* does not explain the instrumental success of the Higgs model *that describes electroweak interactions* because the GL model and Higgs model are related by purely formal analogies. The fact that the analogies are purely formal means that there are no relevant physical similarities between the condensed matter and particle physics systems (as described by the respective models). Neither the horizontally related elements nor the mapped vertical relations are physically similar. Thus getting something approximately right about the superconductor that is captured by theoretical statements in the GL model does not give us any reason to believe that we have thereby gotten the same something right about the analogue particle physics system that is captured by theoretical statements in the Higgs model.

Notice that the Higgs model case study undermines the REDR—an argument offered in support of scientific realism—and not scientific realism directly. The formal analogies presented in Table 13.1 do not rule out giving both the GL and Higgs models physical interpretations that include specifications of causal connections. What is ruled out are specifications of causal connections such that the analogical mappings in Table 13.1 map causal relations in the GL model to causal relations in the Higgs model. This leaves open the possibility of interpreting other relations in the Higgs model as causal relations. However, independent realist interpretations of the GL and Higgs models would not rescue the REDR. Since the causal structures in the GL and Higgs models would not be related by the formal analogies in Table 13.1, appeal to the sets of causal relations in each model would not explain the instrumental success of analogies.

Again, my thesis is not that the instrumental success of the method of formal analogical reasoning is miraculous. There is, I submit, a perfectly satisfactory explanation for the success of formal analogies: that formal analogies served to correct misconceptions about the mathematically possible models of SSB in particle physics. For this purpose, purely formal analogies are sufficient. Ironically, it is the REDR that makes a miracle out of the success of science in this case by insisting that success is to be explained by approximate truth.

13.4.1 *Possible responses*

The scientific realist who wants to hold on to this version of the NMA has several options for responding to case studies of purely formal analogical reasoning. One option is to rule out cases of this sort by considering them to fall outside the scope of scientific methods covered by the REDR. This would require principled grounds for exclusion. It would not be sufficient to exclude the method of formal analogical reasoning merely on the basis that it is used in the context of discovery rather than justification. Even supposing that a satisfactory distinction can be drawn between the contexts of discovery and justification, the case has been made that a variety of heuristic methods for formulating new theories are legitimate inductive methods (e.g., Post, 1971;[14] Thagard, 2012). Furthermore, Boyd and Psillos are explicit that scientific methodology in general is the target of the argument and list examples of methods that fall squarely in the context of discovery (e.g., choice of research problems) (Boyd, 1990, 361).

An apparently more promising strategy would be to exclude the method of purely formal analogical reasoning in particular from the scope of the argument on the grounds that it is not an instrumentally reliable method. Methods such as literally dreaming up new hypotheses and trial and error could be used to discover new theories, but fall outside the class of instrumentally reliable methods covered by the argument. However, this strategy for defending the REDR only appears to be more promising; ultimately, it seems unlikely to pan out. There are some clear differences between these unreliable methods and formal analogical reasoning. For instance, the method of formal analogical reasoning is more systematic and principled than either dreaming or trial and error. Formal analogical mappings are constrained by the requirement that they map elements that play similar formal roles in the theories for source and target domains. But the systematic and principled nature of the method does not automatically translate into instrumental reliability.

A number of philosophical accounts of analogical reasoning offer criteria for evaluating the strength of arguments from analogy (e.g., Hesse, 1966; Holyoak and Thagard, 1995; Bartha, 2010). In each of these accounts, sufficiently strong arguments from analogy confer plausibility on a hypothesis. Unfortunately, all of these accounts base their conclusions on physical analogies, not purely formal analogies, and the arguments do not carry over straightforwardly.

The reason that that formal analogical reasoning in the Higgs case is not tantamount to lucky guessing is that the success of the method is explicable. In the early 1960s, particle physicists mistakenly believed that it was not possible to construct models with SSB and the desired physical features (e.g., massive bosons). Formal analogies to models of superconductivity served as a corrective. Formal analogical reasoning was an appropriate method to use in this case because it allowed particle physicists to probe the space of mathematically possible models. This explanation of the success of formal analogical reasoning in this case undermines the charge that the method is instrumentally unreliable.

A stronger argument in support of the instrumental reliability of formal analogical reasoning would involve establishing that formal analogical reasoning has also led to instrumental success in other cases and that this instrumental success is also explicable. This is a large project, but a plausibility argument can be made for the first conjunct. Analogical reasoning has been a widely used method in the development of quantum theories. Of course, each of these cases needs to be analyzed individually to determine whether the analogies are formal, physical, or both. However, there is a suggestive pattern to these cases. Consider the two domains of condensed

[14] Post notes that the list of heuristic criteria covered in his paper is not exhaustive, and then cites formal analogies as an example (248).

matter physics (e.g., phase transitions in superconductors) and particle physics (e.g., scattering of a few particles). There are many examples of concepts or mathematical frameworks that were passed back and forth between theories for these two domains. Dirac's 'hole theory' of the electron was possibly inspired by ionic crystal models of conductors constructed by Frenkel in the 1920s, and Dirac's idea was certainly picked up in solid state physics in the 1930s (Kojevnikov, 1999). Renormalization techniques developed for QED in the 1940s and the associated concept of dressed electrons were exported from QED to solid state physics in the 1950s (e.g., Bohm and Pines' electron gas model for metals introduces an effective heavy electron and plasmons) (Blum and Joas, 2016). During the same period, Feynman diagrams were borrowed from QED to solve formally analogous perturbative expansions in quantum statistical mechanics (QSM) in which the formal analogue of time in QED is imaginary inverse temperature in QSM (Matsubara, 1955, Abrikosov, Gorkov, and Dzyaloshinski, 1975). In the early 1970s, Kenneth Wilson and collaborators developed renormalization group (RG) methods for condensed matter physics and particle physics by pushing analogies between classical statistical mechanical models of phase transitions and quantum field theoretic models of interactions (Wilson and Kogut, 1974). In this case, the analogue of time t in QFT is imaginary space ix. (Fraser, 2018, offers an account of these analogies and argues that the analogies are purely formal.) The pattern is that in each of these cases the analogies are drawn between a non-relativistic quantum or classical model and a relativistic quantum field theoretic model. In the GL–Higgs case study, the root cause of the analogical mappings failing to respect temporal, causal, and modal structure is that a non-relativistic model is mapped to a relativistic model. Plausibly, the analogical mappings in the other cases carry similar implications.

There is also a practical consideration that makes excluding formal analogical reasoning from philosophical consideration seem unappealing. If formal analogical reasoning is indeed as prevalent a tool in the development of recent and contemporary quantum theories as it seems to be, then disregard for this scientific method in philosophy of science carries the cost of making philosophy less relevant to live foundational issues in physics. Carefully tracing complex patterns of analogical reasoning and attending to the interpretive consequences is the the kind of project that philosophers are well placed to undertake.

13.5 The Argument from the History of Science for Structural Realism

A shift from variants of realism committed to continuity of reference of theoretical terms (e.g., Psillos' position) to variants of realism committed to continuity at the level of physical structure (e.g., variants of structural realism) is accompanied by a shift in the formulation of the NMA. The Higgs case undermines the structural realist version of the NMA in a different way because structural realism's attention to the mathematical structure of theories emphasizes the same aspect of theories that informs formal analogies. In their 2011 review paper on structural realism, Roman Frigg and Ioannis Votsis survey Poincaré's and Worrall's arguments from the history of science for what has come to be known as epistemic structural realism. They extract the following argument incorporating a version of the NMA:

(4a) Only two elements of a theory get preserved through theory change: (a) the theory's mathematical formulation, and (b) the empirical interpretation of the theory's terms.
(4b) A theory's mathematical formulation 'encodes' the structure of that theory's target domain.
(4c) Preservation of an element is a reliable guide to its (approximate) truth.
(4d) Non-preservation of an element is a reliable guide to its (approximate) falsity.

Therefore, the preservation of structural elements through theory change is a reliable guide of their (approximate) truth. The non-preservation of non-structural elements is a reliable guide of their (approximate) falsity. (243)

They note that premises (4c) and (4d) "incorporate an instance of the NMA" (243). In contrast to Psillos' REDR, the explanandum is not the instrumental reliability of scientific methodology, but the instrumental success of a given theory. The explanans appeals not to specific causal connections or mechanisms, but to the physical structure of the theory's target domain (which may or may not include causal relations).

Cases of formal analogical reasoning undermine this argument by presenting a counterexample to premise (4c). The instantiation of (4c) that is important for the structural realist is that in which the element is the theory's mathematical formulation: *Preservation of a theory's mathematical formulation is a reliable guide to the (approximate) truth of a theory's mathematical formulation* where (by 4(b)) truth means " 'encoding' " the (physical) structure of the theory's target domain. The use of purely formal analogical reasoning in the the GL–Higgs case study constitutes a counterexample to this inference because the theory's mathematical formulation is preserved but this does not supply any indication about whether the theory's mathematical formulation is approximately true. In this case, the theory is the GL model. Core aspects of the mathematical formulation of the GL model are (approximately) preserved in the Higgs model; this is what the formal analogical mappings establish. However, contrary to (4c), the mathematical formulation that is preserved cannot be taken to be an approximately true 'encoding' of the shared physical structure of the superconductor and electroweak systems because the systems do not share a physical structure. There are no physical analogies; the formal analogies do not map physical relations to physical relations of the same type. In particular, neither causal nor modal relations are preserved by the mappings, which precludes two prominent structural realist strategies for characterizing physical structure.

13.5.1 Possible responses

The structural realist may object that cases of formal analogical reasoning are irrelevant because they do not relate versions of one theory which are diachronically related. Premise (4a) concerns "elements of a theory [that] get preserved through theory change." An example would be a nineteenth-century version of electromagnetism that includes ether and a twentieth-century version of electromagnetism that posits classical fields and does not include ether. In contrast, the GL model of superconductivity and the Higgs model are not diachronically related versions of the same theory; they apply to different domains of phenomena and are contemporaneous. However, this objection does not address the argument in Section 13.5. The argument is an argument against premise (4c), which states that preservation of mathematical structure is a reliable guide to its (approximate) truth. Cases of formal analogical reasoning are counterexamples: cases of successful theory development in which there is shared mathematical structure between theories and not shared physical structure. The Higgs case demonstrates that the preservation of mathematical structure by intertheoretic relations is not an indicator of shared physical structure. Why couldn't the same situation arise in the special case in which the theories in question are theories for the same domain?

The problem posed by formal analogical reasoning is familiar to structural realists: in order to constitute a genuine variant of realism, the preserved mathematical structure must represent physical structure. Mathematical structure that does not play a representative role would reduce the position to Pythagoreanism; mathematical structure that represents empirical structure but not underlying physical structure would reduce the position to anti-realist empiricism. (See Ruetsche, this volume, Chapter 15, for further discussion of the latter challenge.) The Higgs case study eliminates some of the structural realist's best resources for steering a course between Pythagoreanism and anti-realist empiricism. For example, French (2014) defends a variant of ontic structural realism according to which physical structure is modal structure, but the GL and Higgs models cannot be interpreted as encoding the same modal structure.

One variant of ontic structural realism may be equipped with a strategy for responding to this challenge posed to the Argument from the History of Science for epistemic structural realism by the Higgs case study. In their review of Wallace's defense of the Everett interpretation of quantum theory in *The Emergent Multiverse*, Guido Bacciagaluppi and Jenann Ismael reflect that

> ... the book provides the most comprehensive and best exemplar of a new—and distinctly modern—way of doing metaphysics. On this way of doing metaphysics, one takes one's fundamental ontology from physical theories at face value and simply does the hermeneutic work of trying to understand the structures implicit in the formalism, connecting them with structures that are most readily manifest in our experience of the world, and seeing what needs to be done to accommodate old ideas (about ourselves and our place in nature) to a new world-view. (2015, 18)

Applying this interpretive approach to the Higgs case, the response would be that none of our familiar notions of physical structure are preserved by the analogies, but this just means that we need to exercise ingenuity in characterizing new kinds of physical structure that are preserved (i.e., accommodating our metaphysics to a new world-view). Of course, there is nothing that stands in the way of adherent of this stripe of ontic structural realism pursuing this research program. But bear in mind that this position is being introduced to rescue the NMA for scientific realism; therefore, scientific realism cannot be assumed. At a minimum, this research program would have to be successfully completed in order to yield an argument for scientific realism.

13.6 Conclusion

The main conclusion defended in this chapter is that getting some fact about the world essentially correct is not always a candidate explanation for success in science, which runs contrary to the spirit and the letter of the NMA. Evidence in support of this conclusion is furnished by the instrumentally successful use of purely formal analogies in the development of the Higgs model. This is a case of successful theory development that is not underwritten by approximately veridical representation. The analogies drawn between the GL model of superconductivity and the Higgs model for the electroweak interaction are purely formal. They are not accompanied by physical analogies. In particular, the analogical mappings do not map temporal structure, causal structure, or modal structure in the GL model to structure of the same physical kind in the Higgs model. As a result, this case study undermines both Psillos' Refined Explanationist Defense of Realism and the NMA-inspired premise of the Argument from the History of Science for structural realism. In the former case, the success of the method of purely formal analogical reasoning cannot be explained by appeal to causal relations or even to approximate truth. In the latter case, the Higgs case study blocks the inference from the preservation of formal structure to the formal structure approximately encoding physical structure. Nevertheless, the construction of the Higgs model is not an example of the miraculous success of science. The explanation for the successful use of purely formal analogies in this case is that this was a suitable method for solving the problem that particle physicists had mistakenly ruled out mathematical models that in fact were mathematically or logically possible. The target of the arguments in this chapter is the NMA for scientific realism, not scientific realism itself. The adoption of separate (i.e., not related by the analogies) realist interpretations of the GL and Higgs models is not precluded, as long as the NMA is not invoked.

In response to these arguments, a scientific realist could concede the point and give up on the NMA. This would of course entail reliance on other arguments to support scientific realism. There are several other potentially viable lines of response for scientific realists who wish to hang on to the NMA. One response, mooted in Section 13.5.1, would be to adopt a variant of ontic structural realism which takes as its starting point the commitments that the formal structures in a theory represent the structure of the world and that one aspect of the project of interpreting the theory is to (if necessary) revise our ontology to accord with the formal structures. Of course, how compelling this approach is would hinge on the details of how the ontological structure gets spelled out.

A third response to this counterexample to the NMA would be to concede that the blanket intuition behind the NMA does not hold universally, and to revise the argument accordingly. That is, the scientific realist could concede that approximate truth only explains the success in science in a restricted set of cases. However, this would be a substantial concession. The worry is that weakening the NMA in this way would leave the NMA vulnerable to other lines of objection. The argumentative strategy in this chapter is to raise a counterexample in which the proposed best explanation for the instrumental success of science is not even a candidate explanation. This is in contrast to the more popular strategy for arguing against the NMA, which is to contend that rival candidate explanations for the instrumental success of science are actually superior to the best explanation proposed by the NMA. For example, Ruetsche (this volume, Chapter 15) draws on analysis of the use of renormalization group methods in particle physics to argue for humble empiricism, which denies that approximately true representation of the world (at all energy scales) is the best explanation for the instrumental success of renormalization group methods in particle physics (at low energy scales). If the NMA were modified to include a restriction on its scope, then it would become more difficult to fend off arguments such as this. If it is possible to have instrumental success without approximate truth in some cases, why should the best explanation for instrumental success be approximate truth in other cases (i.e., cases within the intended scope of restricted NMA)?

The use of purely formal analogies in the Higgs case—and plausibly more widely in the development of quantum theories—reveals that quantum theories carry new implications for the scientific realism–anti-realism debate. The underlying reason is that the debate has been informed by historical case studies, such as the use of analogies in the development of electromagnetism, which use different methods. For scientific realists, an additional consequence of the Higgs case study is that the use of purely formal analogies affects how one fixes an interpretation for a theory. For example, in the Higgs case, the fact that the analogies to superconductivity are purely formal means that it is not licit to export the physical interpretation of SSB from superconductivity to particle physics (e.g., genuine mass in the Higgs model does not have the same physical interpretation as effective mass in the GL model, there is no causal process in which mass is gained in electroweak systems). It is tempting to export the physical interpretation from the superconductor model to the electroweak model because there is a clearer physical picture behind the superconductor model. However, this temptation needs to be resisted—not only in the Higgs case, but also in other cases in which the analogies linking models are purely formal. Scientific realists need to be alert to the possibility of purely formal analogies in order to properly interpret theories.

Another broader moral for the scientific realism–anti-realism debate is that quantum theories exhibit different patterns of development than their precursors, which affects where one should look for relevant case studies. Participants in the debate have primarily been interested in the history of science as a source of evidence for either continuity in theories over time (approximately, in some respects) or else discontinuity in theories over time. This has focused attention on diachronic sequences of theories for a single domain. For example, in the domain covered by what is now known as condensed matter physics (which includes, e.g., phase transitions in superconductors), a relevant diachronic sequence of theories is classical statistical mechanics, non-relativistic quantum mechanics (including many-body quantum theory), and quantum statistical mechanics. To take another example, in the domain covered by what is now known as particle physics (which includes, e.g., scattering phenomena) one of the diachronic sequences of theories that is of interest is classical particle mechanics, non-relativistic quantum mechanics, and relativistic quantum field theory. While case studies of theoretical change in a single domain are interesting and important for the scientific realism debate, this focus excludes a prominent pattern of historical development in quantum theories in the twentieth century: the formulation of new theories (or models) based on *synchronic* relations between theories that apply to *different domains*. Reasoning by analogy has served as a mechanism for transferring concepts and frameworks from one domain to another.

This pattern of theory development deserves more attention in the scientific realism–anti-realism debate.

Acknowledgements

Thank you to Steven French and Juha Saatsi for the invitation to contribute to this volume, for organizing a stimulating conference on this topic, and for helpful comments. For feedback on earlier versions of this material, thank you to participants of the Scientific Realism and the Quantum Final Conference at Leeds; members of the Rotman Institute philosophy of physics reading group (Mike Cuffaro, Marie Gueguen, Adam Koberinski, Wayne Myrvold, Stathis Psillos, Chris Smeenk, and Martin Zelko); and anonymous reviewers. Support from a Social Sciences and Humanities Research Council of Canada Standard Research Grant is gratefully acknowledged.

References

Abrikosov, A. A., L. P. Gorkov, and I. E. Dzyaloshinski (1975). *Methods of Quantum Field Theory in Statistical Physics*. Dover Publications, Inc., New York. Republication, with corrections, of the revised English edition (1963), translated from the Russian and edited by Richard A. Silverman.

Bacciagaluppi, G. and J. Ismael (2015). Essay Review: David Wallace, *The Emergent Multiverse*. *Philosophy of Science 82*(1), 129–48.

Bardeen, J., L. N. Cooper, and J. R. Schrieffer (1957). Theory of superconductivity. *Phys. Rev. (2) 108*, 1175–204.

Bartha, P. F. (2010). *By Parallel Reasoning: The Construction and Evaluation of Analogical Arguments*. Oxford: Oxford University Press.

Blum, A. S. and C. Joas (2016). From dressed electrons to quasi-particles: the emergence of emergent entities in quantum field theory. *Studies in the History and Philosophy of Modern Physics* 53, 1–8.

Boyd, R. (1990). Realism, approximate truth, and philosophical method. In C. W. Savage (Ed.), *Scientific Theories*, pp. 355–91. Minnesota: University of Minnesota Press.

Fraser, D. (2018). The development of renormalization group methods for particle physics: formal analogies between classical statistical mechanics and quantum field theory. *Synthese*. https://link.springer.com/article/10.1007%2Fs11229-018-1862-0#

Fraser, D. and A. Koberinski (2016). The Higgs mechanism and superconductivity: a case study of formal analogies. *Studies in the History and Philosophy of Modern Physics* 55, 72–91.

French, S. (2014). *The Structure of the World: Metaphysics and Representation*. Oxford: Oxford University Press.

Frigg, R. and I. Votsis (2011). Everything you always wanted to know about structural realism but were afraid to ask. *European Journal for Philosophy of Science 1*(2), 227–76.

Ginzburg, V. L. and L. D. Landau (2009). On the theory of superconductivity. In *On Superconductivity and Superfluidity: A Scientific Autobiography*, pp. 113–37. Berlin, Heidelberg: Springer.

Harman, P. (1998). *The Natural Philosophy of James Clerk Maxwell*. New York: Cambridge University Press.

Hesse, M. B. (1966). *Models and Analogies in Science*. Notre Dame, IN: University of Notre Dame Press.

Higgs, P. W. (2014). Nobel lecture: evading the Goldstone theorem. *Rev. Mod. Phys. 86*, 851–3.

Holyoak, K. J. and P. Thagard (1995). *Mental Leaps: Analogy in Creative Thought*. Cambridge, MA: MIT Press.

Jona-Lasinio, G. (1964). Relativistic field theories with symmetry-breaking solutions. *Nuovo Cimento* 34, 1790–4.

Jona-Lasinio, G. (2002). Cross fertilization in theoretical physics: the case of condensed matter and particle physics. In A. Fokas, J, Halliwell, T, Kibble, and B. Zegarlinski (Eds), *Highlights of Mathematical Physics*, pp. 143–52. Providence, RI: Americal Mathematical Society.

Kojevnikov, A. (1999). Freedom, collectivism, and quasiparticles: social metaphors in quantum physics. *Historical Studies in the Physical and Biological Sciences* 29(2), 295–331.

Laudan, L. (1981). A confutation of convergent realism. *Philosophy of Science* 48(1), 19–49.

LeBellac, M. (1996). *Thermal Field Theory*. New York: Cambridge University Press.

Matsubara, T. (1955). A new approach to quantum-statistical mechanics. *Progress of Theoretical Physics* 14(4), 351–78.

Maxwell, J. C. ([1861] 1890). On physical lines of force. In W. Niven (Ed.), *The Scientific Papers of James Clerk Maxwell*, Volume I, pp. 452–513. Cambridge: Cambridge University Press.

Maxwell, J. C. ([1873] 1954). *A Treatise on Electricity and Magnetism* (3rd edn). New York: Dover.

Peskin, M. E. and D. V. Schroeder (1995). *An Introduction to Quantum Field Theory*. Reading, MA: Addison-Wesley Publishing Company Advanced Book Program.

Post, H. R. (1971). Correspondence, invariance and heuristics: in praise of conservative induction. *Studies in History and Philosophy of Science Part A* 2(3), 213–55.

Psillos, S. (1999). *Scientific Realism: How Science Tracks Truth*. London: Routledge.

Psillos, S. (2011). The scope and limits of the no miracles argument. In D. Dieks, W. Gonzalo, T. Uebel, S. Hartmann, and M. Weber (Eds), *Explanation, Prediction, and Confirmation*, pp. 23–35. Dordrecht: Springer.

Ruetsche, L. (2011). *Interpreting Quantum Theories*. New York: Oxford University Press.

Saatsi, J. (2005). Reconsidering the Fresnel–Maxwell theory shift: how the realist can have her cake and eat it too. *Studies in History and Philosophy of Science Part A* 36(3), 509–38.

Stein, H. (1989). Yes, but... some skeptical remarks on realism and anti-realism. *Dialectica* 43(1–2), 47–65.

Struyve, W. (2011). Gauge invariant accounts of the Higgs mechanism. *Studies in the History and Philosophy of Modern Physics* 42, 226–36.

Tegmark, M. (2008). The mathematical universe. *Foundations of Physics* 38(2), 101–50.

Thagard, P. (2012). *The Cognitive Science of Science: Explanation, Discovery, and Conceptual Change*. Cambridge, MA: MIT Press.

Wells, J. D. (2018). Beyond the hypothesis: theory's role in the genesis, opposition, and pursuit of the Higgs boson. *Studies in History and Philosophy of Modern Physics* 62, 36–44.

Wilson, K. G. and J. Kogut (1974). The renormalization group and the ϵ expansion. *Physics Reports* 12, 75–200.

Worrall, J. (1989). Structural realism: the best of both worlds? *Dialectica* 43(1–2), 99–124.

14

Towards a Realist View of Quantum Field Theory

James D. Fraser

14.1 Introduction

Quantum field theories (QFTs) seem to have all of the qualities that typically motivative scientific realism. Besides general relativity, the QFTs that make up the standard model of particle physics are our most fundamental physical theories. They have also produced some of the most accurate predictions in the history of science: quantum electrodynamics (QED) famously gives a value for the anomalous magnetic moment of the electron that agrees with experiment at precisions better than one part in a trillion. When it comes to actually articulating a realist reading of these theories, however, we run into serious problems.

This chapter puts forward what I take to be the most promising strategy for developing a realist epistemology of QFT. I set up the discussion by highlighting the difficulty of making sense of QFTs in orthodox realist terms if we restrict our attention to perturbative and axiomatic treatments of the theory (Section 14.2). I then introduce the renormalization group (Section 14.3), and argue, drawing on previous work by Wallace (2006; 2011) and Williams (2019), that it points to a way of rescuing a realist view of QFT (Section 14.4). I close by considering some objections to this programme raised by Ruetsche (2018) (Section 14.5). Besides some brief remarks in this final section, I will mostly be bracketing interpretive puzzles inherited from non-relativistic quantum mechanics which continue to challenge the realist in the QFT context.

14.2 Realism and Quantum Field Theory: A First Look

Scientific realists disagree about how their position should be formulated in detail. We can discern some general commitments that lie behind many of these statements of the realist creed, however. Two, in particular, will be my focus here. The first is an explanatory thesis: realists take the empirical success of the sciences to be explained by the fact that they are getting something right about the way the world is. This contrasts with anti-realists, who either deny that the empirical success of

James D. Fraser, *Towards a Realist View of Quantum Field Theory* In: *Scientific Realism and the Quantum*. Edited by: Steven French and Juha Saatsi, Oxford University Press (2020). © James D. Fraser.
DOI: 10.1093/oso/9780198814979.003.0014

science stands in need of explanation, or else adopt an alternative explanatory strategy that does not invoke the representational veracity of our theories.[1] The second idea concerns the epistemic achievements of current scientific theories. Realists typically take at least some extra-empirical beliefs about the world to be supported by the predictive success of the sciences; that is, they take science to furnish knowledge of the unobservable. Again, this contrasts with anti-realist scepticism towards scientific claims about unobservable entities and properties. The constructive empiricist, for instance, takes an agnostic attitude towards the extra-empirical content of successful theories.

While these two claims do not strictly imply one another, connections between them are often posited. The dominant approach to developing the realist position in recent decades has been to identify the theoretical claims we ought to commit ourselves to using an explanatory criterion. Rather than taking successful theories as a whole to be the locus of realist commitment, this 'selective' form of realism advocates belief in a subset of their descriptive content; namely, those parts of the theory which essentially contribute to, and therefore explain the success of, its empirical predictions. This way of cashing out the epistemic achievement component of realism is explicitly deployed in Psillos (1999) and Kitcher (2001) influential discussions of historical theory change, but the basic selectivist intuition arguably animates many of the other variants of realism put forward in the recent literature, such as structural realism. On this line of thought rolling out a realist epistemology across the various branches of science ought to proceed as follows: start with a theory's empirical predictions and trace them back to theoretical claims that underwrite their success; these are the parts of science that constitute knowledge of the unobservable.

How does this programme fare in the case of QFT? While I will ultimately claim that there is scope for harmonizing QFT with this selective realist scheme, I'll start with a more pessimistic reading of the situation. Ignoring advances in renormalization theory since the 1950s, and restricting our attention to the perturbative and axiomatic approaches to QFT, it is possible to paint a fairly bleak picture of realism's prospects in high energy physics.

As has already been mentioned, QFTs are wildly empirically successful. The problems start when we look at how these predictions are obtained. The most important source is the perturbative approach to QFT developed by Feynman, Tomonaga, Schwinger, and Dyson in the late 1940s.[2] This is where the famous calculation of the anomalous magnetic moment of the electron, and the bulk of the standard model predictions currently being tested at the Large Hadron Collider, come from. I'll sketch how this formalism works with reference to the ϕ^4 theory, a well-understood model

[1] Van Fraassen's analogy between theory selection and natural selection can perhaps be read in either of these two ways (van Fraassen, 1980, 40). See also Saatsi (2017) for a discussion of anti-realist-style explanations of the empirical success of particular theories.

[2] This is less true today than it was some decades ago. Non-perturbative calculation methods, especially Monte Carlo simulations in lattice QFT, are becoming increasingly powerful sources of empirical predictions. A complete epistemological study of high energy physics would need to look at these methods in detail but I leave this as a project for another time.

that shares most of the relevant features of the gauge theories that make up the standard model. The classical action of this model is given by:

$$S = \int d^4x \left[\frac{1}{2}(\partial_\mu \phi)^2 - \frac{1}{2} m^2 \phi^2 - \frac{\lambda_4}{4!} \phi^4 \right], \tag{14.1}$$

where ϕ is a scalar field, m is its associated mass, λ_4 is a positive constant parameterizing the strength of the interaction and the integral ranges over Minkoski space-time. There is little hope of finding exact solutions of this theory, or any realistic interacting QFT. The perturbative approach aims to get around this problem by setting up a power series expansion for quantities in an interaction coupling, in this case, λ_4. The coefficients of this series can, in principle, be calculated using what we know about the corresponding exactly solvable free theory (with $\lambda_4 = 0$), though increasingly difficult computations are required at each order. The hope is that, if λ_4 is small, the first few terms of the series will provide an accurate approximation of the behaviour of a weakly interacting model.

This approximation strategy is, of course, ubiquitous in applied mathematics, but QFT perturbation series have their own peculiarities. Carrying out the series expansion naively, in this case, produces ill-defined results. Computing the series coefficients, for both the ϕ^4 theory and more realistic models, requires one to evaluate integrals over momentum space that diverge in the high momentum, or ultraviolet, region.[3] In order to get meaningful predictions out of perturbative QFT the so-called renormalization procedure must be employed. This works roughly as follows. First, the offending integrals are rendered convergent; in the simplest case, this is achieved by imposing a cutoff—a large but finite upper limit on the momentum integrals. The expansion parameter is then redefined so as to remove the perturbative coefficient's dependence on the cutoff. It turns out to be possible to completely 'absorb' the divergent part of the coefficients at each order into a finite number of parameters in the action if the theory contains only interaction terms parameterized by couplings with a zero or positive mass dimension—so-called renormalizable interactions.[4] The ϕ^4 interaction has this property, as do the electroweak and strong interactions of the standard model. Indeed, during the years the standard model was being developed renormalizability was often viewed as a necessary condition for taking a theory seriously and was demanded a priori. At the end of the renormalization process the cutoff is taken to infinity. Following this recipe produces a series with finite coefficients, and it is truncations of these renormalized perturbation series which are compared to data gathered from collider experiments.

Despite its extraordinary empirical success, this approach to QFT has long been seen as conceptually suspect. Some have suggested that the perturbative formalism is insufficiently mathematically rigorous to engage with philosophically, and even

[3] QFT perturbation series can also contain infrared divergences—integrals which blow up in the region of very low momentum. More generally, the low energy/long distance structure of QFT models raises distinctive problems of its own, which will not be discussed here.

[4] A brief note on dimensional analysis in QFT for the uninitiated. High energy physicists typically work with so-called natural units, which set $\hbar = c = 1$. This has the effect of drastically simplifying dimensional analysis, such that the dimensionality of any quantity can be expressed as a power of energy/mass/momentum, known as its mass dimension.

inconsistent owing to its apparent clash with Haag's theorem. I have argued in previous work that the situation is not quite this bad (J. Fraser, 2017). Perturbative QFT is, at least, internally coherent. There is a problem here, however, which should be especially worrying for the would-be scientific realist. Many of the manipulations that go into perturbative calculations, and the renormalization procedure, in particular, seem to be completely ad hoc. What licences us to remove the infinities that appear in naive QFT perturbation series? The best we can really say within the confines of the perturbative approach itself is that we have to do this for the series to be well behaved and yield sensible predictions. But this is a purely instrumental rationale. Furthermore, it doesn't give us any reason to think that the resulting renormalized series ought to produce accurate predictions. The realist, remember, was committed to there being an explanation for the success of scientific predictions which ultimately rests on the way the world is. On the face of it at least, perturbative QFT violates this demand.

The prospects of extracting knowledge of the unobservable from the perturbative formalism also appear to be grim. Arguably, the root of the perturbative formalism's ad hoc character is that it fails to provide a characterization of the mathematical structure, and descriptive content, of QFT models. I'll push this point in the language of the functional, or path integral, approach to quantization here (in part because it lays the ground for the discussion of the renormalization group in Section 14.3). The key quantity in this approach is the partition function Z, which (in the case of a single scalar field ϕ) is associated with the integral:

$$Z = \int \mathcal{D}\phi e^{iS}. \tag{14.2}$$

Informally, the measure $\mathcal{D}\phi$ indicates that a sum is being taken over all possible configurations of the field (S is, again, the classical action). Once we have the partition function of a QFT model we can, in principle at least, derive all of its observable consequences and, more importantly, construct it as a mathematical object.

It turns out to be very difficult to precisely define this integral for fields that live on continuous space-times, however. Owing to the infinite number of degrees of freedom that exist in any space-time region we need to define a measure over an infinite dimensional space, once again leading to divergences in the ultraviolet region. In the perturbative treatment of QFT this issue was dodged rather than solved. In essence, the perturbative method allows us to set up expansions for Z, and thus for scattering cross sections, without really telling us what it is. The one strategy that might suggest itself for extracting a definition of the partition function from the perturbative formalism is to identify it with the sum of the series, but it turns out that, even after the divergences in the coefficients have been removed, realistic QFT perturbation series do not converge. In the absence of a clear characterization of the physical content of QFT, it is hardly surprising that we struggle to find justifications for the moves made in perturbative calculations. But this situation is also clearly bad news for the epistemic achievement component of realism. How can the empirical success of the standard model possibly support beliefs about unobservable aspects of reality if we can't even specify what the theory is saying about the world?

There is another strand of the QFT programme the realist might turn to here which does directly address the question of where QFTs live in the mathematical universe, namely the axiomatic approach to the theory. In the 1950s and 1960s mathematical physicists dissatisfied with the limitations of the perturbative approach tried to put the theory on a firm non-perturbative footing by writing down sets of axioms that any relativistic quantum field could reasonably be expected to satisfy. In fact, two mathematical treatments of QFT came out of this tradition: one based on the machinery of operator-valued distributions (Streater and Wightman, 1964) and the other on von Neumann algebras (Haag, 1996; Halvorson and Müger, 2007). Mathematically rigorous work in these frameworks succeeded in showing that one can precisely define path integrals for field configurations that take arbitrarily large momenta, at least in the case of some toy models in space-times with a reduced number of dimensions. Unfortunately, the axiomatic tradition ultimately doesn't offer much solace to the aspiring realist about high energy physics. The crucial rub is that, as of yet, neither the standard model nor any interacting QFT in four-dimensional Minkowski space-time has been constructed as a model of these axiomatic systems. While axiomatic treatments of QFT give us a clear set of theoretical principles, and therefore at least potential targets for realist commitment the connection with empirical predictions that the realist needs is missing.

In sum then, both the explanatory and epistemic achievement components of scientific realism seem to run into serious trouble in the QFT context. I should stress that I don't take this to be an existential threat to a realist view of science as a whole. An obvious response to this situation is to weaken the strength of one's epistemic commitments in this context, perhaps citing the theoretical immaturity of the QFT programme. One could admit, for instance, that we cannot yet explain the success of QFT's empirical predictions, or make precise claims about the nature of the unobservable world on their basis while remaining optimistic that future scientific progress will eventually come up with the goods.[5] Furthermore, some philosophers of science have recently been arguing that realism needs to be weakened anyway in the face of anti-realist critique. Saatsi (2016; 2017) advances a position he calls minimal realism which abandons the epistemic achievement component entirely and focuses instead on a stripped back version of the explanatory thesis. The minimal realist is committed to there being an explanation of the success of current theories in terms of the way they latch onto the world but admits that we may not be able to say how this story goes in our current epistemic position. This weakened explanatory thesis is clearly compatible with the state of play in high energy physics as I have just characterized it.

I will be suggesting in what follows, however, that there is scope for rescuing the traditional explanatory and epistemic achievement theses in the QFT context. The preceding discussion may be a fair assessment of the situation as it stood before the 1970s, but developments in renormalization theory open up the possibility of a more full-blooded realist reading of the theory.

[5] Doreen Fraser (2009) can be read as advocating a position like this. According to her, the axiomatic approach should be viewed as a work in progress that we have good reason to hope will eventually solve the puzzle surrounding the physical content of the standard model.

14.3 The Renormalization Group

In the decades following the invention of the perturbative QFT formalism the notion of renormalization broadened in scope and ultimately underwent fundamental conceptual changes. A fruitful exchange of ideas between condensed matter and high energy physics culminated in the emergence of the so-called renormalization group in the 1970s. Nowadays, renormalization methods enjoy applications in many areas of physics and beyond, taking on different forms in different theoretical contexts. I focus here on the incarnation of the renormalization group which is most significant for the philosophy of QFT: the momentum space approach developed by Kenneth Wilson and John Kogut (1974).

The core idea underlying the renormalization group, in all of its guises, is the study of coarse-graining transformations: operations which take us from an initial system of interest to a new one that lacks some of the degrees of freedom associated with high energies and small length scales but shares its large-scale properties. One reason why renormalization group methods are so diverse is that there are many ways of implementing a transformation of this kind. One approach employed in the study of lattice spin systems in statistical physics, for instance, is to replace groups of neighbouring spins with a single 'block spin' degree of freedom and tune the dynamics of the new system so as to reproduce the same (or, in practice, similar) macroscopic behaviour (Figure 14.1). In some cases, it may be possible to invert the transformation, forming a group structure—hence the name. But this is not always possible; blocking transformations of the kind just described are typically not invertible. In any case, group theory seldom plays an important role in renormalization group methods, so the terminology is always misleading in one way or another. The real significance of these transformations is that they can be understood as inducing a 'flow' on a space of possible theories. Studying this flow turns out to be a powerful source of information about the behaviour of physical systems at different scales.

Rather than working in real space, as in the blocking approach, Wilson pioneered the idea of implementing a coarse-graining transformation in momentum space. To see how this works we need to return to the path integral expression for the partition

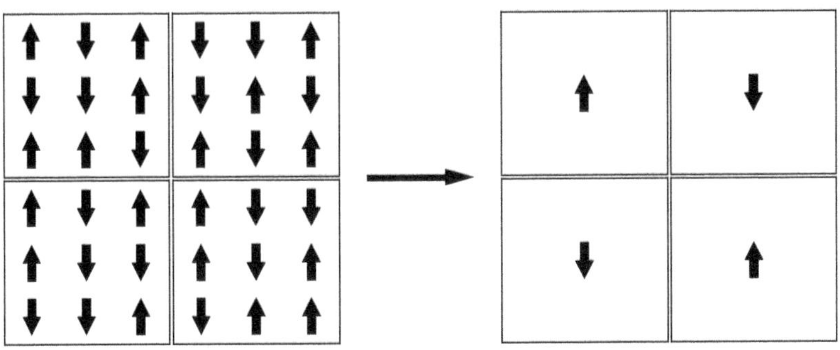

Figure 14.1 A 'majority rule' blocking procedure on a two-valued spin system.

function, Z. As we already discussed, defining integrals over field configurations is a project fraught with difficulty. Wilson's key insight was that, rather than considering all configurations at once, we can evaluate the integral in sections. To get started with this approach we need to explicitly remove from the path integral the configurations associated with momenta above some very high value Λ:

$$Z_\Lambda = \int_{|p| \leq \Lambda} \mathcal{D}\phi e^{iS}. \tag{14.3}$$

We'll see how this can be rationalized in detail shortly, but for now we can think of Λ as the energy scale at which new fields, or physics beyond the scope of the QFT framework entirely comes into play (in the most extreme case this would be the so-called Planck scale at which quantum gravity effects are expected to become important). This is somewhat analogous to the cutoff imposed in perturbative renormalization, but we are now operating in a completely non-perturbative context. What results from this cutoff procedure then is a well-defined physical system that lacks degrees of freedom associated with arbitrarily large momentum, and arbitrarily small length scales. In contrast to the models sought in the axiomatic approach to QFT, these systems have a finite number of degrees of freedom in any finite space-time region. I'll call these structures cutoff QFT models.

The Wilsonian renormalization group sets up a coarse-graining transformation on these cutoff QFTs as follows. We isolate the contribution to the integral due to the highest remaining field configurations, whose Fourier transforms have support above some value μ; this part of the integral is then computed separately and absorbed into a shift in the action. In symbols:

$$\int_{|p| \leq \mu} \mathcal{D}\phi \int_{\mu \leq |p| \leq \Lambda} \mathcal{D}\phi e^{iS} = \int_{|p| \leq \mu} \mathcal{D}\phi e^{i(S+\delta S)}. \tag{14.4}$$

This defines a transformation that takes us from an initial cutoff QFT model to a new one, which has a lower cutoff and a modified dynamics but behaves like the original.

This transformation will not only alter the values of coupling parameters in the initial action but also give rise to new interaction terms. In general, we need to consider all possible terms which are not ruled out by initially demanded symmetries and constraints, including non-renormalizable interactions. For scalar fields, this means going beyond the ϕ^4 theory and considering interactions like $\lambda_6\phi^6$, $\lambda_8\phi^8$, and so on. The renormalization group transformation can then be seen as inducing a flow in a space of possible theories, spanned by the couplings $(\mathrm{m}, \lambda_4, \lambda_6, \ldots)$. The way that these parameters change as μ is lowered, and more and more high momentum degrees of freedom are 'integrated out', is described by the so-called beta functions:

$$\beta_i = \mu \frac{\partial}{\partial \mu} \lambda_i. \tag{14.5}$$

In most interesting contexts the equations governing the renormalization group flow are highly non-linear and cannot be exactly solved. Probing them via various approximation methods, however, has proven to be a powerful tool for answering many important questions about QFTs.[6]

What is most significant for the scientific realism debate is how the renormalization group flow behaves in the low energy regime, far below the cutoff Λ. It turns out to be possible to say something surprisingly general about this. As μ decreases the renormalization group flow is thought to be attracted to a finite dimensional surface spanned by the parameters associated with renormalizable terms in the action. Polchinski (1984) gives a rigorous demonstration of this behaviour in the case of scalar field theories by linearizing the renormalization group equations in the region of small couplings. If we do this it is possible to show that variations in the non-renormalizable couplings at initial cutoff scale Λ are almost completely absorbed into variations in renormalizable couplings as we coarse-grain to a much lower scale $\mu \ll \Lambda$, up to powers of μ/Λ. While this is a somewhat special result, and rests on the assumption that the couplings are small, this sort of behaviour is believed to hold generally and apply to realistic theories like QED and the standard model.

What this means is that QFT models that differ dramatically in their high energy dynamics manifest very similar physics at lower energies; a phenomenon known as universality in the physics literature. Basically any scalar field theory will look like the familiar ϕ^4 model at sufficiently low energies, for instance. This leads to a new perspective on QFT, and on the cutoff imposed at the beginning of the renormalization group analysis in particular. The renormalization group results just discussed demonstrate that removing the high energy degrees of freedom of a field theory leaves its low energy behaviour more or less unaffected. Since a model's low energy physics can be almost completely parameterized by renormalizable couplings, all varying the value of Λ (and the details of how it is imposed) can do is move it around this finite dimensional surface. Fixing the values of the renormalizable couplings via a finite number of experimental measurements absorbs almost all of the cutoff dependence of a QFT model's physical quantities. Cross sections for scattering events at an energy scale E, for instance, will only depend on the cutoff through powers of E/Λ. The cutoff, then, allows us to bracket the question of what the world is like at the fundamental level while accurately modelling its low energy properties.

In the post renormalization group era, QED and the standard model have come to be regarded as 'effective field theories': models that are valid in some limited energy regime but should not be trusted beyond them. It is this shift in the outlook and methodology of the QFT programme which opens new paths for the scientific realist, as I will now argue.

14.4 Brighter Prospects for Realism

I claim that these advances in renormalization theory have improved the situation for both of the components of the realist doctrine in the QFT context. Firstly, the renormalization group framework provides a physical understanding of the original

[6] The renormalization group also turns out to be relevant to the project of constructing QFT models without cutoffs which satisfy the sets of axioms devised in the axiomatic approach to QFT, for instance. See Hancox-Li (2015).

perturbative renormalization procedure. It thus opens up the possibility of achieving the sort of explanation of the empirical success of perturbative QFT the realist seeks. And secondly, it leads to a new perspective on the problem of characterizing QFT's physical content and ultimately suggests a way of articulating claims about the world that are supported by the success of the standard model.

How does the renormalization group help us make sense of perturbative QFT? One important upshot that has been emphasised in the recent philosophical literature is that it transforms our understanding of the notion of renormalizability (Butterfield and Bouatta, 2015). As I mentioned in Section 14.2, renormalizability was traditionally viewed as a property that any viable QFT model must possess. There was always something mysterious about this way of thinking, however. Why should the world be structured in such a way as to be amenable to perturbative approximation? Without the renormalization group results just described, the fact that the actions of empirically successful theories are renormalizable would seem to be a lucky coincidence. We can now see, however, that limiting one's attention to renormalizable interaction terms is a very reasonable thing to do. If Λ is taken to be a very high energy scale at which new physics comes into play, we should expect physics at currently accessible energies to be very well described by a renormalizable action, as the effects of non-renormalizable interaction terms will be heavily suppressed by inverse powers of the cutoff.

This does not quite get to the heart of how the renormalization group illuminates the perturbative approach, however. After all, the fundamental puzzle about renormalized perturbation theory was why it produces accurate approximations at all, even granting the renormalizability of the interactions under consideration. I suggested above that the perturbative approach is incapable of answering this question on its own because it lacks a cogent characterization of the systems and quantities it is supposed to be approximating. The renormalization group framework arguably fills this lacuna, providing a non-perturbative framework that can justify the various steps that go into the perturbative renormalization procedure.

The first step of the perturbative renormalization procedure, you will recall, was to replace the divergent momentum integrals in the perturbative coefficients with finite cutoff expressions. Since the cutoff is removed at the end of the calculation this was typically understood in purely formal terms in the original perturbative treatment of QFT. The renormalization group framework provides a physical interpretation of what is going on here, however. We have seen that it is possible to simulate the effects of high momentum degrees of freedom not included in a cutoff model by tuning the system's dynamics. We can thus legitimize the perturbative cutoff on momentum space integrals in the same terms, on the understanding that the effects of physics beyond the cutoff can be absorbed into an 'effective' action.

The second step of the procedure was to redefine the expansion parameter so as to remove the diverging dependence on the cutoff in the perturbative coefficients. In the original incarnation of the renormalization method, this was viewed as a necessary step to extract sensible predictions from the perturbative formalism. The renormalization group analysis of the low energy regime provides a physical justification for removing the hypersensitivity to the cutoff, however. We saw that low energy physical quantities, and in particular, the scattering cross sections that are typically the targets of perturbative calculations, actually are weakly dependent on the cutoff at low energies. This amounts to a non-perturbative demonstration that the logarithms and powers of Λ that appear in naive perturbative expansions are artefacts of an inappropriate choice of expansion parameter. Once the expansion parameter has been renormalized in the manner described in Section 14.2, the only dependence on the cutoff that remains in perturbative approximations takes the correct form of powers of E/Λ. The perturbative renormalization procedure, on this reinterpretation, is fundamentally about ensuring that our approximations have the right scaling behaviour, not about ensuring that they are mathematically well behaved as the cutoff is removed. Choosing the expansion parameter so as to minimize the dependence on the cutoff can simply be understood as a matter of ensuring that truncations of the series mimic the behaviour of the physical quantity that they are supposed to approximate.

The final step of taking the cutoff to infinity also finds a natural justification in the renormalization group setting. Assuming that the cutoff scale is much higher than the energy scale we are trying to describe, the E/Λ cutoff dependence of renormalized perturbative approximations, and the actual physical quantities they are supposed to approximate, will be very small. In many contexts, they will be much smaller than expected experimental error and can consequently be justifiably ignored. What we are doing when we take the cutoff to infinity in perturbative calculations is neglecting the residual dependence on the cutoff. Since removing the cutoff in the perturbative context has significant computational benefits, and the renormalization group gives us a handle on the kind of errors that result from doing so, it is pragmatically justified.

This is only a sketch of how the renormalization group illuminates the original perturbative QFT formalism and more work is clearly needed to develop this story in detail.[7] Still, the pieces seem to be in place for an explanation of the success of perturbative QFT predictions that should satisfy the realist.

We also find ourselves in a better position regarding the epistemic achievement component of realism. The shift towards an effective field theory perspective on QFT points to a way of extracting knowledge of the unobservable from empirically successful QFTs.

The central problem here was the lack of a clear answer to the question of what empirically successful QFTs are—both mathematically and physically. We saw in Section 14.3, however, that it is possible to precisely define the path integral for the partition function and explicitly construct realistic QFTs as mathematical models if the degrees of freedom of the field associated with arbitrarily large energies and momenta are removed via the cutoff. This provides a non-perturbative characterization of QFT which has a crucial advantage over the axiomatic systems discussed in Section 14.2; we can explicitly write down cutoff formulations of empirically successful QFTs, and the standard model in particular. Furthermore, we have seen good reasons

[7] One potential worry here is that, since perturbation theory itself is often used to analyse the renormalization group flow—and the Polchinski results mentioned in Section 14.3 turns on a small coupling assumption—there is a danger of circularity in appealing to the renormalization group to rationalize the perturbative formalism. I think this objection can be rebutted, but discussing this point in detail is beyond the scope of this chapter. Thanks to Laura Ruetsche for raising this issue.

to regard these cutoff models as conceptually respectable, and empirically successful, theories in their own right. As Wallace (2006; 2011) has argued, this suggests that we should be looking to cutoff QFT models when it comes to the question of what we ought to believe about the world given the empirical successes of high energy physics.[8]

What would it mean exactly to be realist about a cutoff QFT? Following Williams (2019), we can make sense of this along the lines of the selective realist programme introduced in Section 14.2.[9] The basic idea behind this approach was that the realist ought to take a differentiated attitude towards the content of a successful theory, saving their optimism for those theory constituents that underwrite its predictive success. The renormalization group comes into its own again here, providing a powerful tool for developing this sort of selective realist reading of a QFT.

On the one hand, it allows us to identify features of cutoff QFT models that we should not take representationally seriously. Much of the empirical success of the standard model takes the form of predictions of cross sections for scattering events produced in particle colliders, with the current energy limit being at the order of 10^{13} electron volts. The renormalization group results just discussed reveal that many features of current QFTs do not really make a difference to these empirical successes, in the sense that they can be varied without much affecting scattering cross sections at relevant energy scales. For one thing, it establishes that these quantities are highly insensitive to the imposition of a cutoff at some much higher energy scale, as well as the details of how this is done. We can also vary the dynamics of a model at the cutoff scale without affecting its predictions; adding small non-renormalizable interactions to the standard model action, for instance, does not undermine its empirical adequacy. What this tells us is that many of the claims QFTs make about the world at the fundamental level do not contribute to, and consequently are not supported by, the predictive successes of modern particle physics.

On the other hand, the renormalization group helps us articulate positive theoretical commitments that are supported by the success of the standard model. As well as sharing empirical content, the classes of QFT models that flow to the same surface under the action of the renormalization group transformation arguably make common claims about relatively large-scale, non-fundamental, aspects of the world. They agree, for instance, about low energy correlation functions—expectation values of products of field operators associated with well-separated space-time regions. These quantities are preserved by the renormalization group coarse-graining transformation and encode the long distance structure of a QFT model. They are also directly connected to its successful predictions—unlike the theoretical features mention above, you cannot vary the long distance correlation functions of a theory without drastically

[8] Taking cutoff models to provide an adequate characterization of the mathematical and physical content of QFT is controversial. Doreen Fraser (2011), in particular, has argued against this move and defended the superiority of axiomatic formulations of the theory. Note, however, that taking cutoff QFTs to be appropriate objects of realist commitment does not imply that the axiomatic approach has nothing to offer philosophically. There are arguably other philosophical issues raised by the QFT programme that are most naturally addressed in the context of the axiomatic tradition. For more on this pluralist approach to the dispute surrounding the formulation of QFT see J. Fraser (2016).

[9] See J. Fraser (2018) for a discussion of how this proposal fits into, and sheds light on, broader debates about how a selective form of scientific realism ought to be formulated.

affecting its low energy scattering cross sections. Furthermore, the renormalization group tells us that these quantities are extremely insensitive to the details of physics at very high, currently inaccessible, energy scales. It thus demonstrates that they are, at least in one sense of the term, robust, another quality which selective realists often take to motivate belief in a theoretical claim (Wimsatt, 2007; Williams, 2019).

The idea then is that the information the renormalization group provides about the dependencies that hold between theoretical claims at different scales allows us to sort those we should take representationally seriously from those we should not. The picture that emerges from this analysis is that QFTs enjoy a kind of coarse-grained representational success, capturing some (relatively) long distance, low energy, features of the world without limning its fundamental structure. This accords with the effective field theory methodology of much contemporary high energy physics, but, crucially, the claim that the standard model is an effective theory is not taken to mean that it is purely phenomenological. It furnishes genuine extra-empirical knowledge on this view—just not about the fundamental.

14.5 Challenges

Drawing on renormalization group resources in the way I have just outlined is, I think, our best hope for assimilating QFT into a traditional realist view of science. Many questions remain about how this proposal should be fleshed out and defended, however. In the interest of sharpening the position, or, more accurately, identifying those areas where it needs sharpened, I want to conclude by addressing some objections recently posed by Laura Ruetsche (2018). I will touch, in particular, on two worries raised in her discussion that point to the need for further work in an epistemological and metaphysical/semantic direction respectively.[10]

The first objection targets my claim at the end of the last section that results about the low energy behaviour of the renormalization group flow give us grounds to be confident in some coarse-grained properties of QFT models. The worry is that this move falls foul of familiar anti-realist arguments concerning historical theory change. Anti-realists have long pointed to the plethora of predictively successful yet false scientific theories in the historical record as a challenge for realism. A recent incarnation of this sort of argument, which is particularly pertinent in the present context, is Kyle Stanford's problem of unconceived alternatives (Stanford, 2006). When we look at the history of science, according to Stanford, we find that scientists have repeatedly failed to conceive of rival theories that were just as well supported by the available empirical evidence as the theories they accepted. Inductively then, we should expect present scientists, and scientific theories, to be in the same boat, undermining our confidence in current extra-empirical scientific claims.

The renormalization group results discussed above might seem to offer respite from this problem. The standard model will likely someday be replaced by a new theory that describes the physics of currently inaccessible energy regimes. Since the low energy

[10] Another line of objection, which will not be dealt with here, is Doreen Fraser's (2011) claim that renormalization group results reveal widespread underdetermination in QFT and therefore push against realism, rather than coming to its aid. I hope to address this worry in future work.

features of the standard model are highly insensitive to what is going on at these higher energy scales, however, one might think that they are very likely to be retained by these more fundamental theories, whatever they end up looking like. As Ruetsche points out, however, this line of response rests on the assumption that future theories can be situated in the space on which the renormalization group transformation acts and this is not obvious a priori. Rigorous renormalization group results, such that there are, deal with rather circumscribed spaces of theories. Furthermore, the QFT framework itself is expected to break down and be replaced by a radically new quantum gravity theory as the Planck scale is approached. Who is to say this theory can be treated within the renormalization group framework? In fact, if we accept the moral of Stanford's induction, it seems that we should actively expect future theory change to outstrip the reach of renormalization group considerations.

How to respond? We should concede, I think, that renormalization group results do not defeat historically generated scepticism on their own.[11] Indeed, it is hard to see how local scientific arguments could ever sway a thoroughgoing anti-realist who is already convinced of their unreliability. It is important to distinguish two tasks facing the scientific realist here, however: spelling out the content of their epistemology and defending it. This chapter has focused on the former; I presupposed a realist perspective and advanced a strategy for articulating two central realist theses in the QFT context. When it comes to the project of arguing for a realist view of science more general epistemological issues that have not been touched on here need to be considered, the implications of historical theory change being prominent among them. A flat-footed response to this sort of attack then is to admit that the plausibility of the programme set out in this chapter is conditional on a successful rebuttal of historical arguments for anti-realism.[12]

Having said that, it is possible that the renormalization group might play some role in the broader project of defending realism. Here is a different way it might feature in a response to Stanford's unconceived alternatives problem, for instance. One supposed advantage of Stanford's argument is that, whereas traditional incarnations of the pessimistic induction generalize over scientific theories, and are therefore vulnerable to the response that false historical theories were unlike currently accepted ones in

[11] One might try to resist this conclusion by pointing to further theoretical considerations which suggest that the renormalization group apparatus can encompass the class of relevant alternatives to the standard model. We might point, for instance, to Weinberg's (2004) 'folk theorem', which states that any theory that satisfies some general principles, like cluster decomposition and (approximate) Lorentz invariance, will have to recover the basic theoretical structure of QFT at low energy scales. This suggests that, no matter how radically novel a future theory of quantum gravity is, its low energy limit should be a QFT that is amenable to the kind of renormalization group analysis described in Section 14.3. Again though, this argument rests on a posteriori scientific premises, so unrestrained historically generated scepticism will also throw it into doubt. Thanks to Porter Williams for discussions on this point.

[12] Godfrey-Smith (2008), Chakravartty (2008), Ruhmkorff (2011) and Devitt (2011) offer responses to Stanford's unconceived alternatives problem which the aspiring realist about QFT might draw on. I should note that if it turns out that the epistemic achievement component of realism has to be abandoned in light of arguments like Stanford's, the renormalization group might still have something to offer a more minimal realism of the kind advocated by Saatsi (2016). Saatsi suggests that the most promising strategy for articulating the realist position is to point to exemplars, from contemporary science and historical record, of how a theory's latching onto the world could explain its empirical success. We could thus treat the story told in Section 14.4 as an exemplar of this sort without committing ourselves to its truth.

relevant respects, his induction generalizes over scientific theorists. Stanford suggests that it is more difficult to drive a wedge between the inferential capacities of past and present scientists—surely current scientists are no better than luminaries like Darwin and Maxwell at conceiving of relevant alternatives to their theories (Stanford, 2006, 43)? One might reject this assumption, however, claiming that theoretical and methodological progress in the sciences has improved our ability to probe the space of possible theories. The renormalization group is arguably an instance of this phenomenon. Even if it is not exhaustive in its scope, the information the renormalization group provides about how modifications of a theories dynamics at one scale affect its behaviour at others seems to be a novel epistemological resource that was not available to previous generations of physicists. While Ruetsche is right to flag the danger of exaggerating the power and generality of the renormalization group approach then, I do not think her discussion gives us reason to abandon the realist programme advanced in this chapter. It does point to a need for a more careful examination of the epistemological mileage that can really be extracted from renormalization group results, however.

The second worry I want to consider concerns the nature of the representational success allegedly enjoyed by effective field theories. Even granting that some of the low energy claims of current QFTs are supported by renormalization group considerations, Ruetsche (2018) suggests, it is not obvious that they really concern the unobservable. This line of attack exploits the fact that renormalization group enthusiasts are often rather vague about the features of the world they take effective field theories to be latching onto. Furthermore, when they do try to make more explicit commitments they seem to be vulnerable to reinterpretation in empiricist friendly terms. In Section 14.4, I identified low energy correlation functions as plausible examples of the sort of theoretical quantities we ought to be realist about. But correlation functions are intimately related to the scattering cross sections measured in particle colliders. If correlation functions can be interpreted as merely encoding information about the empirical signatures of scattering events measured at collider experiments, then the view of QFT I have been developing threatens to collapse into a form of empiricism.

Again, I do not think this line of objection is fatal, but it does add urgency to a key challenge facing my programme. In order to substantiate the claim that effective field theories are capturing unobservable aspects of the world, in accordance with the epistemic achievement thesis, we need a precise characterization of the non-fundamental descriptive content of QFTs. This is a difficult problem for, at least, two reasons.

First of all, the status of non-fundamental entities and properties is a controversial issue in its own right. As Williams (2019) points out, pervasive methodological assumptions about the nature of the interpretive project have led philosophers of physics to focus on the fundamental ontology posited by physical theories. As a result, non-fundamental physical ontology is undertheorized and some may even doubt that there is such a thing, adding fuel to the suspicions about the realist credentials of my programme. In philosophy more generally, there is no agreed-upon framework for regimenting claims about the non-fundamental, with different approaches gaining currency in different sub-disciplines. Philosophers of science in the structural realist tradition (Ladyman and Ross, 2007; Wallace, 2012) have appealed to the Dennetian notion of 'real patterns', for instance, while analytical metaphysics has seen an explosion of interest in notions of metaphysical dependence and grounding (Bliss and Trogdon, 2014). There are certainly frameworks on the table that the effective field theory realist might turn to in order to clarify their position then, but doing so will only be as reassuring as these approaches are well motivated.[13]

A second obstacle is the perennial controversy surrounding the physical content of quantum theories. I have skirted around the measurement problem thus far but the would-be realist about QFT clearly cannot put it aside indefinitely. In order to rebut suggestions that low energy correlation functions have no extra-empirical content we seem to need a physical interpretation of quantum operators and states, and different approaches to the measurement problem give different answers to the question of what the quantum world is basically like.[14] Furthermore, in addition to the puzzles inherited from non-relativistic quantum mechanics, QFT raises interpretive problems of its own. As Ruetsche (2011) sets forth in detail, the existence of unitarily inequivalent Hilbert space representations of a quantum theory with infinitely many degrees of freedom arguably poses a novel interpretive challenge.[15]

All this suggests that the strategies advanced in this chapter will have to be re-examined alongside debates about the interpretation of quantum theory and non-fundamental ontology before the prospects of a realist view of QFT can be fully assessed.

Acknowledgements

Thanks to participants of the Pittsburgh Center for Philosophy of Science reading group (Pablo Acuña, Colin Allen, Geoffrey Gorham, Karen Kovaka, Edouard Machery, Matthias Neuber, Daniel Wilkenfeld, and Porter Williams) as well as an anonymous referee for useful comments on earlier versions of this chapter.

References

Bliss, R. and K. Trogdon (2014). Metaphysical grounding. In Edward N. Zalta (Ed.), *The Stanford Encyclopedia of Philosophy* (Winter 2014 Edition). https://plato.stanford.edu/entries/grounding/.

Butterfield, J. and N. Bouatta (2015). Renormalization for philosophers. *Metaphysics in Contemporary Physics 104*, 437–85.

[13] One problem here is that many discussions of the fundamental/derivative distinction in the metaphysics literature are sufficiently far removed from scientific practice that it is difficult to see how to apply them to actual physics. McKenzie's (2013) application of Fine's notion of grounding to quantum theory shows that this gap can be bridged, however.

[14] Wallace's realism about QFT is wedded to his Everetianism. It is worth noting that Everetian interpretations of quantum theory are perhaps the most natural fit with the approach to realism about QFT described in this chapter, since approaches to the measurement problem which modify the basic structure of quantum mechanics, such as Bohmian and dynamical collapse models, cannot currently be extended to empirically successful QFTs.

[15] Indeed, she claims that unitarily inequivalent representations ultimately pose a problem for a realist view of QFT. See J. Fraser (2016, ch. 6), for a response to this challenge.

Chakravartty, A. (2008). What you don't know can't hurt you: realism and the unconceived. *Philosophical Studies 137*(1), 149–58.
Devitt, M. (2011). Are unconceived alternatives a problem for scientific realism? *Journal for General Philosophy of Science 42*(2), 285–93.
Fraser, D. (2009). Quantum field theory: underdetermination, inconsistency, and idealization. *Philosophy of Science 76*(4), 536–67.
Fraser, D. (2011). How to take particle physics seriously: a further defence of axiomatic quantum field theory. *Studies In History and Philosophy of Science Part B: Studies In History and Philosophy of Modern Physics 42*(2), 126–35.
Fraser, J. D. (2016). What is quantum field theory? Idealisation, explanation and realism in high energy physics. PhD thesis, University of Leeds.
Fraser, J. D. (2017). The real problem with perturbative quantum field theory. *British Journal for the Philosophy of Science*. https://doi.org/10.1093/bjps/axx042.
Fraser, J. D. (2018). Renormalization and the formulation of scientific realism. *Philosophy of Science 85*(5), 1164–75.
Godfrey-Smith, P. (2008). Recurrent transient underdetermination and the glass half full. *Philosophical Studies 137*(1), 141–8.
Haag, R. (1996). *Local Quantum Physics: Fields, Particles, Algebras*. Dordrecht: Springer.
Halvorson, H. and M. Müger (2007). Algebraic quantum field theory. In J. Butterfield and J. Earman (Eds), Handbook of the Philosophy of Science, *Philosophy Of Physics, Part A*, pp. 731–864. Amsterdam: Elsevier
Hancox-Li, L. (2015). Coarse-graining as a route to microscopic physics: the renormalization group in quantum field theory. *Philosophy of Science 82*(5), 1211–23.
Kitcher, P. (2001). Real realism: the Galilean strategy. *The Philosophical Review 110*(2), 151–197.
Ladyman, J. and D. Ross (2007). *Every Thing Must Go: Metaphysics Naturalized*. Oxford: Oxford University Press.
McKenzie, K. (2013). Priority and particle physics: ontic structural realism as a fundamentality thesis. *British Journal for the Philosophy of Science 65*(2), 353–80.
Polchinski, J. (1984). Renormalization and effective Lagrangians. *Nuclear Physics B 231*(2), 269–95.
Psillos, S. (1999). *Scientific Realism: How Science Tracks Truth*. London: Routledge.
Ruetsche, L. (2011). *Interpreting Quantum Theories*. Oxford: Oxford University Press.
Ruetsche, L. (2018). Renormalization group realism: the ascent of pessimism. *Philosophy of Science 85*(5), 1176–89.
Ruhmkorff, S. (2011). Some difficulties for the problem of unconceived alternatives. *Philosophy of Science 78*(5), 875–86.
Saatsi, J. (2016). What is theoretical progress of science? *Synthese*. https://doi.org/10.1007/s11229-016-1118-9.
Saatsi, J. (2017). Replacing recipe realism. *Synthese 194*(9), 3233–44.
Stanford, P. K. (2006). *Exceeding Our Grasp: Science, History, and the Problem of Unconceived Alternatives*. Oxford: Oxford University Press.
Streater, R. and A. Wightman (1964). *PCT, Spin and Statistics, and All That*. W. A. Benjamin.
Van Fraassen, B. C. (1980). *The Scientific Image*. Oxford: Oxford University Press.
Wallace, D. (2006). In defence of naiveté: the conceptual status of Lagrangian quantum field theory. *Synthese 151*(1), 33–80.
Wallace, D. (2011). Taking particle physics seriously: a critique of the algebraic approach to quantum field theory. *Studies In History and Philosophy of Science Part B: Studies In History and Philosophy of Modern Physics 42*(2), 116–25.
Wallace, D. (2012). *The Emergent Multiverse: Quantum Theory According to the Everett Interpretation*. Oxford: Oxford University Press.

Weinberg, S. (2004). What is quantum field theory, and what did we think it is? In T. Y. Cao (Ed.), *The Conceptual Foundations of Quantum Field Theory*, pp. 241–51. Cambridge: Cambridge University Press.

Williams, P. (2019). Scientific realism made effective. *British Journal for the Philosophy of Science*, *70*(1), 209–37.

Wilson, K. G. and J. Kogut (1974). The renormalization group and the ϵ expansion. *Physics Reports* *12*(2), 75–199.

Wimsatt, W. C. (2007). *Re-Engineering Philosophy for Limited Beings: Piecewise Approximations to Reality*. Cambridge, MA: Harvard University Press.

15

Perturbing Realism

Laura Ruetsche

15.1 Introduction: A New Development?

In soft-focus but at breakneck speed, some of the more interesting moves in the debate over scientific realism unfold as follows. Take a stupendously successful contemporary scientific theory T. A scientific realist contends that T is (at least approximately) true, where 'true' is understood in some heartily non-epistemic, non-pragmatic way. The realist's opponent observes that, as a matter of historical fact, every single physical theory celebrated as successful in its heyday was subsequently overthrown in favor of successor theories offering radically different pictures of the world. Executing what's known as the pessimistic meta-induction, the opponent asks: Why think our currently beloved theory T will meet a different fate? Psillos (2005) dubs the by-now standard realist response *divide et impera*: announce a selection principle that identifies those aspects of T worthy of belief. A symptom of worthiness is the persistence of these aspects in T's successors. The diagnosis of this persistence is that those aspects are essential to theoretical success. They're latching on to 'hidden springs,' not merely cataloguing phenomena. Selection principles proposed to date—they include 'structure,' 'essential explanatory elements,' and 'causal features'—strike non-realists as suspiciously vague, absent augmentation by accounts of how to individuate structure, distinguish essential explanatory elements from idle wheels, or identify causal features. 'What's preserved in all future theories' may be an apt functional characterization of the intended locus of commitment—but it's not one we can use now. In the absence of a concrete selection principle that *right now* identifies theoretical elements worthy of commitment, realism's detractors are unmoved by *divide et impera*.

Some antirealists advocate an empiricist attitude toward T: T's success warrants belief not in its truth but in its empirical adequacy, its capacity to save (and to continue to save) the phenomena. The realist indicts this attitude for committing 'selective skepticism.' The empiricist believes that T extends successfully to the not-yet-observed but does not believe that T extends successfully to the unobservable. Such selective skepticism lands the empiricist in an epistemic predicament realism avoids. By the agency of its hidden springs and mechanisms, T explains why the phenomena it saves *have to* unfold as they do. Believing in the operation of those springs and mechanisms, realists have reason to believe the phenomena will continue

Laura Ruetsche, *Perturbing Realism* In: *Scientific Realism and the Quantum*. Edited by: Steven French and Juha Saatsi, Oxford University Press (2020). © Laura Ruetsche. DOI: 10.1093/oso/9780198814979.003.0015

to conform to expectations guided by *T*. The agnosticism about underlying operations empiricism insists upon delivers no analogous grounds for faith in the persistence of the phenomena *T* saves. This is the predicament of empiricism, which requires us to believe in the persistence of regularities it denies us underlying reason to expect to continue.[1]

Thus both the realist and the empiricist face problems of circumscription. For realists, the challenge is to circumscribe defensible knowledge claims. For empiricists, the challenge is to circumscribe sustainable skepticisms.

And there the matter might unsteadily rest but for a new development: a contribution to the realism debate its advocates call *Effective Realism*. Effective Realism offers a strategy for pursuing selective realism, a strategy inspired by both technologies and ideologies of modern physics. Ideologically, Effective Realism takes seriously what is increasingly a commonplace among working physicists: our current best physical theories, including the interacting Quantum Field Theories (QFTs) making up the Standard Model of contemporary particle physics, are *merely effective*. Effective theories aren't absolutely true at all orders. However, they don't need to be, because they are applied only to limited regimes—for instance, the regimes of phenomena accessible at energies modern particle accelerators can reach.[2] *In those regimes*, effective theories approximate the implications of fundamental physics (whatever it is!).

The technology inspiring Effective Realism is the technology of *Renormalization Group* (RG) approaches to effective theories. Effective Realism marshals RG resources to announce a *committal* and *physically motivated* selection principle for *divide et impera* realism. Thus Effective Realism extracts from recent developments in cutting-edge physics a refined and resilient realism. Along the way, Effective Realists often criticize as misguided and stultifying an established tradition in philosophy of physics: the tradition of 'Standard Interpretation' (about which more soon) (J. Fraser, 2016 and this volume, Chapter 14; Miller, 2017; Williams, 2019; see also Hancox-Li, 2015).

My aim here is, having explicated and motivated Effective Realism (Section 15.2), to subject it to some friendly criticism (Section 15.3), culminating in an attempt to appropriate RG resources for empiricism. I will suggest that resisting empiricist appropriation entangles Effective Realists in something similar to the very project of Standard Interpretation they regard as misguided. Understanding why this apparent irony isn't an inconsistency illuminates how the matters of interpretation, commitment, and explanation intersect in the scientific realism debate. I hope to convince Effective Realists to join their empiricist appropriators in disavowing, not the project of Standard Interpretation itself, but its involvement in a further project that I'll call 'fundamentalism.'

15.2 Motivating Effective Realism

15.2.1 *Perturbative QFT's 'real problem'*

Typically, non-interacting (aka *free*) QFTs—QFTs describing particles persisting in lofty isolation—are under good mathematical control. Their Lagrangians determine

[1] I am grateful to Lina Jansson for emphasizing this point.

[2] For those of you keeping score at home, the Large Hadron Collider is designed to operate at 13 TeV—that is, 13 *trillion* electron volts.

equations of motion that are exactly soluble; we know how to use their solutions to build Hilbert spaces, and in the setting of those Hilbert spaces we conduct quantum physics (see Wald, 1994 for details). The free QFTs we thereby build can be shown to satisfy standard sets of axioms codifying desiderata for reasonable relativistic quantum theories. That free QFTs admit an axiomatic characterization reinforces our sense that we have a grip on their contents.

Interacting QFTs—QFTs describing the eventful particle physics explored in accelerators such as the Large Hadron Collider—are far less well understood mathematically. Their Lagrangians generate equations of motion we don't know how to solve. Ergo we can't use explicit solutions to construct the standard apparatus of a quantum theory. Nor do axiomatic approaches afford insight into interacting QFTs. No non-trivial example of an interacting QFT satisfying standard axioms has been found—a circumstance often cited as a reason to expand philosophical attention to QFTs beyond the confines of those we can characterize axiomatically (Wallace, 2011).

One way to manage interacting QFTs is to resort to perturbative methods. To approach an interacting QFT *perturbatively* is to build *approximations* of key theoretical structures from the counterparts of those structures exactly realized in a free theory to which the interacting QFT is supposed to be closely related. The construction unfolds roughly as follows.

First, assume that the full Lagrangian of the interacting theory contains free and interacting parts. A self-interacting mass m scalar field $\phi(x)$, where x ranges over points in a two-dimensional spacetime, may be the simplest case. Schematically, the interacting theory's *full* Lagrangian

$$\mathcal{L}(\phi, m, \lambda) = \mathcal{L}_0 + \lambda \mathcal{L}_I \qquad (15.1)$$

depends on the field ϕ, its mass m, and a 'coupling' λ. The free part of $\mathcal{L}(\phi, m, \lambda)$ is the Lagrangian of the free scalar field[3]

$$\mathcal{L}_0 = \frac{1}{2}\partial^2\phi^2 - \frac{1}{2}m^2\phi^2 \qquad (15.2)$$

(15.2) defines equations of motion we know how to solve *exactly*; let $\phi_0(x)$ denote an exact solution.

The term $\mathcal{L}_I$ gives the interacting part of $\mathcal{L}(\phi, m, \lambda)$ and describes the field's self-interaction. For a mass m scalar field interacting with itself via the so-called ϕ^4 potential, $\mathcal{L}_I = -\frac{\phi^4}{4!}$. $\mathcal{L}_I$ appears in $\mathcal{L}(\phi, m, \lambda)$ multiplied by a coefficient λ. When this *coupling constant* is small, the full Lagrangian represents a small disturbance to the free dynamics. In all its glory, the full Langrangian for the ϕ^4 theory is

$$\mathcal{L}(\phi, m, \lambda) = \frac{1}{2}\partial^2\phi^2 - \frac{1}{2}m^2\phi^2 - \lambda\frac{\phi^4}{4!} \qquad (15.3)$$

It is high energy physics custom to identify particles with excitations of fields appearing in the Lagrangian.[4] Following this custom, take $\mathcal{L}(\phi, m, \lambda)$ to describe a field

[3] Cognoscenti will recognize that I am observing the convention of natural units ($\hbar = c = 1$) and a $(+ - - -)$ signature for the Minkowski metric.

[4] In philosophy of physics, even for free theories, it is just as customary to question what a particle interpretation is and whether it's warranted. For details, see Ruetsche (2011, chs. 9–11) and references therein.

ϕ whose excitations correspond to particles of mass m interacting weakly with themselves.

Because the coupling λ is small, the perturbative approach assumes solutions to the equations of motion arising from (15.3) to be well-approximated by free solutions. Expressed as a *perturbative expansion* around a free solution ϕ_0, a full solution ϕ becomes:

$$\phi(x) = \phi_0(x) + \lambda\phi_1(x) + \lambda^2\phi_2(x) + \ldots \tag{15.4}$$

In this expansion, the (as yet unknown) corrections to the free solution induced by the interaction are sorted by magnitude (aka order): the term $\phi_1(x)$ tracks corrections of order λ, $\phi_2(x)$ tracks those of order λ^2, and so on. Because λ is small, each successive term in the expansion matters less to the full solution. If only we can determine a few low-order contributions $\phi_1(x), \phi_2(x) \ldots$ to this perturbative expansion, we can use the resulting approximation to calculate physically interesting quantities accurately enough to do empirical work.

Such quantities include n-point functions, S-matrix elements, and experimental cross sections. An n-point function is a vacuum expectation value of n (time ordered) field operators. The two point function $\langle\phi(x)\phi(y)\rangle$ is an example. In free theories, such expressions are perfectly well defined. Indeed, they define the theory: according to the Wightman Reconstruction Theorem, a QFT satisfying the Wightman axioms is equivalent to (and can be reconstructed from) a full set of n-point functions (see Strocchi, 2013). No interacting QFTs in four spacetime dimensions have been shown to satisfy the Wightman axioms. So as far as we know, realistic interacting QFTs fall outside the scope of this result.

Textbooks typically gloss n-point functions as encoding probabilities for creation, annihilation, and propagation of particles associated with the fields in question: e.g., $\langle\phi(x)\phi(y)\rangle$ is said to determine the probability that the scalar field particle propagates from spacetime point x to spacetime point y. n-point functions enable us to calculate the S-matrix, which encodes probabilities for transitions between incoming and outgoing states of particles/fields scattering off one another. The S-matrix enables us to define cross sections for those interactions, which can be compared to data from particle accelerators. Thus with n-point functions in hand, we can equip our QFT with testable predictions.

Alas, if our QFT is interacting, we haven't got n-point functions in hand. The best we can do is approximate them perturbatively. Feynman diagrams direct these approximations. But enigmatically. Following the Feynman rules, we discover that *individual terms* in the perturbative expansion diverge, that is, go meaninglessly to infinity. For example, for the ϕ^4 self-interaction in four dimensions, starting at second order, integrals like

$$\int_0^\infty \frac{k^3}{k^2 + m^2} dk$$

(k momentum) occur. As the momentum $k \to \infty$, this integral diverges, apparently quashing all hope of wresting empirical meaning from our perturbative expansion.

Because high momentum corresponds to short (ultraviolet) wavelength, this fiasco is known as an *ultraviolet divergence*. Perturbative QFT also suffers from infrared divergences—terms in the perturbative expansion that diverge as the momentum goes to 0. And even if each term of the perturbative expansion could be rendered finite, there remains the question of whether the infinite sum (15.4) itself converges. I'll focus here on how perturbative QFT treats ultraviolet/high momentum/high energy divergences.

The treatment has two major phases. The first phase is *regularization* wherein the problematic integral is rendered finite by (for instance) imposing an upper limit, an ultraviolet cutoff Λ_{UV}, on the momentum modes over which integration is performed:

$$\int_0^{\Lambda_{UV}} \frac{k^3}{k^2 + m^2} dk$$

(Other ways to achieve regularization are to introduce fictitious particles and to execute the integral over non-integer dimensions; see Peskin and Schroeder, 1995 for details.) The phenomena we're using the interacting QFT to describe occur at momenta much lower than the cutoff momentum Λ_{UV}. Although imposing the cutoff renders the integral finite, regularization has an unsettling side-effect: the value of the no-longer-divergent integral looks like it depends on the choice of Λ_{UV}, a choice left to the discretion of the calculating physicist. Such *cutoff-dependence* threatens to hold the theory's empirical predictions hostage to apparently arbitrary features of our calculational techniques.

The second phase of the treatment, *renormalization*, relieves this side-effect. Through art and craft evolved over a period of almost thirty years, physicists learned how to re-write regularized Lagrangians using new coefficients and new fields artfully tuned so that as $\Lambda_{UV} \to \infty$, terms in the perturbative expansion converge at each order. Much finesse, including delicate constructions of counter-terms fostering cascades of cancellations, is required. This hard-won good behavior delivers the theory's empirical predictions from the embarrassment of cutoff dependence.

The renormalized ϕ^4 Lagrangian $\mathcal{L}_R$ assumes the same form as the original Lagrangian, with the 'bare' field amplitude ϕ, bare mass m, and bare coupling λ parameterizing that Lagrangian replaced by renormalized counterparts ϕ_r, m_r, λ_r. Provided a theory's ultraviolet divergences can be tamed by reparameterizing a finite set of coefficients appearing in its Lagrangian, that theory is said to be *(perturbatively) renormalizable*. In addition to the self-interacting scalar field theory just discussed, perturbatively renormalizable interacting QFTs include Quantum Electrodynamics (QED), Yukawa theory, and Yang–Mills theory for compact groups. In practice, the values of the renormalized masses and couplings specifying the renormalized Lagrangian aren't obtained by calculations presupposing specific (and possibly maladroit) renormalization schemes. They are *determined experimentally*. Once these parameters are supplied, perturbatively renormalized QFTs like QED enjoy stunning predictive success.

The actual history of efforts to renormalize theories like QED is littered with poignant tales of renormalization strategies that succeeded for the first dozen or so orders but came to ruin at the thirteenth (see Schweber, 1994). So it is not surprising that those who witnessed the saga of perturbative renormalization are prone to a measure of ambivalence about the technique.

There exists a celebrated procedure, called *renormalization*, which removes the short distance divergences from the field theoretic Perturbative Expansion. Born as partly magic, partly suspicious manipulations on formal series with infinite coefficients, it led, when applied to QED, to finite results which were in spectacular agreement with experiment.

(Gawedzki, 1986, 1279–80)

The ambivalence reflects what James Fraser (2017) has called perturbative QFT's "real problem." Consider an interacting QFT whose misbehaving perturbative expansions can be rendered finite by standard renormalization techniques—an interacting QFT whose consequent empirical predictions are confirmed. We've got to ask: why? Why does this theory, evidently resting on apparently suspicious manipulations, work as well as it does? Perturbative expansions are approximations, but renormalization techniques for taming them tell us nothing about what they approximate.

The success of the perturbative approach is mysterious, I suggest, precisely because it dodges the question of what an interacting QFT is.... [There is] an absence of any non-perturbative characterization of the system of interest.... [This] undercuts the possibility of telling a physical story which could explain its success. (Fraser, 2017, 17–18)

15.2.2 *Renormalization group resolution*

Perturbative QFT's real problem is that it offers no clues about how to construct a physical picture that would explain its success. A number of authors have mustered Renormalization Group considerations to address this problem (Wallace, 2006; Hancox-Li, 2015; J. Fraser, 2016 and this volume Chapter 14; Williams, 2019). Their crucial move is to abandon hope of extracting from an interacting QFT itself an explanation of its success. Instead, they develop a different sort of explanation. Mirroring an attitude increasingly prevalent among physicists who work with such theories (see Wells, 2012), these authors cast interacting QFTs as 'merely effective' theories. T is merely effective just in case T, while not itself a complete and accurate account of physical reality, approximates that account *whatever it is(!)* within a restricted domain of application. Let's suppose there is such a thing as a complete and accurate account of physical reality, and let's use T_{final} to denote that account. To say that T is effective is to say that T succeeds not because it *is* T_{final} but because, in regimes to which we apply T, T *impersonates* T_{final}.

The bare idea that perturbative QFTs are effective theories is not on its own an especially articulate explanation of their empirical success. For one thing, the bare idea engages none of the particulars of perturbative QFTs. We can entertain the impersonation hypothesis about *any* successful theory. We might hope for an explanation of the success of perturbative QFTs that involves features specific to those theories. Renormalization Group (RG) considerations are thought to secure this hope. The following is a very rough sketch of RG approaches to interacting QFTs—see Williams (forthcoming) for a guide aimed at philosophers or Duncan (2012) for one aimed at cognoscenti.

Section 15.2.1 presented the ϕ_4 theory by specifying its Lagrangian

$$\mathcal{L}(\phi, m, \lambda) = \frac{1}{2}\partial^2\phi^2 - \frac{1}{2}m^2\phi^2 - \lambda\frac{\phi^4}{4!}$$

Such introductions are standard etiquette: a physical theory (assumed to operate in a spacetime setting) is identified with its Lagrangian $\mathcal{L}$, where $\mathcal{L}$ is a function of some collection of fields and their spacetime derivatives. Indeed, once we've organized all candidate fields and their derivatives into a list, we can identify a Lagrangian with a list of coefficients (most of which are 0 in Lagrangians we actually calculate with!): the coupling coefficients with which the corresponding fields and spacetime derivatives on the master list appear in that Lagrangian.

Suppose we're interested in how a theory defined by a Lagrangian $\mathcal{L}$ governs phenomena accessible at some scale ℓ. ℓ might be the energy scales attainable by our best current experimental technologies, and we might want to know what $\mathcal{L}$ implies about experiments we can conduct using those technologies. Introduce the notation $\mathcal{L}^\ell$ for the Lagrangian that encapsulates $\mathcal{L}$'s implications for ℓ-scale physics. $\mathcal{L}^\ell$ is an element of a space I'll call $\mathcal{T}$. $\mathcal{T}$ has the form *Lagrangians* × *scales*—it's a space of various Lagrangians, understood as governing physics at various scales.

$\mathcal{T}$ teems with elements such as the Lagrangian $\mathcal{L}^\Lambda$, where Λ is an energy scale higher than 'our' scale ℓ. It is natural to wonder what such relatively fundamental $\mathcal{L}^\Lambda$s imply about experiments we can conduct. Consider a map $R_{\ell\Lambda}$ that acts on $\mathcal{T}$ to track these implications:

$$\mathcal{L}^\ell = R_{\ell\Lambda}\mathcal{L}^\Lambda$$

$R_{\ell\Lambda}$ carries the underlying Lagrangian $\mathcal{L}^\Lambda$ to a Lagrangian $\mathcal{L}^\ell$ governing physics at our scale.[5] $\mathcal{L}^\ell$ is the *effective* (at scale ℓ) Lagrangian induced by the underlying (at scale Λ) Lagrangian $\mathcal{L}^\Lambda$. This effective Lagrangian encapsulates what the underlying Lagrangian implies about physics at our scale.

$R_{\ell\Lambda}$ is a member of a family of transformations acting on $\mathcal{T}$ to extract, from Lagrangians at higher energy scales, the implications for physics at lower energy scales. Put picturesquely, $R_{\ell\Lambda}$ induces a 'flow' on the space of Lagrangians, a flow connecting physics at different scales. To the horror of mathematicians everywhere, the family of transformations is known as the '*Renormalization Group*' (RG).[6]

Let's place perturbative QFTs in the RG framework, and see if that helps with the real problem. Suppose that the perturbatively renormalized QFTs obtained by means of 'suspicious manipulations on formal power series' are low-energy effective theories induced by (unknown) higher energy theories, and connected to them by an RG flow. This supposition on its own hardly solves the problem. If we don't know the high energy theory $\mathcal{L}^{final}$, RG analysis won't help us identify the effective theory $\mathcal{L}^\ell = R_{\ell final}\mathcal{L}^{final}$ governing physics at experimentally accessible scales. In particular, RG analysis won't help us identify our favorite perturbative QFTs as *the* effective theories induced by underlying higher energy physics. What's more, the RG picture

[5] One strategy for constructing such a map is to coarse-grain—for instance, to start with the Lagrangian at higher energy scale Λ and 'integrate out' over contributions from momentum modes lying between Λ and the effective scale ℓ. See Hancox-Li (2015) for details.

[6] The mathematicians are horrified because the map, at least when obtained by coarse-graining, isn't invertible. You can't 'de-integrate' to extract from the low energy encapsulation *the* high energy physics inducing it. So the map is properly a *semi*group. For another way of regarding the RG flow, as well as another route to suspicion of the particle commitments discussed in Section 15.4, see Rosaler and Harlander (forthcoming).

is consistent with a distressing possibility: the possibility that the effective Lagrangian $\mathcal{L}^\ell$, whatever it is, depends intractably on *infinitely* many fields. An $\mathcal{L}^\ell$ depending on infinitely fields is evidently *not* among the *perturbatively renormalizable* QFTs we've been using for particle physics, for those theories require only *finitely* many reparameterizations of coefficients to specify.

It would alleviate these concerns if the RG's action on $\mathcal{T}$ were such that

(i) *All* high energy theories flow to the *same* subspace of $\mathcal{T}$—the subspace $\mathcal{T}^\ell_{\mathit{eff}}$ of effective (at ℓ) theories.
(ii) This *surface of attraction* $\mathcal{T}^\ell_{\mathit{eff}}$ is *finite dimensional*: only finitely many parameters are required to specify a Lagrangian in $\mathcal{T}^\ell_{\mathit{eff}}$.

(i) would make the details of the underlying theory irrelevant to the shape of the theory at scale ℓ—even if we're ignorant of underlying details, we know what signature they leave on physics at our scales. (ii) would ensure that that signature reduces to a tractably finite set of couplings.

Let's call effective theories residing in finite dimensional surfaces of attraction *effectively specifiable.*[7] Section 15.2.1 introduced the notion of 'perturbatively renormalizable' to apply to theories whose ultraviolet divergences could be tamed by finitely many reparameterizations of terms appearing in their original Lagrangians. Perturbatively renormalizable theories are hailed as such because they're susceptible to a certain family of divergence-taming strategies. By contrast, effectively specifiable theories are hailed as such because of where in a space of theories with a remarkable topography they're located.

In principle, when a theory is effectively specifiable, the contribution of unknown physics to physics at the scales we care about can be determined experimentally, simply by measuring coefficients (masses and couplings and the like) defining (that is, locating in $\mathcal{T}^\ell_{\mathit{eff}}$) the effective Lagrangian governing actual physics at experimentally accessible scales. Empirically successful interacting QFTs are widely regarded to be examples of effectively specifiable theories:

> It is a remarkable property of local QFTs that for a certain subset of theories, ... the low energy amplitudes can be parameterized by just a finite set of parameters—namely, those needed to locate the theory on the finite-dimensional attractive submanifold, and which can in principle be determined by making an equal number of independent experimental measurements.
>
> (Duncan, 2012, 587)

Polchinski (1984) demonstrates that at least one interacting QFT exhibits the "remarkable property" Duncan celebrates. Polchinski considers a space of theories whose Lagrangians that are polynomials in a four- (spacetime) dimensional scalar field and its derivatives. Theories in the space describe the free field, as well as that field perturbatively disturbed by a wide variety of *weakly coupled* interactions. The RG flow $R_{\ell\Lambda}$ is required to be such that whenever $\mathcal{L}^\ell = R_{\ell\Lambda}\mathcal{L}^\Lambda$, $\mathcal{L}^\ell$ and $\mathcal{L}^\Lambda$ both

[7] They are also known as 'non-perturbatively renormalizable'—a locution I avoid because it's so easy to hear as 'NOT perturbatively renrmalizable,' which isn't what's meant.

determine the same n-point functions for phenomena at scale ℓ.[8] Such agreement about n-point functions is sufficient for the scale ℓ empirical predictions of theories $\mathcal{L}^{\ell}$ and $R_{\ell\Lambda}\mathcal{L}^{\Lambda}$ to agree within experimental tolerances.

Polchinski shows that the RG flow so defined has a three-dimensional surface of attraction, where this surface is parameterized by couplings m^{*} (the effective mass), λ^{*} (the effective coupling) and ϕ^{*} (the effective field amplitude renormalization).[9] Effectively specifiable theories in this surface all share the form of the *perturbatively renormalized* ϕ^4 Lagrangian obtained by 'suspect' conventional renormalization methods of ultraviolet divergence taming.

Although there aren't at present equally rigorous demonstrations that more complicated interacting QFTs exhibit the remarkable property, a host of considerations suggest that they do. Some illustrate the possibility that effective Lagrangians depend on a roster of fields different from the roster of fields appearing in the higher energy Lagrangians inducing them. Quantum Chromodynamics (QCD) and its low-energy effective realization at atomic scales affords an example. The high energy QCD Lagrangian $\mathcal{L}_{QCD}$ features fields corresponding to quark and gluon degrees of freedom. These are absent from the effective Lagrangian encoding $\mathcal{L}_{QCD}$'s implications for atomic scales. The effective Lagrangian features instead hadronic fields corresponding to protons and neutrons and their ilk. The absence from the effective Lagrangian of quark degrees of freedom illustrates 'quark confinement'—isolated quarks aren't encountered at atomic scales.

Perturbative QFT's real problem was: why should the approximation technique of perturbative renormalization work at all? RG analyses suggest an answer: perturbatively renormalized Lagrangians, astonishingly successful when applied to experimentally accessible phenomena, correspond to finitely specifiable effective theories. As such, they encapsulate the low-energy implications of unknown higher energy physics. The lower energy physics is tractable because $\mathcal{T}^{\ell}_{\mathit{eff}}$ is finite dimensional: effective Lagrangians at scale ℓ require only finitely many parameters to specify. These parameters correspond to the finite set of renormalized couplings, masses, and so on emerging from perturbative renormalization techniques. Reassuringly, despite our innocence of high energy details, we can completely specify the physics at scale ℓ by *measuring* these couplings. We'll get the effective physics right even if our renormalization schemes are opportunistic and our cutoff procedures imperfectly understood.

There's no a priori reason to expect susceptibility to domestication by standard perturbative renormalization techniques to be either necessary or sufficient for the "remarkable property" to obtain. Another way RG analysis figures significantly in the project of modern physics is by making theories that aren't perturbatively renormalizable tractable.

[8] The proof proceeds by constructing an appropriate $R_{\ell\Lambda}$, an operation that turns out to be far more delicate than suggested by the naive coarse-graining recipe sketched in fn. 5.

[9] More precisely: only three terms in the effective Lagrangian contribute meaningfully to the physics. For other terms in the effective Lagrangian, their contributions to perturbative expansions of n-point functions are suppressed by factors of ℓ/Λ_{UV}, where Λ_{UV} is a high energy cutoff. This codicil will matter later.

15.2.3 *Effecting realism*

15.2.3.1 THE POSITION STATED

A number of philosophers bold enough to engage interacting QFTs have marshaled the considerations just presented to refine and reinforce scientific realism (J. Fraser, 2016 and this volume, Chapter 14; Miller, 2017; Williams, 2019[10]). Porter Williams has branded the resulting position 'Effective Realism.' Effective Realism is an attitude toward empirically successful effective theories in the ambit of RG techniques. It is an attitude whose virtues emerge on the field of deeds. Recall the plight of the aspiring realist about actual successful theory T. Confronted with the pessimistic metainduction, she mounted a *divide et impera* defense. Her detractors complain that because she neither offers nor defends principles for restricting T's domain or content, her position is neither well-articulated nor justified.

Effective Realism would neutralize this complaint. Effective Realism casts successful physical theory T as effective, and thereby "accept[s] that it *does not* provide... a true and complete description of the world in all respects" (Williams, 2019, 212). Indeed, "incorporating explicitly into their mathematical framework a short-distance length [high energy] scale beyond which the theory is known to be empirically inadequate" (212), effective theories offer domain restrictions on their own behalf. An interacting QFT presented as effective at scale ℓ—a presentation that warrants studying how the RG flow connects it to QFTs governing physics at much higher energies—exemplifies a theory with an avowedly limited domain, a theory whose mathematical analysis proceeds by acknowledging this domain restriction.

As for content restrictions, Williams suggests "using the RG to provide a means of identifying elements of Effective Field Theories that are invariant across independent and distinct choices about how to model the physics at the short distances [high energies] where the theory is empirically inapplicable" (215). The distinguished elements are "stable and 'robust'"; as such they "can be expected to survive future episodes of theory change" (14). *Robust* is a notion borrowed from Wimsatt, who glosses it as "accessible (detectable, measurable, derivable, definable, producible, or the like) in a variety of independent ways" (2007, 95, as quoted in Williams, 2019, 218). The kind of variety most directly surveyed by RG analysis of effective field theories is variety in underlying physics/ways of managing cutoffs.

To illustrate the principled and motivated content restriction effected by RG considerations, Williams considers the case of mirror fermions in lattice QCD. QCD describes strong interactions, fields involving quark and gluon degrees of freedom, in Minkowski spacetime. Regarding QCD as effective requires acknowledging some length scale below which it ceases to apply. One way to express this regard is to dislodge QCD from continuous Minkowski spacetime and reformulate it instead on a $4d$ spacetime lattice, choosing a spacing a between lattice sites shorter than the characteristic length scale at which QCD ceases to apply. It is know that a latticized QCD hosts a variety of unwelcome guests, including many more varieties of fermion than one set out to describe, many more varieties than one needs to accommodate the

[10] See also Bricmont and Sokal (2004, section 3.2) for what may be an anticipation of the large-scale structure of the view. (I'm obliged to Jeremy Butterfield for bringing this to my attention.)

phenomena.[11] These *mirror fermions* are denigrated as 'lattice artifacts.' They are also elements of successful physics. How can the consistent realist avoid endorsing them when she endorses QCD?

If the consistent realist is an Effective Realist, Williams suggests, she stands committed only to those elements of a successful effective theory that are robust under changes in the underlying physics. Mirror fermions fail this test. The term in the dynamics of lattice QCD that describes mirror fermions is proportional to the lattice spacing a. In the limit as this spacing a goes to 0, the theory reverts to its continuous form, whose dynamics describe only the physical fermions we know and love. Thus *how* mirror fermions appear in lattice QCD depend on *how* the theory is latticized—on both the choice of the lattice spacing a and the details of the latticization. How physical fermions appear is independent of these choices. "The mirror fermions then depend on an arbitrary choice of modeling scheme in a way that genuinely representational quantities do not" (2019, 224). Mirror fermions fail, and physical fermion satisfy, the Effective Realist's demand for robustness. "By interpreting EFTs," Williams concludes, "one has resources to make fine-grained distinctions between mathematical artifice and physical significance, which supports a 'divide and conquer' approach to ontological commitment" (227).

15.2.3.2 AGAINST INTERPRETATION

Williams articulates Effective Realism in concert with a criticism of what he calls "Standard Interpretation," particularly as it is pursued in the grip of the idea that our best physical theories are "fundamental" insofar as they provide a "true and complete description of the world in all respects" (2019, 217). Expressions of the Standard Account Williams cites (212) include:

> Whatever else it means to interpret a scientific theory, it means saying what the world would have to be like if the theory is true. (Earman, 2004, 1234)
>
> *Hence we come to the question of interpretation*: Under what conditions is the theory true? What does it say the world is like? These two question are the same. (van Fraassen, 1991, 242)

A "Standard Interpretation" of T is an account of T's truth conditions. Williams takes Standard Interpretation to further the quest for an underlying ontology: "the goal of interpreting physical theory is to identify and characterize its *fundamental* ontological structure" (2019, 212). Thus, confronted with an effective theory T, a theory acknowledged to break down at some length/energy scale—a theory avowedly false—there is a natural question to ask about its Standard Interpretation: Why bother?[12] After all, if T is false, its truth conditions aren't going to tell us what *our world* is like. A contingent further complication: our best present effective QFTs, which include the lion's share of empirically applicable QFTs, lack the sort of axiomatic formulation that some (e.g., D. Fraser, 2011 and Halvorson and Müger, 2007) regard as a congenial, even essential, setting for Standard Interpretation.

The foregoing considerations suggest that pursuing a Standard Interpretation of interacting QFT could be pointless and difficult. Williams thinks there is worse news

[11] For details, see Williams (2019, 227–32) and references therein.
[12] See Belot (1998) for further development of the point.

still: "[a]dhering to Standard Interpretational Methodology (that the description of the world offered by the theory be true and exhaustive)" (2019, 216–17) issues in absurd and counterproductive conclusions. Consider the case of quark confinement. Recall that at high energy scales, the QCD Lagrangian $\mathcal{L}_{QCD}$ uses only quark and gluon degrees of freedom. Only in the effective Lagrangian induced for lower (atomic) energy scales do hadronic degrees of freedom (such as protons and neutrons) appear. Insisting on a Standard Interpretation of QCD, Williams argues, we'd draw on the underlying theory $\mathcal{L}_{QCD}$ to articulate an ontology of quarks and gluons. Because explanations of atomic behavior rely on protons and neutrons, this ontology would hamstring explanations of atomic behavior. Phenomena eluding our explanatory reach include hadronic jets central to particle physics data confirming QCD.

Insisting upon a Standard Interpretation of $\mathcal{L}_{QCD}$ as fundamental physics, then, we'd suppress the theory's capacity to explain atomic physics, as well as its account of ontology at atomic scales. We'd also ignore the possibility, dramatized by RG considerations, that—because $\mathcal{T}^{\ell}_{eff}$ is a catchment for *all* higher energy Lagrangians in $\mathcal{T}$—some underlying physics *other than QCD* induces hadronic physics effective at scale ℓ. We'd "be dramatically misled about the genuine ontological information provided by a quantum field theory by interpreting only its empirically *unreliable* 'fundamental' structure and conflating that with its structure at all scales" (226), Williams concludes. His verdict is unsparing

> The central vice of Standard Interpretation... is that it declares essentially all empirically applicable quantum field theories to be unfit for interpretation... An alternative response... is to conclude that it may be the approach to interpretation that is unfit for interpretive work, not the theory. (215, 217)

15.3 Perturbing Realism

Effective Realism is a genuinely new position in the scientific realism debate. It is a position propelled by considerations internal to our best contemporary physics. It is a subtle, even cagey, position, one that turns underdetermination and agnosticism, standard empiricist weapons, into resources for the realist.

Promising views deserve critical scrutiny. This section offers some. Section 15.3.1 questions how *effective* Effective Realism is. That is, it asks how resistant to antirealist affronts, such as the Pessimistic Metainduction, Effective Realism is. Section 15.3.2 questions how *realist* Effective Realism is. That is, it asks what qualifies the commitments RG considerations circumscribe as *distinctively realist* commitments. The pursuit of these questions motivates an empiricist appropriation of the RG apparatus, which Section 15.3.3 develops and the conclusion (fleetingly) defends.

15.3.1 *Resisting the pessimistic metainduction?*

Briefly put, the Effective Realist's response to the Pessimistic Metaindution is: that *no matter what the future of physics,* specific features of our best current physics will persist in that physics. RG analysis not only helps us identify those features, it also gives us reason to expect their persistence as elements of physics applicable at effective scales. The Effective Realist asserts features robust in this sense to be reliable guides to the way the world is. Also briefly put (and developed at more length in Ruetsche, 2018), a criticism of this response is that it trades on a false generality. The action of the RG is defined on a specific space of theories $\mathcal{T}$ (if it's defined at all). In the case of the Polchinski result sketched in Section 15.2.2, $\mathcal{T}$ is the space of free scalar field theory and small perturbative corrections thereto. The most we can conclude on the basis of rigorous and explicit RG analysis is: *provided the true underlying theory* T_{final} *lies in* $\mathcal{T}$, features robust under the action of the RG will persist in future physics. But if true and complete theory T_{final} lies outside that space, then all bets are off. Our best current theories could be dramatically incompetent guides to T_{final}. Put another way, our best current theories could be dramatically incompetent guides to how the physical world is.

This isn't an idle possibility. Newton's Law of Universal Gravitation (LUG) is a historical example of how overconfidence in the comprehensiveness of a theoretical space could lead us into grave ontological error. LUG isn't exactly true: famously, it fails to accommodate the perihelion advance of Mercury and other planets. However, other elements of a $\mathcal{T}$-like space, a space including LUG and perturbative corrections to LUG, can save these phenomena, provided finitely many free parameters are fixed by experiment (see Wells, 2012, for details). This might inspire optimism that, no matter what the details of underlying fundamental physics, its implications for physics at orbital radii we care about can be captured by a finite number of small empirically tractable corrections to LUG. Inspired by this optimism, we might endorse those features of LUG's picture of the world that it shares with these corrections. These include central gravitational forces acting instantaneously at a distance across Euclidean three-space.

From the perspective of General Relativity (GR), our confidence is misplaced. The equations of motion determined by phenomena-saving corrections to LUG are low velocity, low eccentricity, large radius limits of the equations of motion arising from the Schwarzschild solution to the field equations of GR. GR dispenses with central forces acting instantaneously across flat Euclidean three-space. Instead, GR attributes gravitational phenomena to the curvature of a four-dimensional spacetime. Reassuring considerations of features it shares with an apparently encompassing space of theories notwithstanding, *LUG gets the ontology dead wrong*. If we had taken LUG to represent how the world is, we would have been led astray.[13]

15.3.2 *An alternative to empiricism?*

This section waives the skeptical concerns of the last section to ask how *realist* Effective Realism is. Conceding for the sake of argument that the commitments circumscribed by the Effective Realist are worth endorsing, a further question is whether there's anything distinctively realist about the endorsement. More dramatically put: what does the Effective Realist believe that garden variety empiricists would reject?

To identify, in spite of our ignorance, aspects of current theories that represent the way the world is, Effective Realism appeals to *robustness*, explicated as invariance

[13] Those inclined to regard LUG as 'approximately true' won't feel the pull of these considerations. Given how wrong LUG is, I'm puzzled about what exactly the unmoved are praising LUG for, when they hail it as approximately true.

under the RG flow. Several strong candidates for endorsement by this robustness criterion emerge from effective theories RG analysis locates in a surface of attraction $\mathcal{T}^{\ell}_{eff}$. The strong candidates include the finite set of couplings that parameterize $\mathcal{T}^{\ell}_{eff}$; the n-point functions defined by an effective Lagrangian located in $\mathcal{T}^{\ell}_{eff}$; and the particles correlate to the fields appearing in effective Lagrangians located in $\mathcal{T}^{\ell}_{eff}$. (Or, for that matter, the fields themselves—I'll focus on the associated particles in what follows.) Let us consider each in turn.

Couplings. For instance, m_r, λ_r and ϕ_r for the perturbatively renormalized and effectively specifiable ϕ^4 Lagrangian. These are invariant under the RG flow in the sense that, no matter what the underlying physics is, the Lagrangian effective at scale ℓ will be an element of $\mathcal{T}^{\ell}_{eff}$ and so will incorporate these couplings. They will have different values for different elements of $\mathcal{T}^{\ell}_{eff}$, but the quantities corresponding to the couplings are omnipresent elements of the physics on that surface of attraction. No matter what the underlying physics, quantities corresponding to the couplings will feature in effective physics. Thus these quantities invite endorsement by the robustness criterion.

But, I surmise, as close as those quantities live to the empirical surface of the theory, most empiricists would be happy to share in the endorsement. One key to relieving interacting QFT of an unseemly dependence on details of poorly understood renormalization schemes is that *we can measure the couplings* specifying the effective theory. Understanding the couplings as phenomena enables empiricists to join Effective Realists in commitment to couplings without compromising their empiricism.

Correlation functions. Correlation/n-point functions—for example the two point function $\langle\phi(x)\phi(y)\rangle$ for the scalar field—are another strong candidate for endorsement by the robustness criterion, provided that the correlation extends over ranges longer than the length scale at which the effective theory is supposed to break down (see J. Fraser, this volume, Chapter 14). Correlation functions are adamantly invariant under the action of the RG: the RG is defined to act in such a way that that $\mathcal{L}^{\ell}$ and $R_{\ell\Lambda}\mathcal{L}^{\Lambda}$ define the same correlation functions at scale ℓ. The challenge for the Effective Realist is how to understand this endorsement as distinctively realist. Fueling the challenge is a natural way to interpret n-point functions as bookkeeping devices for particle physics data: by way of the LSZ reduction formula, n-point functions determine S-matrix elements and scattering cross sections. *Were* the QFTs under consideration free QFTs, the Effective Realist could justify commitment to n-point functions as commitment to an underlying theory by appeal to the Wightman Reconstruction Theorem, according to which a full set of n-point functions is equivalent to a theory satisfying the Wightman axioms. Interacting QFTs in 4 spacetime dimensions are not known to satisfy the axioms. So we've stepped outside the context where appeal to the Wightman Reconstruction Theorem is known to be warranted, and the Effective Realist needs other ways to resist the empiricist's attempt to assimilate n-point functions to phenomena.

One way to meet this need is to embark on the project of interpreting interacting QFTs. Viewed through an appropriate interpretation, expressions such as $\langle\phi(x)\phi(y)\rangle$ describe subatomic goings-on—'a photon of momentum k is created here while one of momentum k' is annihilated there'—and aren't just a *façon de parler* for scattering cross sections. Earnest ontological commitment to $\langle\phi(x)\phi(y)\rangle$ *so understood* would distinguish Effective Realism from empiricism. But notice that in this case what makes Effective Realism realist isn't the mere circumscription of contents satisfying the robustness criterion but also the interpretation of those contents as representational. In cases, such as n-point functions and couplings, where there are also empiricist interpretations of those contents, there is something Effective Realist could use: considerations favoring the realist over the phenomenal readings.

Section 15.3.3 will have more to say about the apparent irony that, its suspicion of Standard Interpretation notwithstanding, Effective Realism requires something like interpretive work to count as realism at all. There remain candidates for endorsement by the robustness criterion that seems less susceptible to empiricist deflation.

Fields and/or entities corresponding to fields in the effective Lagrangian. Every Lagrangian in $\mathcal{T}^{\ell}_{\mathit{eff}}$ incorporates the same list of fields and their spacetime derivatives; what alters as one moves around $\mathcal{T}^{\ell}_{\mathit{eff}}$ are the coupling coefficients with which those fields appear. The roster of fields is robust. Invoking the customary association between the fields appearing in a particle physics Lagrangian and elementary particles, Effective Realists might endorse *particles* corresponding to fields robustly present in a Lagrangian at scale ℓ. The effective Lagrangian that QCD induces at atomic scales, for example, incorporates fields correlate to protons, neutrons, and other hadrons. Endorsing these particles, the Effective Realist undertakes commitments apparently unpalatable to the empiricist.

I say 'apparently' because the commitment to particles just described is complicated in a number of respects, some of which are elaborated in Section 15.4, and each one of which draws Effective Realism further into debates typical of Standard Interpretation, typical even of metaphysics. Yet another reason Effective Realism merits and rewards engagement is that engaging it entangles us in questions such as these.

15.3.3 *An alternative empiricism?*

Our elaboration of Effective Realism has encountered a hitch that might be an irony. To present the commitments circumscribed by RG considerations as distinctively *realist* commitments, Section 15.3.2 suggests, the Effective Realist must engage in interpretive projects. The apparent irony arises because Effective Realists repudiate Standard Interpretation. I think the irony is only apparent. This section explains why. It begins with a mildly idiosyncratic characterization of realism, one which makes apparent that Effective Realists needn't disavow interpretation per se. The section closes by describing a mildly idiosyncratic (but to my mind quite attractive) variation of empiricism.

Presented with an extraordinarily successful scientific theory T, here are two questions we might ask about it:

(i) What does T say the world is like?; and
(e) Why does T work as well as it does?

One way to characterize and distinguish positions in the scientific realism debate is to compare their attitudes about the interdependence, or lack thereof, of these questions.

The first is a question about how to *interpret* T, where 'interpretation' is understood in what Williams called its 'standard sense.' To interpret a theory is to characterize the ways the world might be, according to that theory. That is to say, it is to characterize the possible worlds at which the theory is true. Supposing that to understand an assertion is to grasp its truth conditions, an interpretation of *T* enables *understanding* of *T* by affording truth conditions to grasp.

The second question is a question about how to *explain T*'s success. An adequate answer furnishes an understanding (not of *T* proper but) of why *T* works as well as it does. That answer may or may not be mediated by the claim that *T* is true. That is, it may or may not be mediated by an interpretation of *T*.

According to a position I'll label *fundamentalism*, and that I take to be a variety of realism, the interpretive and the explanatory questions are intimately related. A proper understanding of why *T* succeeds requires understanding *T*, for the simple reason that *T* succeeds as well as it does *because* the world is the way *T* says it is. An interpretation of *T* tells us what that way is. The fundamentalist regards any response to the explanatory demand that prescinds from *T*'s truth, as articulated by an answer to the interpretive question, shallow by comparison. Some shallow answers include: *T* works as well as it does because the world is *as if T were true*; *T* works as well as it does because we like our scientific theories to work and we picked *T* (van Fraassen, 1980); *T* works as well as it does because it abounds in pragmatic virtues (one way to put this is: *T* is true, where 'true' is construed pragmatically (Fine, 1986)).

Fundamentalism endows *T* with a sort of explanatory *self-sufficiency*. For a fundamentalist about *T*, no (empirical) features exogenous to *T* are required to explain *T*'s success. This is why *understanding T* and *understanding why T works as well as it does* are the same project. The properties Williams attributes to fundamental physics—*generality* and *rectitude*—are near-corollaries of self-sufficiency. *T* is explanatorily self-sufficient only if *T*'s domain is unrestricted and *T* describes everything in that domain accurately. Otherwise, something exogenous to *T* is required to explain why *T* works in some regimes but not others, and why things aren't quite the way *T* says they should be.

Regarding our current most successful scientific theories, a collection which adamantly includes the perturbatively renormalizable QFTs making up the Standard Model of particle physics, fundamentalists abound. Champions of "naturalistic metaphysics," which Wallace describes as "the thesis that we have no better guide to metaphysics than the successful practice of science" (2011, 58), are among them. Take Maudlin:

> Metaphysics is ontology. Ontology is the most generic study of what exists. Evidence for what exists, at least in the physical world, is provided solely by empirical research. Hence the proper object of most metaphysics is the careful analysis of our best scientific theories (and especially of fundamental physical theories) with the goal of determining what they imply about the constitution of the physical world. (2007, 104)[14]

Supposing 'metaphysics' is an account of the way the world is, positions like these assume that making sense of scientific success, answering the explanatory question,

[14] See also the authors cited in Williams (2019, n.14). This is not to say fundamentalism is universal: Wallace's (2011) naturalism leads him away from fundamentalism; Myrvold (this volume, Chapter 12) is another instance at hand of a non-fundamentalist.

requires understanding what successful science says the world is like, answering the interpretive question.

We can understand T without believing it. And we can be moved by reasons to interpret T that aren't reasons to believe T (e.g., identifying GR's observables as a prolegomenon to quantizing GR). So simply engaging in the interpretive project doesn't commit us to implicating it in the explanatory project as fundamentalists would. And if T is effective, it's a mistake to prosecute the explanatory project as fundamentalists would. This is because if T is effective, something empirical and exogenous to T explains T's success: the underlying physics T_{final} that T approximates in T's domain of application.

Effective Realists' suspicion of Standard Interpretation derives from its allegiance with fundamentalism. Fundamentalism's sin isn't to interpret T. It's to decide that, having interpreted T, no further explanatory work remains—and thereby suppress the possibility that T is merely effective. Interpretation itself isn't in tension with Effective Realism. What's inconsistent with the core insight of Effective Realism, the insight that our best physical theories are merely effective, is involving interpretation in fundamentalist explanatory projects.

But having noticed this, we are in a position to appreciate just how much of Effective Realism is compatible with, even supportive of, a variety of empiricism. The effective theory idea alerts us that *we can construct explanations of T's success which make no appeal to T's truth.* Such explanations lie at the core of a position I'll call *Humble Empiricism*, and which I beg you not to identify with more notorious varieties of empiricism. For the humble empiricist, successful T *isn't* latching on to hidden springs and *shouldn't* be plumbed for an account of those springs. Granting (for the sake of argument) that T's truth explains its success, here's a *humbler explanation*:

> In experimentally-accessible regimes, T approximates T_{final}'s predictions.

Here T_{final} is the true, fundamental theory. For the humble empiricist, successful T is approximating fundamental T_{final}'s *experimentally-accessible predictions*, not T_{final}'s account of physical reality. Section 15.3.1 offered LUG and GR as a near example of a T/T_{final} pair.[15]

Presented with a successful scientific theory, the fundamentalist believes the right way to answer the explanatory question (why does the theory work as well as it does?) is to answer the interpretive question (what does the theory say the world is like?) and believe (that is, accept as at least approximately true, where 'true' is in construed in some non-epistemic, non-pragmatic sense apt for realism) the theory as interpreted. The humble empiricist has nothing against answering the interpretive question. Indeed, she recognizes that taking our best theories literally—something we need to interpret them to do—promotes core scientific enterprises, including theory development. However, the humble empiricist prefers an answer to the explanatory question mediated neither by an interpretation of the theory nor by a commitment

[15] Only a near-example because not even General Relativity is regarded as a fundamental theory of gravity. That role, if it is occupied at all, will be played by some future theory of quantum gravity. The humble empiricist position is stable if we remove the assumption that there is a T_{final}, provided that for any successful physical theory T, there is an underlying theory T' such that (i) T's success consists in mimicking T' in a limited domain, and (ii) T and T' differ dramatically in the accounts they offer of physical reality.

to its truth. The humble empiricist's answer is that our theory succeeds as well as it does where it does, not because it's true, but because, whatever the true theory is, our theory approximates it in its domain of application.

The humble empiricist differs in significant ways from more notorious empiricists. To my mind, these differences all redound to the favor of the humble empiricist.

Some nonrealists (e.g., Cartwright, 1999) contend that the physical world is irremediably untidy or irreconcilably disjointed—not the sort of thing to afford the kind of truth conditions that would vindicate fundamentalism. The humble empiricist, by contrast, entertains the possibility of a true, fundamental theory T_{final}. For her, the question isn't whether it's coherent or plausible to posit such a theory. The real question occurs downstream from that posit. It's: how much can/does our best current theory T tell us about T_{final}? A lot, thinks the fundamentalist. Only that T_{final} is roughly empirically equivalent to T in T's domain of application, thinks the humble empiricist. That we still have a lot to learn about T_{final} is one of the things we have to be humble about.

Empiricists are often characterized as believing that the success of science requires no explanation. This wrongs the humble empiricist, who offers an explanation for T's success. It's not the explanation the fundamentalist favors. Still, both offer explanations. The core issue dividing them is whose explanation we should accept. The nature of explanation and norms for acceptable explanations have been core concerns of philosophy of science since Hempel—concerns animated by a rich history of scientific practice trading in (prima facie) explanations. Whereas philosophy of science may be ill-equipped to settle whether or not the universe is fundamentally tidy and domain-cohesive, perhaps progress can be made on the question of whose explanation—the humble empiricist's or the fundamentalist's—is preferable.

The humble empiricist is clearly stealing pages from the Effective Realist's playbook. Here is one more. Recall that empiricism, charged with committing selective skepticism, faced a problem of circumscription: to explain what justifies faith in the continued empirical success of our best physical theories—faith in what they say about what's not yet observed—while allowing agnosticism about the accuracy of their representation of hidden springs—agnosticism about what they say about the unobservable. RG considerations deliver the circumscription required. *Many* distinct underlying theories flow to the surface of attraction $\mathcal{T}^{\ell}_{eff}$ containing all effective theories at the scale ℓ within our experimental reach—a surface that includes T. Such underdetermination makes agnosticism about T's account of hidden springs permissible: T succeeds not because it is the unknown truth T_{final} but because it mimics the truth at scale ℓ. T_{final} is what warrants belief that T's successful mimicry will continue. Whatever the hidden springs are, they're out there, and RG considerations reassure us that T adequately describes the signature they leave at scale ℓ.

15.4 Conclusion: Effective Empiricism

Section 15.3.2 identified a promising *distinctively realist* candidate for endorsement by Effective Realism. It applies to effective (at scale ℓ) theories RG analysis locates in a surface of attraction $\mathcal{T}^{\ell}_{eff}$. Deploying the same set of fields ψ_i^{eff}, Lagrangians in this surface differ only in the coupling coefficients with which those fields appear. The presence in effective Lagrangians ψ_i^{eff} of the fields is in this sense robust. Hence the candidate:

[CRITERION] The particles corresponding to the fields ψ_i^{eff} exist.

Subjecting the candidate to somewhat closer (but still too rapid) scrutiny, this section suggests that it's an even more promising candidate for a sort of endorsement congenial to the humble empiricist: pragmatic endorsement. Motivating the suggestion are three puzzles about the candidate that, I'll suggest, evaporate if we construe the endorsement pragmatically.

First puzzle: mirror fermions. One puzzle is that CRITERION appears to commit the Effective Realist to the existence of mirror fermions! Although the details of their appearance vary with the details of the lattice spacing and model, the *fact* of their appearance does not. Regarding the theory as effective requires latticizing, and latticizing means mirror fermions will be present in the effective Lagrangian. So mirror fermions satisfy CRITERION. But it is clear that the Effective Realist wants to join the rest of us in rejecting mirror fermions. (While I confess that this puzzle feels sophistical to me, I suspect an account of *why* it's sophistical would clarify the Effective Realist position.)

Second puzzle: 'relevance.' Section 15.2.2's discussion of Polchinski's result skated over a subtlety: what parametrizes $\mathcal{T}_{eff}^{\ell}$ are coefficients of terms in the effective Lagrangian that contribute *meaningfully* to the physics at scale ℓ. I'll abuse terminology by calling the fields associated with these terms 'relevant.'[16] The contrast class are terms whose contributions are negligible, in the sense that they are suppressed by factors of ℓ/Λ_{UV} (Λ_{UV} is a high energy cutoff scale). *Other* interactions, including interactions involving fields that aren't relevant at at scale ℓ, typically figure in the effective Lagrangian; they just matter less to physics at scale ℓ. So either CRITERION endorses all these fields, or just the relevant ones. If it endorses them all, it's failing to act as a selection principle. But if it endorses just the relevant fields, we have a puzzle. On the one hand, the distinction between relevant fields and others is graded—fields are more or less relevant, and their status as such changes as a function of both the effective scale ℓ and the cutoff scale Λ_{UV}. On the other hand, the distinction between existence and non-existence is traditionally reckoned to be sharp. The puzzle is that we might have expected CRITERION to effect a distinction of kind—but CRITERION delivers instead a distinction of degree.

Third puzzle: scale relativity. The commitments circumscribed by CRITERION are scale-relative. Properly understood, CRITERION declares hadrons (not real *tout court* but) real at atomic scales. Tied as they are to phenomena we can achieve experimentally, the scales to which effective commitments are indexed are manifestly tied to us and our limitations. (A similar feature is sometimes called as a strike against van Fraassen's notion of *observable.*) In the saga of perturbative renormalization, renormalization vanquished the threat that interacting QFT's empirical content was cutoff scale dependent. Why should scale-dependence be less of a threat to legitimate ontological commitment?

[16] See Duncan (2012, section 16.3) for the proper use of 'relevant,' 'irrelevant,' and 'marginal.'

Each one of these puzzles evaporates if we construe the commitments circumscribed by CRITERION as pragmatic, rather than realist, commitments. On the reconstrual, the commitment is commitment to utility for organizing phenomena at scale ℓ. It is a commitment to what fosters coping rather than to what achieves representational success. Empirically inert, mirror fermions fail to qualify for such commitment. This addresses the first puzzle.[17] Pragmatic commitment comes in degrees, just as utility does—which addresses the second puzzle. Utility is utility *for us*—so it would be mysterious if pragmatic commitment *weren't* keyed to our abilities and limitations—which addresses the third puzzle. Regarding the commitments satisfying CRITERION as pragmatic makes better sense of them than realist renderings do. It also makes them commitments the empiricist can live with.

Solid state physicists sometimes draw an ontologically invidious distinction between two sorts of *quasi-particles*. A beloved text describes situations calling out for quasi-particles as follows: a "system composed of *strongly* interacting *real* bodies acts *as if* it were composed of *weakly* interacting (or non-interacting) *fictitious* bodies" (Mattuck, 1992, 3). The quasi-particles are the "fictitious bodies." One sort is exemplified by the 'dressed' electron. The intuitive picture is that as a negatively-charged 'bare' electron travels through a semi-conductor, it accretes a 'cloud' of positive charges. Effective semi-conductor physics results by treating the electron and its cloud as a single unit, propagating freely—a unit with a mass different from the bare electron mass.

Another sort of quasi-particle is exemplified by the phonon. A crystal lattice constituted by (let us agree for the sake of argument) protons, neutrons, and electrons exhibits collective excitations whose energy is quantized. Happily, these collective excitations behave in tractable ways—indeed, like sound waves. A phonon is an elementary such excitation. Just as with the dressed electron, there are many ways, beyond present calculational resources to comprehend, the crystal's atomic constituents might realize a phonon. For treatment of crystal-level questions concerning phenomena such as heat transfer, differences between different atomic realizations don't matter. They all induce the same effective physics, expressed in terms of phonons.

Phonons, mediators of the effective physics of crystals, are clear candidates for endorsement by CRITERION. So too are dressed electrons. But not all solid state physicists are eager to hail phonons and dressed electrons as ontological peers. Here is more from the beloved text:

> The [electron] quasiparticle consists of the original real, individual particle, plus a cloud of disturbed neighbors. It behaves very much like an individual particle, except that it has an effective mass and a lifetime. But there also exist other kinds of fictitious particles in many-body systems, i.e. 'collective excitations.' These do not center around individual particles, but instead involve collective, wavelike motion of *all* the particles in the system simultaneously.
>
> (Mattuck, 1992, 10)

[17] Williams (2019) also emphasizes the roles of physics practice, and particularly of "having empirical applications" (212), in singling out what's worthy of commitment. In the case of mirror fermions, that component of Williams' view may be doing more work than the robustness criterion.

Physicists so-minded regard dressed electrons as *better* candidates for endorsement as 'real particles' than phonons. Mattuck doesn't elaborate the basis of this distinction—but here's a guess: in the case of phonons, according to our best picture of the underlying physics, there are no particles *there*; in the case of dressed electrons, there are particles there, just not quite the particles featuring in our effective physics. Drawing the distinction, so understood, presupposes a lucid grip on underlying physics. Drawing the distinction also destabilizes "real but not fundamental" as a cohesive ontological category. The 'real but not fundamental' category comprehends both electron quasi-particles and phonons; if (as Mattuck suggests) there is an ontological distinction to be drawn between them, indiscriminate use of the category threatens to efface it.

CRITERION, particularly when deployed with respect to interacting QFTs, is ill-equipped to respect Mattuck's distinction. The Effective Realist disavows knowledge of the physics underlying the effective Lagrangians that engage the gears of CRITERION. But it's by appeal to that physics that Effective Realists might defend particles endorsed by CRITERION against demotion to the status of phonons, worthy of pragmatic commitment, but wholly fictitious.

This underscores the theme that whether the commitments of Effective Realism count as realist depends on how they're understood. To that theme it adds the variation: there are possible phonon-like understandings of the commitments endorsed by CRITERION in terms of underlying physics that attenuate their claim to be realist. What is more, there is bound to be something revisionary and/or deflationary about *any* understanding of existence claims in terms of underlying physics. Taking 'there are hadrons' to be true only if paraphrased/corrected/ . . . to 'there are fundamental string modes vibrating hadronwise' is a far more tepid commitment to hadrons than it is to strings. But the lesson of Effective Realism was supposed to be we shouldn't reserve our warmest commitments for what's fundamental.

Acknowledgments

I am grateful to Doreen Fraser, Wayne Myrvold, Juha Saatsi, and Porter Williams, as well as to audiences at the University of California-Berkeley and Rutgers University, for feedback on earlier incarnations of this work. I am also grateful to the editors of this volume for their heroic patience and unreasonable encouragement.

References

Belot, G. (1998). Understanding electromagnetism. *British Journal for the Philosophy of Science* 49, 531–55.

Bricmont, J. and Sokal, A. (2004). Defense of a modest scientific realism. In M. Carrier, J. Roggenhofer, G. Küppers, and Ph. Blanchard (Eds), *Knowledge and the World: Challenges beyond the Science Wars*, pp. 17–53. Berlin: Springer.

Butterfield, J. and J. Earman (2007). *Philosophy of Physics, Part A*. Amsterdam: Elsevier.

Cartwright, N. (1999). *The Dappled World: A Study of the Boundaries of Science*. Cambridge University Press.

Duncan, A. (2012). *The Conceptual Framework of Quantum Field Theory*. Oxford: Oxford University Press.

Fine, A. (1986). Unnatural attitudes: realist and instrumentalist attachments to science. *Mind*, 95(378), 149–79.
Fraser, D. (2011). How to take particle physics seriously: a further defence of axiomatic quantum field theory. *Studies In History and Philosophy of Science Part B: Studies In History and Philosophy of Modern Physics* 42(2), 126–35.
Fraser, J. D. (2016). What is Quantum Field Theory? Idealisation, Explanation and Realism in High Energy Physics. PhD thesis, University of Leeds.
Fraser, J. D. (2017). The real problem with perturbative quantum field theory. *The British Journal for the Philosophy of Science*. https://core.ac.uk/download/pdf/160113683.pdf.
Gawedzki, K. (1986). Renormalization group, from magic to mathematics, talk delivered at the International Congress on Mathematics (1986, Berkeley).
Halvorson, H. and M. Müger (2007). Algebraic quantum field theory. In J. Butterfield and J. Earman (Eds), *Philosophy of Physics Part A*, Volume 2, pp. 731–922. Amsterdam: Elsevier.
Hancox-Li, L. (2015). Coarse-graining as a route to microscopic physics: the renormalization group in quantum field theory. *Philosophy of Science* 82(5), 1211–23.
Mattuck, R. D. (1992). *A Guide to Feynman Diagrams in the Many-Body Problem*. New York: Dover.
Maudlin, T. (2007). *The Metaphysics within Physics*. Oxford: Clarendon Press.
Miller, M. E. (2017). The Structure and Interpretation of Physical Theories. PhD thesis, University of Pittsburgh.
Peskin, M. E. and D. V. Schroeder (1995). *An Introduction to Quantum Field Theory* (Frontiers in Physics). Boulder: Westview Press Incorporated.
Polchinski, J. (1984). Renormalization and effective lagrangians. *Nuclear Physics B 231*(2), 269–95.
Psillos, S. (2005). *Scientific Realism: How Science Tracks Truth*. London: Routledge.
Rosaler, J. and R. Harlander (forthcoming). Naturalness, Wilsonian renormalization, and "fundamental parameters" in quantum field theory. *Studies in the History and Philosophy of Modern Physics*. http://philsci-archive.pitt.edu/15810/1/naturalnessI.pdf.
Ruetsche, L. (2011). *Interpreting Quantum Theories*. Oxford: Oxford University Press.
Ruetsche, L. (2018). Renormalization group realism: the ascent of pessimism. *Philosophy of Science* 85(5), 1176–89.
Schweber, S. S. (1994). *QED and the Men who Made it: Dyson, Feynman, Schwinger, and Tomonaga*. Princeton: Princeton University Press.
Strocchi, F. (2013). *An Introduction to Non-Perturbative Foundations of Quantum Field Theory*, International Series of Monographs on Physics, Volume 158. Oxford: Oxford University Press.
Van Fraassen, B. C. (1980). *The Scientific Image*. Oxford: Oxford University Press.
Wald, R. M. (1994). *Quantum Field Theory in Curved Spacetime and Black Hole Thermodynamics*. Chicago: University of Chicago Press.
Wallace, D. (2006). In defence of naiveté: the conceptual status of Lagrangian quantum field theory. *Synthese 151*(1), 33–80.
Wallace, D. (2011). Taking particle physics seriously: a critique of the algebraic approach to quantum field theory. *Studies In History and Philosophy of Science Part B: Studies In History and Philosophy of Modern Physics* 42(2), 116–25.
Wells, J. D. (2012). *Effective Theories in Physics: From Planetary Orbits to Elementary Particle Masses*. Amsterdam: Springer Science & Business Media.
Williams, P. (2019). Scientific realism made effective. *The British Journal for the Philosophy of Science 70*(1), 209–37.
Williams, P. (forthcoming). Renormalization group methods. In A. Wilson and E. Knox (Eds), *Routledge Companion to the Philosophy of Physics*. London: Routledge.

Name Index

Subject Index